London Mathematical Society Lecture Note Series. 72

Commutative Algebra: Durham 1981

Edited by
R.Y. SHARP
Reader in Pure Mathematics
University of Sheffield

CAMBRIDGE UNIVERSITY PRESS
Cambridge
London New York New Rochelle
Melbourne Sydney

Published by the Press Syndicate of the University of Cambridge
The Pitt Building, Trumpington Street, Cambridge CB2 1RP
32 East 57th Street, New York, NY 10022, USA
296 Beaconsfield Parade, Middle Park, Melbourne 3206, Australia

First published 1982

Printed in Great Britain at the University Press, Cambridge

Library of Congress catalogue card number: 82-12781

British Library Cataloguing in Publication Data

Commutative algebra - (London Mathematical Society
Lecture note series, ISSN 0076-0552; 72)
1. Mathematics - Congresses
I. Sharp, R.Y. II. Series
510 QA3

ISBN 0 521 27125 8

LONDON MATHEMATICAL SOCIETY LECTURE NOTE SERIES

Managing Editor: Professor I.M. James,
Mathematical Institute, 24-29 St Giles,Oxford

1. General cohomology theory and K-theory, P.HILTON
4. Algebraic topology, J.F.ADAMS
5. Commutative algebra, J.T.KNIGHT
8. Integration and harmonic analysis on compact groups, R.E.EDWARDS
9. Elliptic functions and elliptic curves, P.DU VAL
10. Numerical ranges II, F.F.BONSALL & J.DUNCAN
11. New developments in topology, G.SEGAL (ed.)
12. Symposium on complex analysis, Canterbury, 1973, J.CLUNIE & W.K.HAYMAN (eds.)
13. Combinatorics: Proceedings of the British Combinatorial Conference 1973, T.P.McDONOUGH & V.C.MAVRON (eds.)
15. An introduction to topological groups, P.J.HIGGINS
16. Topics in finite groups, T.M.GAGEN
17. Differential germs and catastrophes, Th.BROCKER & L.LANDER
18. A geometric approach to homology theory, S.BUONCRISTIANO, C.P. BOURKE & B.J.SANDERSON
20. Sheaf theory, B.R.TENNISON
21. Automatic continuity of linear operators, A.M.SINCLAIR
23. Parallelisms of complete designs, P.J.CAMERON
24. The topology of Stiefel manifolds, I.M.JAMES
25. Lie groups and compact groups, J.F.PRICE
26. Transformation groups: Proceedings of the conference in the University of Newcastle-upon-Tyne, August 1976, C.KOSNIOWSKI
27. Skew field constructions, P.M.COHN
28. Brownian motion, Hardy spaces and bounded mean oscillations, K.E.PETERSEN
29. Pontryagin duality and the structure of locally compact Abelian groups, S.A.MORRIS
30. Interaction models, N.L.BIGGS
31. Continuous crossed products and type III von Neumann algebras, A.VAN DAELE
32. Uniform algebras and Jensen measures, T.W.GAMELIN
33. Permutation groups and combinatorial structures, N.L.BIGGS & A.T.WHITE
34. Representation theory of Lie groups, M.F. ATIYAH et al.
35. Trace ideals and their applications, B.SIMON
36. Homological group theory, C.T.C.WALL (ed.)
37. Partially ordered rings and semi-algebraic geometry, G.W.BRUMFIEL
38. Surveys in combinatorics, B.BOLLOBAS (ed.)
39. Affine sets and affine groups, D.G.NORTHCOTT
40. Introduction to Hp spaces, P.J.KOOSIS
41. Theory and applications of Hopf bifurcation, B.D.HASSARD, N.D.KAZARINOFF & Y-H.WAN
42. Topics in the theory of group presentations, D.L.JOHNSON
43. Graphs, codes and designs, P.J.CAMERON & J.H.VAN LINT
44. Z/2-homotopy theory, M.C.CRABB
45. Recursion theory: its generalisations and applications, F.R.DRAKE & S.S.WAINER (eds.)
46. p-adic analysis: a short course on recent work, N.KOBLITZ
47. Coding the Universe, A.BELLER, R.JENSEN & P.WELCH
48. Low-dimensional topology, R.BROWN & T.L.THICKSTUN (eds.)

49. Finite geometries and designs, P.CAMERON, J.W.P.HIRSCHFELD & D.R.HUGHES (eds.)
50. Commutator calculus and groups of homotopy classes, H.J.BAUES
51. Synthetic differential geometry, A.KOCK
52. Combinatorics, H.N.V.TEMPERLEY (ed.)
53. Singularity theory, V.I.ARNOLD
54. Markov processes and related problems of analysis, E.B.DYNKIN
55. Ordered permutation groups, A.M.W.GLASS
56. Journées arithmétiques 1980, J.V.ARMITAGE (ed.)
57. Techniques of geometric topology, R.A.FENN
58. Singularities of smooth functions and maps, J.MARTINET
59. Applicable differential geometry, F.A.E.PIRANI & M.CRAMPIN
60. Integrable systems, S.P.NOVIKOV et al.
61. The core model, A.DODD
62. Economics for mathematicians, J.W.S.CASSELS
63. Continuous semigroups in Banach algebras, A.M.SINCLAIR
64. Basic concepts of enriched category theory, G.M.KELLY
65. Several complex variables and complex manifolds I, M.J.FIELD
66. Several complex variables and complex manifolds II, M.J.FIELD
67. Classification problems in ergodic theory, W.PARRY & S.TUNCEL
68. Complex algebraic surfaces, A.BEAUVILLE
69. Representation theory, I.M.GELFAND et. al.
70. Stochastic differential equations on manifolds, K.D.ELWORTHY
71. Groups - St Andrews 1981, C.M.CAMPBELL & E.F.ROBERTSON (eds.)
72. Commutative algebra: Durham 1981, R.Y.SHARP (ed.)
73. Riemann surfaces: a view toward several complex variables, A.T.HUCKLEBERRY
74. Symmetric designs: an algebraic approach, E.S.LANDER

CONTENTS

PART III: MULTIPLICITY THEORY, HILBERT AND POINCARÉ SERIES, ASSOCIATED GRADED RINGS, AND RELATED TOPICS

PREFACE

A Symposium on Commutative Algebra was held at the University of Durham during the period 15-25 July, 1981, under the auspices of the London Mathematical Society and with financial support from the Science and Engineering Research Council. There were 71 participants.

The academic programme was built round a series of invited one hour lectures; in addition, many participants volunteered lectures at sessions of short talks. It was decided, on account of limitations of space, to restrict this volume to articles by the invited speakers related to lectures given at the Symposium, although all participants were welcome to contribute to the section of 'Further problems' at the end of the book. The articles have been grouped together in sections in the hope that the result will reflect the flavour of the main themes of the Symposium. The first group of papers is concerned with the local homological conjectures, big Cohen-Macaulay modules, and related topics and applications; the second group consists of articles related to determinantal ideals and finite free resolutions; and the third group is concerned with various topics in local algebra, including multiplicity theory, Hilbert and Poincaré series, and associated graded rings. Within each section, the papers are arranged in alphabetical order of authors' names.

Participants at the Symposium were invited to submit open problems in commutative algebra for inclusion in a Problem Section in these proceedings, and the response to that invitation is contained in the final section of the book, entitled 'Further problems'. I am grateful to the contributors for their efforts.

I am also very grateful to Mrs. Elsie Benson and Mrs. Janet

Williams for their beautiful typing of the camera-ready copy for this book.

It is a pleasure to record my gratitude, and that of my co-organizer, D.G. Northcott, to the numerous members of staff of Durham and Sheffield Universities who contributed so much to the smooth-running arrangements and friendly atmosphere of the Symposium: we are particularly grateful to Dr. L.M. Woodward of the Durham University Mathematics Department, Mrs. M.O. Lund of the Durham University Finance Department, Captain G.R.T. Duffay, the Bursar of Grey College, and our Sheffield colleagues Drs. T.B. Cruddis, A.J. Douglas and D.W. Sharpe.

Finally, we should like to record our gratitude to the London Mathematical Society, under whose auspices the Symposium took place and this book will appear, and the Science and Engineering Research Council, without whose financial support the Symposium could not have taken place.

R.Y. Sharp

ADDRESSES OF CONTRIBUTORS

KAAN AKIN and DAVID A.BUCHSBAUM, Department of Mathematics, Brandeis University, Waltham, Mass. 02154, U.S.A.

WINFRIED BRUNS, Fachbereich 3, Naturwissenschaften, Mathematik, Universität Osnabrück-Abteilung Vechta-Driverstrasse 22, D-2848 Vechta, West Germany.

E.G.EVANS and PHILLIP GRIFFITH, Department of Mathematics, University of Illinois, Urbana, Illinois 61801, U.S.A.

HANS-BJØRN FOXBY, Matematisk Institut, Københavns Universitet, Universitetsparken 5, DK 2100 København Ø, Denmark.

SHIRO GOTO, Department of Mathematics, Nihon University, Sakurajosui 3-25-40, Setagaya-ku, Tokyo, Japan.

J.HERZOG, Universität Essen-Gesamthochschule, Fachbereich 6-Mathematik, Universitätsstrasse 3, Postfach 103764, D-4300 Essen 1, West Germany.

MELVIN HOCHSTER, Department of Mathematics, University of Michigan, Ann Arbor, Michigan 48109, U.S.A.

G.HORROCKS, School of Mathematics, University of Newcastle upon Tyne, Newcastle upon Tyne NE1 7RU, U.K.

D.G.NORTHCOTT, Department of Pure Mathematics, University of Sheffield, Hicks Building, Sheffield S3 7RH, U.K.

D.REES, Department of Mathematics, University of Exeter, North Park Road, Exeter EX4 4QE, U.K.

JAN-ERIK ROOS, Department of Mathematics, University of Stockholm, Box 6701, S-113 85 Stockholm, Sweden.

JUDITH D.SALLY, Department of Mathematics, Northwestern University, Evanston, Illinois 60201, U.S.A.

R.Y.SHARP and H.ZAKERI, Department of Pure Mathematics, University of Sheffield, Hicks Building, Sheffield S3 7RH, U.K.

L.SZPIRO, École Normale Supérieure, 45, Rue d'Ulm, 75230 Paris Cedex 05, France.

LIST OF PARTICIPANTS

Barnard, A.D.	(King's, London)	Macdonald, I.G.	(Q.M.C., London)
Bartijn, J.	(Utrecht)	McLean, K.R.	(Liverpool)
Bijan-Zadeh, M.H.	(Tehran)	MacRae, R.E.	(Boulder)
Bøgvad, R.	(Stockholm)	Massaza, C.	(Siena)
Boratyński, M.	(Warsaw)	Merriman, J.R.	(Canterbury)
Brodmann, M.	(Zürich)	Moore, D.J.	(Glasgow)
Brown, M.L.	(Coventry)	Nastold, H.-J.	(Münster)
Bruns, W.	(Osnabrück/Vechta)	Northcott, D.G.	(Sheffield)
Buchsbaum, D.A.	(Brandeis)	O'Carroll, L.	(Edinburgh)
Chatters, A.W.	(Bristol)	Orbanz, U.	(Köln)
Cruddis, T.B.	(Sheffield)	Orecchia, F.	(Napoli)
Douglas, A.J.	(Sheffield)	Porter, T.	(Bangor)
Evans, E.G.	(Urbana)	Pragacz, P.	(Toruń)
Faltings, G.	(Münster)	Qureshi, M.A.	(Edinburgh)
Flenner, H.	(Osnabrück)	Ragusa, A.	(Catania)
Foxby, H.-B.	(Norman, Oklahoma)	Ratliff, L.J.	(Riverside)
Fröberg, R.	(Stockholm)	Rayner, F.J.	(Liverpool)
Gotô, S.	(Tokyo)	Rees, D.	(Exeter)
Greco, S.	(Torino)	Rhodes, C.P.L.	(Cardiff)
Hajarnavis, C.R.	(Warwick)	Riley, A.M.	(Sheffield)
Herzog, J.	(Essen)	Robbiano, L.	(Genova)
Hochster, M.	(Ann Arbor)	Roberts, P.C.	(Salt Lake City)
Horrocks, G.	(Newcastle)	Roos, J.-E.	(Stockholm)
Huneke, C.	(Ann Arbor)	Rotthaus, C.	(Münster)
Józefiak, T.	(Toruń)	Sally, J.D.	(Evanston)
Kirby, D.	(Southampton)	Schenzel, P.	(Halle)
Lascoux, A.	(Paris)	Sharp, R.Y.	(Sheffield)
Lech, C.	(Stockholm)	Sharpe, D.W.	(Sheffield)

Shimoda, Y. (Tokyo)
Simis, A. (Rio de Janeiro)
Stafford, J.T. (Cambridge)
Strano, R. (Catania)
Strooker, J.R. (Utrecht)
Szpiro, L. (Paris)
Valla, G. (Genova)
Vámos, P. (Sheffield)
Vasconcelos, W.V. (New Brunswick)
Vetter, U. (Osnabrück/Vechta)
Weyman, J. (Toruń)
Wilson, P.M.H. (Cambridge)
Wiseman, A.N. (Sheffield)
Woodcock, C.F. (Canterbury)
Zakeri, H. (Sheffield)
71

PART I

THE LOCAL HOMOLOGICAL CONJECTURES, BIG COHEN-MACAULAY MODULES, AND RELATED TOPICS

THE SYZYGY PROBLEM: A NEW PROOF AND HISTORICAL PERSPECTIVE

E.G.EVANS and PHILLIP GRIFFITH

This article is a brief survey of the results that led up to our solution of the syzygy problem [8] as well as a discussion of our solution of that problem as it was generalized during the Durham Symposium following conversations with Bruns, Foxby, Hochster, Huneke, Roberts, Szpiro and others.

From our view the syzygy problem began with three separate and unrelated events in 1969. One was the submission of Hackman's (so far unpublished) Ph.D. Thesis [11] "Exterior powers and homology", which contains on the penultimate page the statement of the problem. Using techniques of his thesis he proved that regular local rings of dimension three are unique factorization domains and writes as follows.

"In order to prove the general UFD Theorem along the same lines, one would need the following theorem: If the projective dimension of M is r and M admits a projective resolution

$$0 \to F_r \to F_{r-1} \to \ldots \to F_0 \to M \to 0,$$

where the F_i, $i < r$, are finitely generated free modules and F_r is an admissible projective module, then

$$\sum_{j=k}^{r} (-1)^{j-k} rk(F_j) \geq k$$

for all $k \leq r-1$; in other words, the k-th syzygy module of the resolution is of rank $\geq k$."

A second event in 1969 was the appearance of Auslander and Bridger's monograph [1] "Stable module theory" which contains among many other results an explicit criterion for a module of finite projective dimension to be a k-th syzygy. Because they were interested in a different circle of ideas and were writing in more generality than we need, it is somewhat difficult to give a concise and explicit reference to their connection with the syzygy problem.

Perhaps the best one is their Theorem 4.25 [1; p.127] which proves that, if M has finite Gorenstein dimension, then seven statements concerning M are equivalent. We need to remark that a module of finite projective dimension has the same finite Gorenstein dimension. Two of the seven equivalent conditions are the following.

(b) There is an exact sequence $0 \to M \to P_{k-1} \to \dots \to P_1 \to P_0$ where the P_i are projective (that is, M is a k-th syzygy).

(f) For each prime ideal P every R_P regular sequence of length $\leq k$ is M_P-regular.

It is interesting to note that this memoir is an outgrowth of Bridger's Ph.D. Thesis (Brandeis).

Thirdly in 1969, Peskine and Szpiro circulated a preprint of what was to become the first two chapters of their remarkable article [17] "Dimension projective finie et cohomologie locale" which was their joint Ph.D. Thesis. This article established their famous intersection property from which they settled many open problems for rings containing a field. The intersection property (cf. [17; p.84]) is defined as follows. Let A be a local ring and M a nonzero A-module of finite projective dimension. Then A has the intersection property if for all finitely generated A-modules N, one has $\dim N_P \leq \text{pd}.M$ for each prime P which is minimal in $\text{Supp}\, M \cap \text{Supp}\, N$. Peskine and Szpiro [17; Theorem 2.1, p.86] prove that, if A is a local ring containing a field, then all finitely generated A-modules of finite projective dimension have the intersection property. It is of some interest to note that the weaker inequality that $\text{depth}\, N_P \leq \text{pd}.M$, for prime ideals P minimal in $\text{Supp}\, M \cap \text{Supp}\, N$, is much easier to prove and, indeed, follows from their Lemme d'Acyclicité [17; p.55] which states as follows.

Let A *be a local ring and let* $0 \to L_s \to \dots \to L_0 \to 0$ *be a complex of finitely generated* A*-modules. Suppose that*

(1) $\text{depth}\, L_i \geq i$ *and*

(2) $\text{depth}\, H_i(L.) = 0$ *or* $H_i(L.) = 0$.

Then $H_i(L.) = 0$ *for all* $i \geq 1$.

The proof of the weaker inequality can be deduced from the lemma quite easily. First one replaces A by A_P, M by M_P and N by N_P, where P is a minimal prime in $\operatorname{Supp} M \cap \operatorname{Supp} N$. This is harmless since the projective dimension of M can only get smaller. Thus P is the only prime in $\operatorname{Supp} M \cap \operatorname{Supp} N$. Therefore all of the homology modules $\operatorname{Tor}_i^A(M,N)$ (including $M \otimes N$) have finite length. Let $0 \to L_d \to \ldots \to L_0 \to M \to 0$ be a minimal projective resolution of M. If $\operatorname{depth} N > \operatorname{pd.} M$, we apply the Lemme d'Acyclicité to the complex

$$0 \to L_d \otimes N \to \ldots \to L_0 \otimes N \to 0$$

to conclude that the complex is acyclic and, further, using a standard depth counting argument for long exact sequences, we conclude that the complex is too short to have its zero-th homology nonzero and of finite length. Thus, the real difficulty Peskine and Szpiro overcame was to be able to establish the stronger inequality one obtains by replacing depth by dimension.

It is interesting to recall that all three of these early contributions were connected with Ph.D. Theses. It is also noteworthy that the clear vision provided by hindsight has shown that the essential ingredient in the final solution came from a better understanding of the Lemme d'Acyclicité. Perhaps we should also remark that by 1970 Hartshorne [12] was developing a theme in algebraic geometry concerning questions on vector bundles of small rank and complete intersections which, as it turns out, was rather closely related to subsequent work on the syzygy problem (cf. Bruns-Evans-Griffith [4] and Hartshorne [12],[13]).

The next collection of results came in the mid 1970's. Lebelt [15], [16] tried to extend Hackman's ideas in order to settle the syzygy problem. Briefly, he wanted to compare a projective resolution of $\wedge^k M$ with that of M and then use it to show that, if M was a module of rank k which was also a "large" syzygy, then $\wedge^k M$ would necessarily be a reflexive ideal. Thus it would follow that both $\wedge^k M$ and M would be free. Lebelt managed to obtain several interesting results along these lines which provided the first explicit and affirmative results on the problem.

In 1976 Bruns [2] established that the bound given by Hackman [11] was the best possible in a very general sense. He showed that, if R is a Cohen-Macaulay domain and if M is a k-th syzygy of finite projective dimension of rank exceeding k, then M contains a free submodule F such that M/F is a k-th syzygy of rank exactly k. The proof consists of two parts. One part uses the basic element results of Eisenbud and Evans [5] to show that, if M has rank exceeding k, then there is an $x \in M$ such that x is a minimal generator of M_P for all prime ideals P of height k. One should note that, if k is less than dim R, then x can be taken in $\underline{m}M$, where $\underline{m}$ denotes the maximal ideal of R. The second part of the proof uses the criterion of Auslander and Bridger [1] previously mentioned to show that, if x is as described above and if M is a k-th syzygy, then M/Rx is again a k-th syzygy. Bruns' proof concludes by descending induction on the rank of M. One should note that, if M is not free, then M/F also is not free and, if M is free with $k < \dim R$, then one can take $F \subset \underline{m}M$ so that M/F has projective dimension one. In particular, non-free k-th syzygies of rank k exist in abundance as soon as non-free k-th syzygies exist. However, there is no evidence at this point to suggest that one can expect non-free k-th syzygies of rank less than k.

One may analyze the second part of Bruns' proof as follows. If M is a k-th syzygy and P is a prime ideal of height k, then M_P is a free R_P-module. If x is a minimal generator of M_P, then there is a map of M to R sending x outside P as a result of M_P being free. Thus the ideal of images of such an x, defined by

$$O_M(x) = \{f(x) \mid f \in \operatorname{Hom}(M,R)\},$$

cannot be contained inside any prime ideal of height k. Conversely, if M is a k-th syzygy and if $O_M(x)$ is an ideal of height greater than k, then M/Rx is again a k-th syzygy. This remark led Eisenbud and Evans [6] to investigate how the height of $O_M(x)$ depends on the rank of M. For if $M = R^k$ is free of rank k and $x \in \underline{m}R^k$, then $x = \langle r_1,\ldots,r_k\rangle$ with $r_i \in \underline{m}$ for each i and $O_M(x) = (r_1,\ldots,r_k)$. It follows from the Krull altitude theorem that the height of $O_M(x)$ is at most k. Eisenbud and Evans [6] established this result for arbitrary M of rank k as long as R was a domain which contained a field. These assumptions on R were later removed by Bruns [3]. Evans [7] further remarked that the preceding result showed that, if M was a k-th syzygy of rank k and if $x \in \underline{m}M$, $x \neq 0$, then M/Rx is never a k-th syzygy. This observation gave further evidence that the bound on rank was the correct one.

On another tack Peskine and Szpiro [18] and, independently, Roberts [20] strengthened the Lemme d'Acyclicité in that depth was replaced by dimension. More exactly they proved [18],[20] that, if R contains a field and if $0 \to L_s \to \ldots \to L_0 \to 0$ is a complex of free R-modules having finite length homology, then s is at least the dimension of R (assuming the zero-th homology is nonzero). This gave added strength to the philosophy that many inequalities using depth as a bound might remain true when dimension is used.

In 1974 Hochster [14] in his remarkable memoir "Topics in the homological theory of modules over commutative rings" validated the above philosophy. He constructed for any local ring R, which contained a field, a maximal Cohen-Macaulay module (not necessarily finitely generated). These modules, by their existence, allowed one to replace depth by dimension in many existing inequalities. In particular, the earlier result of Peskine and Szpiro [17] and Roberts [19] could be improved in this way.

Now we shall examine our original proof of the syzygy problem [8]. Our first version contained various restrictions on the ring R involved. To be precise we needed that R contains a field in order that factor rings of R have maximal Cohen-Macaulay modules (as discussed above). We needed that R is a domain so that rank is well defined and we needed the Cohen-Macaulay property in order to apply the Auslander-Bridger criterion for a module to be a k-th syzygy. During discussions at the Symposium it became clear that the last two assumptions on the ring R could be dropped. Firstly, if M has a finite free resolution, then one can define the rank of M to be the alternating sum of the Betti numbers. Secondly, if R fails to be Cohen-Macaulay, then the depth of R is smaller and it becomes even more difficult for a module to be a k-th syzygy of finite projective dimension. It also became apparent during the course of the Symposium that one could modify the proofs of Peskine and Szpiro [18] and Roberts [20] in order to provide a version of the Lemme d'Acyclicité suitable for one of the crucial steps in our proof. The final deviation in our new proof of the syzygy theorem (as presented below) occurs in the use of our result on order ideals of minimal generators [9]. There is a slight drawback here of a technical nature in that we established our statement concerning the heights of order ideals of minimal generators under the assumption that the residue field is algebraically closed. However we shall state a slight modification of this result which is suitable for our needs here. Except for this our proof is rather easier than our original one in addition to being more general.

Definition. Let R be a local ring and let M be an R-module having a finite free resolution $0 \to F_s \to \dots \to F_1 \to F_0 \to M \to 0$. Then the *rank of* M is defined by $\operatorname{rank} M = \sum_{i=0}^{s} (-1)^i \operatorname{rank} F_i$. Of course this definition agrees with the usual one in case R is a domain or as generalized by Bruns [3].

LEMMA. *Let* $(R,\underline{m},\underline{k})$ *be a* <u>*complete*</u> *local ring and let* M *be a finitely generated non-free* R*-module of finite projective dimension and having rank* r. *Then there is a* <u>*finite*</u> *faithfully flat residue field extension* $(R',\underline{m}',\underline{k}')$ *of* $(R,\underline{m},\underline{k})$ *and a minimal generator* x *of* $R' \otimes M$ *having order ideal of height less than or equal to* r.

Proof. The argument is essentially that given in [9]. We fix a minimal generating set $e_1,\dots,e_t$ of M, noting that $t > r$ since M is not free. Next we form the polynomial ring $S = R[X_1,\dots,X_t]$ and the S-module $N = S \otimes M'$. Considering the element $v = \sum_{i=1}^{t} X_i e_i$ of N we have from Bruns [3] that the height of the order ideal $O_N(v)$ does not exceed $r = \operatorname{rank}_R M = \operatorname{rank}_S N$. Let P be a prime ideal of height r containing $O_N(v)$. As noted in [9], the height of the ideal $P + \underline{m}S$ is at most $r + \dim R$, which is less than $\dim S$. Hence there is a maximal ideal Q (actually infinitely many) of S which contains $P + \underline{m}S$ and which corresponds modulo $\underline{m}$ to a maximal ideal of $\underline{k}[X_1,\dots,X_t]$ other than $(X_1,\dots,X_t)$. The remainder of our proof is exactly that given in [9] provided Q is of the form $Q = (\underline{m}, X_1 - a_1,\dots,X_t - a_t)$ where $a_i \in R$, the element $x = \sum_{i=1}^{t} a_i e_i$ being the desired element. However, this can be achieved after a finite extension of $\underline{k}$, since only a finite number of equations are involved. Since R is complete and local we may extend its residue field (finitely) so that the resulting ring $(R',\underline{m}',\underline{k}')$ is complete and local as well as being a faithfully flat extension of R. Thus we can achieve a maximal ideal Q of the desired form by passing to a suitable finite extension $(R',\underline{m}',\underline{k}')$. □

The more general solution to the syzygy problem now follows.

THEOREM. *Let* $(R,\underline{m},\underline{k})$ *be a local ring containing a field. Let* M *be a finitely generated* k*-th syzygy of rank* r. *If* r *is less than* k, *then* M *is free.*

Proof. We may assume that M is locally free on the punctured spectrum of R since otherwise we may localize to a ring of smaller dimension while keeping M a k-th syzygy of finite projective dimension. We may also assume that R is complete and that M contains a minimal generator x having order ideal $O_M(x)$ of height $\le r$. This follows from the fact that the syzygy problem remains unchanged under faithfully flat change of base as well as the preceding lemma.

Let $0 \to F_s \to \dots \to F_1 \to F_0 \to M \to 0$ be a minimal projective resolution of M. Then the complex $F. \otimes R/I$, where $I = O_M(x)$, has homology $\mathrm{Tor}_i^R(M,R/I)$ of finite length for $i > 0$, since M is locally free on the punctured spectrum of R. Moreover, the element $x + IM$ in M/IM is nonzero, since x is not even in $\underline{m}M$, but generates a submodule of finite length, since $x \in IM$ on the free locus of M. It follows that the zero-th homology of $F.\otimes R/I$, namely M/IM, has depth zero.

It remains to show that s is at least $\dim R/I$. For if $s \ge \dim R/I$, then $\mathrm{pd}.M = s \ge \dim R/I \ge \dim R - r$, since complete local rings are catenary. Hence we obtain the inequality $\mathrm{pd}.M + r \ge \dim R$. On the other hand one has that $\mathrm{pd}.M + k \le \dim R$ which together with the previous inequality gives that $r \ge k$ as desired. The second inequality $\mathrm{pd}.M + k \le \dim R$ actually can be improved to $\mathrm{pd}.M + k \le \mathrm{depth}\, R \le f$, where $f = \min\{\dim R/P \mid P \in \mathrm{Ass}\, R\}$. We isolate this last step as a separate lemma (called by some the "New New" Intersection Conjecture). Note that S plays the role of R/I in the lemma. □

LEMMA. *Let* S *be a local ring containing a field. Let* $\mathbb{F}. : 0 \to F_s \to \dots \to F_0 \to 0$ *be a complex of finitely generated free* S*-modules such that* $H_i(\mathbb{F}.)$ *is of finite length for* $i > 0$ *and* $H_0(\mathbb{F}.)$

has a minimal generator which generates a nonzero submodule of finite length. Then s ≥ dim S.

Remark. Note that by the Lemme d'Acyclicité, if s < depth S, $\mathbb{F}_.$ would be exact and then in turn too short to be a free resolution of the module $H_0(\mathbb{F}_.)$ of depth zero. Our proof merely has to strengthen the inequality in the Lemma d'Acyclicité from depth S to dim S.

Proof of the Lemma. Suppose in fact that s < dim S. Let C be a maximal Cohen-Macaulay module over S. Then we obtain a complex $\mathbb{F}_. \otimes C$ with $H_i(\mathbb{F}_. \otimes C)$ annihilated by an $\underline{m}$-primary ideal for i > 0. Moreover, if c is any element in $C - \underline{m}C$, then $x \otimes c$ represents an element in $H_0(\mathbb{F}_. \otimes C) - \underline{m}H_0(\mathbb{F}_. \otimes C)$ and consequently is nonzero. Moreover, $x \otimes c$ is annihilated by an $\underline{m}$-primary ideal since $x \in H_0(\mathbb{F}_.)$ has this property. Thus $H_0(\mathbb{F}_. \otimes C)$ has depth zero. Now one applies the Lemme d'Acyclicité (see [8] or [10] for an appropriate version) to obtain the contradiction that the complex is too short to have such properties. Thus it must be that s ≥ dim S. □

The final argument here is a bit brief. This is for two reasons. First, the essential details are in our original version [8]. More interestingly, the above lemma was already proved (but not stated) by Roberts [20] in his proof that, if the homology of $\mathbb{F}_.$ has finite length, then s ≥ dim S. There Roberts carefully analyzed what was needed in a separate lemma. That lemma covers our case. This theorem (and the nearly identical one of Peskine and Szpiro) really is obtained from a better understanding of the 1969 Lemme d'Acyclicité. Thus, in some sense, the essential part of the argument was known for some time although the reduction of the question to that case was not apparent (to us) until after our original proof. Perhaps more importantly, this result gives yet another application of this circle of ideas to commutative ring theory. In particular the proof of this case is made no simpler if we start with a Cohen-Macaulay (or even regular) local ring while some of the earlier applications of this technique were already understood in such cases. Thus one is enticed to look for more applications.

References

1. M.Auslander and M.Bridger, Stable module theory, Mem.Amer. Math.Soc. 94 (American Mathematical Society, Providence, 1969).

2. W.Bruns, "'Jede' endliche freie Auflösung ist freie Auflösung eines von drei Elementen erzeugten Ideals", J.Algebra, 39 (1976), 429-439.

3. W.Bruns, "The Eisenbud-Evans generalized principal ideal theorem and determinantal ideals", preprint, University of Osnabrück at Vechta.

4. W.Bruns, E.G.Evans and P.Griffith, "Syzygies, ideals of height two and vector bundles", J.Algebra, 67 (1980), 143-162.

5. D.Eisenbud and E.G.Evans, "Generating modules efficiently: theorems from algebraic K-theory", J.Algebra, 27 (1973), 278-305.

6. D.Eisenbud and E.G.Evans, "A generalized principal ideal theorem", Nagoya Math.J., 62 (1976), 41-53.

7. E.G.Evans, "Position generale et position speciale en algèbre commutative", Colloque d'Algèbre Commutative Rennes 1976.

8. E.G.Evans and P.Griffith, "The syzygy problem", Ann. of Math., 114 (1981), 323-333.

9. E.G.Evans and P.Griffith, "Order ideals of minimal generators", preprint, University of Illinois at Urbana-Champaign.

10. H.-B. Foxby, "On the μ^i in a minimal injective resolution, II", Math.Scand., 41 (1977), 19-44.

11. P.Hackman, "Exterior powers and homology", Ph.D.Thesis, University of Stockholm, 1969.

12. R.Hartshorne, "Varieties of small codimension in projective space", Bull.Amer.Math.Soc., 80 (1974), 1017-1032.

13. R.Hartshorne, "Algebraic vector bundles on projective spaces: a problem list", Topology, 18 (1979), 117-128.

14. M.Hochster, Topics in the homological theory of modules over commutative rings, C.B.M.S.Regional Conference Series in Mathematics 24 (American Mathematical Society, Providence, 1976).

15. K.Lebelt, "Zur homologischen Dimension ausserer Potenzen von Moduln", Arch.Math. (Basel), 26 (1975), 595-601.

16. K.Lebelt, "Freie Auflösungen ausser Potenzen", Manuscripta Math., 23 (1977), 341-355.

17. C.Peskine and L.Szpiro, "Dimension projective finie et cohomologie locale", Publications Mathématiques 42 (Institut des Hautes Études Scientifiques, Paris, 1973), pp.47-119.

18. C.Peskine and L.Szpiro, "Syzygies et multiplicités",C.R.Acad. Sci. Paris Sér.A, 278 (1974), 1421-1424.

19. P.Roberts, "Two applications of dualizing complexes over local rings", Ann.Sci.École Norm.Sup. (4), 9 (1976), 103-106.

20. P.Roberts, "Cohen-Macaulay complexes and an analytic proof of the new intersection conjecture", J.Algebra, 66 (1980), 220-225.

Department of Mathematics,
University of Illinois,
Urbana,
Illinois 61801, U.S.A.

THE THEORY OF HOMOLOGICAL DIMENSIONS OF COMPLEXES

HANS-BJØRN FOXBY

The object of this article is to comment on the theory of homological dimensions of bounded complexes of modules over a Noetherian commutative ring. When a module is thought of as a complex concentrated in degree zero, this theory extends parts of the theory of homological dimensions of modules; thus the idea is just to replace "modules" by "complexes of modules" whenever possible. This idea is not new. For example, it is essential, and used extensively, in the seminar notes "Residues and duality" [Ha] and "Théorie des intersections et théorème de Riemann-Roch" [SGA6].

Now let X be a bounded complex of modules over a ring A. Thus

$$X = 0 \to X^r \to X^{r+1} \to \ldots \to X^s \to 0.$$

The homological dimensions mentioned in the title are the following:

pd X, the projective dimension of X;

fd X, the flat dimension of X;

id X, the injective dimension of X;

depth X, the depth of X.

(In order that we can define the depth the ring must be local.) In addition to these there is another important dimension, namely

dim X, the Krull dimension of X.

Let me point out some reasons why one would (or could) want to study these concepts.

(1) *It is possible*

As an example let me define $\mathrm{id}_A X$. For an integer n we say that $\mathrm{id}_A X \leq n$ if there exists a quasi-isomorphism ϕ: $X \to I$ where I is a bounded complex of injective modules such that $I^\ell = 0$ for all $\ell > n$. (That ϕ is a *quasi-isomorphism* means that ϕ induces isomorphisms $H^\ell(\phi)$: $H^\ell(X) \to H^\ell(Y)$ in cohomology for all ℓ.)

This extends the definition of injective dimension of a module M: if M has an injective resolution I then the inclusion $M \to I^0$ induces a morphism of complexes $M \to I$ which is a quasi-isomorphism; on the other hand if $M \to I$ is a quasi-isomorphism then $H^{\ell}(I) = 0$ for $\ell \neq 0$, and so it is easy to split off the irrelevant injective modules sitting in negative degrees in I and thus obtain an injective resolution of M.

The definitions of the homological dimensions and the Krull dimension can be found in [F_3].

(2) *These dimensions behave very nicely (in fact, just as for modules)*

Let me mention two examples.

(2.1) They can be characterized by the vanishing of the functors Ext and Tor. For example, id $X \leq n$ if and only if $\mathrm{Ext}^{\ell}(M,X) = 0$ for all $\ell > n$ and modules M. Here Ext^{ℓ} is the ℓ-th hyperExt: see [Ha; Chapter I, §6]. (Thus $\mathrm{Ext}^{\ell}(M,X)$ is not the complex obtained by applying the additive (module) functor $\mathrm{Ext}^{\ell}(M, \ )$ to the complex X.)

(2.2) Let the ring A be local (and this includes the requirement that it be commutative and Noetherian; all rings in this article possess non-zero multiplicative identities). For finitely generated modules M and N there are many relations between the various dimensions. The next three examples of such relations are among the most well-known (and oldest).

(2.2.a) pd M + depth M = depth A if pd M $< \infty$.

(2.2.b) id M = depth A if id M $< \infty$.

(2.2.c) $\sup\{\ell \mid \mathrm{Ext}^{\ell}(M,N) \neq 0\}$ = depth A - depth M if id N $< \infty$.

See [K; Theorems 173, 214, 217, 218 and 219]. An important tool in the proofs of these results is the notion of a regular sequence (of ring elements).

These relations, and many others, hold also for complexes, that is, when the modules M and N are replaced by bounded complexes X and Y with $H^{\ell}(X)$ and $H^{\ell}(Y)$ finitely generated for all ℓ: see [F_3].

(3) *The proofs get simpler (in a certain sense)*

Let me mention two examples where the results for complexes are simpler to prove than the corresponding results for modules. Of course, the proofs are only simpler when one knows some basic facts about complexes, such as the definitions and basic properties of the hyperExt and hyperTor. The theory in [Ha; Chapter I] would provide sufficient knowledge, but much less suffices.

(3.1) To prove formulae like (2.2.a,b,c) one could develop a theory of regular sequences for complexes and then generalize the known proofs. However, completely different methods are available. There are three canonical functorial isomorphisms:

(3.1.a) $\mathrm{Hom}(X\otimes Y,Z) \simeq \mathrm{Hom}(X,\mathrm{Hom}(Y,Z))$;

(3.1.b) $X\otimes \mathrm{Hom}(Y,Z) \simeq \mathrm{Hom}(\mathrm{Hom}(X,Y),Z)$;

(3.1.c) $\mathrm{Hom}(X,Y)\otimes Z \simeq \mathrm{Hom}(X,Y\otimes Z)$.

Here X,Y and Z are bounded complexes. For (3.1.b,c) to hold these complexes have to satisfy special requirements: see [Ha] and [F_1]. These isomorphisms, together with a very simple trick (concerning computation of cohomology modules), give many results directly, including (2.2.a,b,c): see [F_1] and [F_3].

(3.2) In [F_2] it was proved that if M is a (not necessarily finitely generated) module over a local ring A such that $M \otimes k \neq 0$ where k is the residue class field, then

(3.2.a) $\dim A \leq \dim M + \mathrm{fd}\, M$,

provided that A contains a field (as a subring).

In the later paper [F_3] it was proved that if X is a bounded complex such that $\mathrm{Tor}_\ell(k,X) \neq 0$ for some ℓ, then

(3.2.b) $\dim A \leq \dim X + \mathrm{fd}\, X$,

provided that A contains a field (and is local).

The latter result is stronger than the first one. In the proof of the latter the theory of homological dimensions of complexes was used. The proof is simpler than the proof of the first result in the sense that [F_2;(6.3)] and [G] have become superfluous.

The reason for the requirement that the ring A contain a field is that Hochster's big Cohen-Macaulay modules from [Ho_1] play a key role in the proof, and these modules are only (so far) available in this case (and some other cases).

(4) *The results get better*

Sometimes it is interesting to know that a particular result holds also for complexes. For example, a special case of (3.2.b) states that if each module F_ℓ in the complex

$$F = 0 \to F_s \to F_{s-1} \to \dots \to F_0 \to 0$$

is finitely generated and free and if there exists an integer t such that $\dim H_\ell(F) \le t + \ell$ for all ℓ, then $\dim A \le s + t$, provided that F is not exact and the local ring A contains a field. (For this complex F we have fd $F \le s$ and

$$\dim F = \sup_{\ell \in \mathbb{Z}}(\dim H_\ell(F) - \ell) \le t$$

by definitions and assumptions.)

This result is the NEW INTERSECTION THEOREM of Peskine, Szpiro, Roberts, and Hochster: see [PS], [R], [Ho_2] or [F_3].

(5) *This theory is ideal for working with dualizing complexes*

Dualizing complexes certainly are important tools. Examples can be found in, for example, [Ha], [R], [S], [F_1], [F_2] and [F_3].

To describe vaguely what dualizing complexes are good for assume that A is local with residue field k. The useful Matlis duality, that is, duality with respect to E(k), has the disadvantage that the dual of a finitely generated module in general is not finitely generated. Dualizing complexes can be thought of as one way to make up for this, but the price that has to be paid is that one has to work with complexes of modules instead of just modules.

Some applications of dualizing complexes are described in [F_4] (in these proceedings). Let me describe another one.

Assume that the ring R has finite left finitistic projective dimension t, that is, left-pd $M \le t$ whenever M is a left R-module with left-pd $M < \infty$. Then Jensen has proved that left-pd $M < \infty$ for all flat modules M: see [J].

Let A be a commutative Noetherian ring. For all non-negative integers $n \le \dim A$ Bass has constructed a module M with pd M = n (see [B]) and Bass stated that he knew of no example where

$\dim A < \text{pd } M < \infty$. Gruson and Raynaud settled this by proving that

$$\text{pd } M < \infty \Rightarrow \text{pd } M \leq \dim A:$$

see [GR]. (Note that M is not supposed to be finitely generated.)

The proof is not easy. However, if A has a dualizing complex (as most rings have) then it is easy to prove even that

$$\text{fd } M < \infty \Rightarrow \text{pd } M \leq \dim A:$$

see [F_1]. From this Gruson and Raynaud's result follows (when A has a dualizing complex) as well as Jensen's (in this restricted case).

Acknowledgement

The author has been supported, in part, by the Danish Natural Science Research Council.

References

[B] H. Bass, "Injective dimension in Noetherian rings, II", Trans. Amer. Math. Soc., 88 (1958), 194-206.

[F_1] H.-B. Foxby, "Isomorphisms between complexes with applications to the homological theory of modules", Math. Scand., 40 (1977), 5-19.

[F_2] H.-B. Foxby, "On the μ^i in a minimal injective resolution, II", Math. Scand., 41 (1977), 19-44.

[F_3] H.-B. Foxby, "Bounded complexes of flat modules", J. Pure Appl. Algebra, 14 (1979), 149-172.

[F_4] H.-B. Foxby, "Complexes of injective modules", Commutative algebra: Durham 1981, London Mathematical Society Lecture Notes 72(ed. R.Y. Sharp, Cambridge University Press, Cambridge, 1982), pp. 18-31.

[G] P. Griffith, "A representation theorem for complete local rings", J. Pure Appl. Algebra, 7 (1976), 303-315.

[GR] L. Gruson and M. Raynaud, "Critères de platitude", Invent. Math., 13 (1971), 1-89.

[Ha] R. Hartshorne, Residues and duality, Lecture Notes in Mathematics 20 (Springer, Berlin, Heidelberg, New York, 1966).

[Ho_1] M. Hochster, "The equicharacteristic case of some homological conjectures on local rings", Bull. Amer. Math. Soc., 80 (1974), 683-686.

[Ho_2] M. Hochster, "Big Cohen-Macaulay modules and algebras and embeddability in rings of Witt vectors", Proceedings of the conference on commutative algebra, Queen's University, Kingston, Ontario, 1975 (Queen's University Papers on Pure and Applied Mathematics No. 42, 1975), pp. 106-195.

[J] C.U. Jensen, "On the vanishing of $\varprojlim^{(i)}$", J. Algebra, 15 (1970), 151-166.

[K] I. Kaplansky, Commutative rings (Allyn and Bacon, Boston, 1970).

[PS] C. Peskine and L. Szpiro, "Syzygies et multiplicités", C. R. Acad. Sci. Paris Sér. A, 278 (1974), 1421-1424.

[R] P. Roberts, "Two applications of dualizing complexes over local rings", Ann. Sci. École Norm. Sup. (4), 9 (1976), 103-106.

[S] P. Schenzel, "Dualizing complexes and systems of parameters", J. Algebra, 58 (1979), 495-501.

Department of Mathematics,
University of Oklahoma,
Norman, Oklahoma 73019,
U.S.A.

and (from 1982)

Matematisk Institut,
Københavns Universitet,
Universitetsparken 5,
DK 2100 København Ø, Denmark.

COMPLEXES OF INJECTIVE MODULES

HANS-BJØRN FOXBY

The object of this article is to describe the modules in a bounded below complex of injective modules

$$I = 0 \to I^i \to I^{i+1} \to \dots \to I^\ell \to \dots .$$

In this complete generality it is of course impossible to formulate reasonable assertions (since the modules could be any injective modules and all the differentials could be zero). So we have to impose restrictions. We assume that the cohomology modules $H^\ell(I)$ vanish for sufficiently large ℓ. For example, I could be an injective resolution of a module.

However we will have to impose further restrictions. For example, take any *exact* complex J of injective modules. Then the complex $I \oplus J$ has also only finitely many non-vanishing cohomology modules. Therefore we shall restrict the study to the case where I is *minimal* in the sense that $\mathrm{Ker}(I^\ell \to I^{\ell+1})$ is an *essential* submodule of I^ℓ for all ℓ. For example, I could be a *minimal* injective resolution of a module.

The ring we are working over is denoted by A and is supposed to be *commutative* and *Noetherian*. (The word "ring" always incorporates the existence of a non-zero multiplicative identity.) Thus the injective module I^ℓ decomposes into the direct sum of indecomposable injective modules, that is,

$$I^\ell \simeq \coprod_{\underline{p} \in \mathrm{Spec}\, A} E(A/\underline{p})^{(\mu_A(\underline{p},I))}$$

for some cardinal numbers $\mu_A(\underline{p},I)$. That is, to describe I^ℓ it suffices to describe the numbers $\mu_A^\ell(\underline{p},I)$. It turns out that

$\mu_A^\ell(\underline{p},I) = \mu_{A_{\underline{p}}}^\ell(\underline{p}_{\underline{p}},I_{\underline{p}})$, and so it suffices to assume that A is local and seek information on $\mu_A^\ell(I) = \mu_A^\ell(\underline{m},I)$ where $\underline{m}$ denotes the maximal ideal in A. The best results are obtained when $H^\ell(I)$ is *finitely generated* (*f. g.*) for all ℓ, because then $\mu^\ell(I) < \infty$ for all ℓ.

Next follows a list of some of the known results about $\mu^\ell(I)$. The references are Bass [B], Roberts [R_1], [R_2], Foxby and Thorup [FT], and Foxby [F_1], [F_2], [F_3], but further comments on the results will be provided in the subsequent sections of the article. To facilitate the presentation of these results here in the introduction let us assume that I is a minimal injective resolution of a module M (although this restriction is not necessary). The complex I is uniquely determined by M, and so we write $\mu_A^\ell(M) = \mu_A^\ell(I)$.

There is a close connection between I and a minimal *free* resolution L of M when M is f. g. If $\beta_\ell^A(M)$ denotes the rank of L_ℓ, then

$$\beta_\ell^A(M) = \sum_{p\in\mathbb{Z}} \mu_A^p(A)\mu_A^{p-\ell}(M)$$

for all ℓ if I is bounded, and

$$\mu_A^\ell(M) = \sum_{p\in\mathbb{Z}} \mu_A^p(A)\beta_{p-\ell}^A(M)$$

for all ℓ if L is bounded: see (2.6).

Let B be a flat A-algebra such that $C = B/\underline{m}B$ is non-trivial. Then

$$\mu_B^\ell(M\otimes_A B) = \sum_{p\in\mathbb{Z}} \mu_A^p(M)\mu_C^{\ell-p}(C)$$

for all ℓ: see (3.1).

If $\underline{p} \in \operatorname{Spec} A$ and $v = \dim A/\underline{p}$, then $\mu_A^\ell(\underline{p},M) \le \mu_A^{\ell+v}(M)$ for all ℓ: see (5.1).

It is well known that $\operatorname{depth}_A M \le \ell \le \operatorname{id}_A M$ if $\mu^\ell(M) > 0$. Here $\operatorname{id}_A M$ denotes the injective dimension of M (which might be infinite). The opposite holds also if M is f. g.. That is, $\mu^\ell(M) > 0$ if $\operatorname{depth}_A M \le \ell \le \operatorname{id}_A M$ ($\le \infty$): see (6.2). In many cases even the following holds: $\mu^\ell(M) \ge 2$ if $\operatorname{depth}_A M < \ell < \operatorname{id}_A M$ ($\le \infty$) (see (6.3)). As a consequence the ring A is Cohen-Macaulay (and therefore Gorenstein) if $\mu^d(A) = 1$ when $d = \dim A$: see (6.4).

Even if M is not f. g. we can say something about the vanishing of $\mu^\ell(M)$. If $\ell \ge \dim M$ and $\mu^\ell(M) = 0$ then $\mu^m(M) = 0$ for $m \ge \ell$, and if $m > \operatorname{depth} A$ and $\mu^m(M) > 0$ then $\mu^\ell(M) > 0$ for $\ell \geq m$: see (8.1).

The *small support*, supp M, of M is the set of prime ideals $\underline{p}$ with $\mu_A^\ell(\underline{p}, M) > 0$ for some ℓ. This is a subset of Supp M, the usual support of M, and supp M = Supp M if M is f. g.. In general supp M has nicer properties than Supp M. For example,

$$\operatorname{supp} M \cap \operatorname{supp} N = \bigcup_{\ell \geq 0} \operatorname{supp} \operatorname{Tor}_\ell(M,N):$$

see (7.1).

1. *Notation and generalities*

The symbols X, Y and Z denote *bounded* complexes of A-modules such that $H^\ell(X)$ and $H^\ell(Y)$ (but not $H^\ell(Z)$) are f. g. for all ℓ. The The complex I is always assumed to be *bounded below* and consisting of *injective* A-modules (as in the introduction). The symbols F and L denote *bounded above* complexes of, respectively, *flat* modules and *f. g. free* modules. The ring A is always supposed to be *commutative* and *Noetherian*. When A is supposed to be local, $\underline{m}$ denotes the maximal ideal and $k = A/\underline{m}$, the field of residue classes.

1.1 LEMMA. *There exist complexes* E *and* J *such that* $I \simeq E \oplus J$, *the complex* E *is minimal, and the complex* J *is exact, that is,* J *is of the form*

$$0 \longrightarrow K^i \xrightarrow{\begin{bmatrix} 1 \\ 0 \end{bmatrix}} \begin{matrix} K^i \\ \oplus \\ K^{i+1} \end{matrix} \xrightarrow{\begin{bmatrix} 0 & 1 \\ 0 & 0 \end{bmatrix}} \begin{matrix} K^{i+1} \\ \oplus \\ K^{i+2} \end{matrix} \xrightarrow{\begin{bmatrix} 0 & 1 \\ 0 & 0 \end{bmatrix}} \cdots .$$

The complex E *is uniquely determined up to isomorphism of complexes.*

Proof. Each I^ℓ decomposes into the direct sum of submodules $I^\ell = E^\ell \oplus K^{\ell-1} \oplus \overline{K}^\ell$ in such a way that each differential $I^\ell \to I^{\ell+1}$ is of the form

$$\begin{bmatrix} \phi^\ell & 0 & 0 \\ 0 & 0 & \psi^\ell \\ 0 & 0 & 0 \end{bmatrix} : \begin{matrix} E^\ell \\ \oplus \\ K^{\ell-1} \\ \oplus \\ \overline{K}^\ell \end{matrix} \longrightarrow \begin{matrix} E^{\ell+1} \\ \oplus \\ K^\ell \\ \oplus \\ \overline{K}^{\ell+1} \end{matrix} ,$$

where $\psi^\ell : \bar{K}^\ell \to K^\ell$ is an isomorphism, and where Ker ϕ^ℓ is an essential submodule of E^ℓ. (Use induction on ℓ.)

The complex E is minimal, and the inclusion $E \to I$ and the projection $I \to E$ are quasi-isomorphisms (that is, they induce isomorphisms in cohomology). If E′ is another minimal direct summand of I with exact complement, then the composite $E \to I \to E'$ is a quasi-isomorphism, and then it is relatively easy to see that $E \to E'$ is actually an isomorphism. □

1.2 *Definitions and Remarks.* The complex I is an *injective resolution* of Z if there exists a quasi-isomorphism $Z \to I$, and it is a *minimal* injective resolution of Z if, in addition, it is minimal (as defined in the introduction). The complex Z always has a minimal injective resolution (see [Ha; Chapter 1]) and this minimal injective resolution is unique up to isomorphism, by (1.1).

The complex F is a *flat resolution of* Z if there exists a quasi-isomorphism $F \to Z$. The complex Z always has a flat resolution: see [Ha].

When A is local, then L is a *minimal free resolution of* X if there exists a quasi-isomorphism $L \to X$ and $\mathrm{Im}(L^{\ell-1} \to L^\ell) \subseteq \underline{m}L^\ell$ for all ℓ. Such always exists and is unique up to isomorphism.

If $H^\ell(I) = 0$ for ℓ large, say for $\ell \geq s$, then I is the injective resolution of the bounded complex

$$U = 0 \to I^i \to \dots \to I^{s-1} \to \mathrm{Ker}(I^s \to I^{s+1}) \to 0,$$

since the inclusion $U \to I$ is a quasi-isomorphism.

1.3 *Definitions and Remarks.* The two complexes U and V are *equivalent*, and we write $U \approx V$, if there exists a third complex W and quasi-isomorphisms $U \to W$ and $V \to W$. This *is* an equivalence relation: see [Ha; Chapter I].

The *supremum* and *infimum* of a complex U are defined by

$$s(U) = \sup\{\ell \mid H^\ell(U) \neq 0\}$$

and

$$i(U) = \inf\{\ell \mid H^\ell(U) \neq 0\}.$$

The complex U is equivalent to a non-zero *module* M if and only if $s(U) = 0 = i(U)$ (and then $M \simeq H^0(U)$). The complex U is *trivial* if it is equivalent to 0, that is, if it is exact.

1.4 *Definitions and Remarks.* $\underline{\underline{\mathrm{Hom}}}(X,Z)$ denotes the equivalence class of the *complex* $\mathrm{Hom}(X,I)$ whenever I is an injective resolution of Z. This makes sense: see [Ha; Chapter I].

$$\mathrm{Ext}^{\ell}(X,Z) = H^{\ell}(\underline{\underline{\mathrm{Hom}}}(X,Z))$$

is then determined up to isomorphism. This is the ℓ-th hyperExt.

$X \underline{\underline{\otimes}} Z$ denotes the equivalence class of the complex $X \otimes F$ whenever F is a flat resolution of Z, and

$$\mathrm{Tor}_{\ell}(X,Z) = H^{-\ell}(X \underline{\underline{\otimes}} Z),$$

the ℓ-th hyperTor.

If A is local then the ℓ-th *Bass number* $\mu^{\ell}_{A}(Z)$ and the ℓ-th *Betti number* $\beta^{A}_{\ell}(Z)$ are the dimension of, respectively, $\mathrm{Ext}^{\ell}_{A}(k,Z)$ and $\mathrm{Tor}^{A}_{\ell}(k,Z)$ considered as vector spaces over k.

The numbers $\mu^{\ell}_{A}(X)$ and $\beta^{A}_{\ell}(X)$ are finite because $H^{\ell}(X)$ is supposed to be f. g. for all ℓ (see [Ha]) and two formal Laurent series (with integral coefficients) are defined as follows:

$$\mathscr{I}^{X}_{A}(t) = \sum_{\ell \in \mathbb{Z}} \mu^{\ell}_{A}(X)t^{\ell},$$

$$\mathscr{P}^{X}_{A}(t) = \sum_{\ell \in \mathbb{Z}} \beta^{A}_{\ell}(X)t^{\ell}.$$

1.5 LEMMA. *If* I *is the minimal injective resolution of* Z *then*

$$I \simeq \coprod_{\underline{p} \in \mathrm{Spec}\,A} E(A/\underline{p})^{(\mu^{\ell}_{A}(\underline{p},Z))}$$

for all ℓ, *where* $\mu^{\ell}_{A}(\underline{p},Z) = \mu^{\ell}_{A_{\underline{p}}}(Z_{\underline{p}})$.

Proof. The complex $\mathrm{Hom}_{A}(A/\underline{p},I)_{\underline{p}}$ of vector spaces over the field $k(\underline{p}) = A_{\underline{p}}/\underline{p}_{\underline{p}}$ is minimal, and so all the differentials are zero. □

The next result is dual to (1.5).

1.6 LEMMA. *If* A *is local and* L *is the minimal free resolution of* X, *then*

$$L^{-\ell} \simeq A^{\beta^{A}_{\ell}(X)}. \quad □$$

2. *Finite homological dimension*

The ring A is supposed to be local in this section.

2.1 *Definitions and Remarks.* The *injective dimension of* Z is denoted by $\mathrm{id}_A Z$, and is at most n ($\in \mathbb{Z}$), if Z has an injective resolution I with $I^{\ell} = 0$ for $\ell > n$. This implies that, if I is the *minimal* injective resolution of Z, then $\mathrm{id}_A Z = \sup\{\ell \mid I^{\ell} \neq 0\}$.

If A is local then the *projective dimension*, $\mathrm{pd}_A X$, of X is at most n if X has a projective (i.e. free) resolution L with $L^{-\ell} = 0$ for $\ell > n$. This implies that if L is the *minimal* free resolution of X, then $\mathrm{pd}_A X = \sup\{\ell \mid L^{-\ell} \neq 0\}$.

The proof of the next result is an easy application of Nakayama's lemma (as for a f. g. module instead of the complex X).

2.2 LEMMA. (a) $\mathrm{id}_A X = s(\underline{\underline{\mathrm{Hom}}}(k,X))$ = *degree of* $\mathcal{I}_A^X(t)$, *and*

(b) $\mathrm{pd}_A X = -i(k \underline{\underline{\otimes}} X)$ = *degree of* $\mathcal{P}_A^X(t)$.

(The degree might be infinite. The degree of 0 is $-\infty$.)

2.3 *Definition.* We set depth X = $i(\underline{\underline{\mathrm{Hom}}}(k,X))$ = order of $\mathcal{I}_A^X(t)$. This extends the definition of depth for a f. g. module. As for f. g. modules there is a connection with regular sequences: see [F_3; (3.18)]. However, this connection will not be used in this article (and seems not to be important).

2.4 *Remark.* $-s(X)$ = order of $\mathcal{P}_A^X(t)$.

2.5 THEOREM. (a) $\mathcal{I}_A^{Y\underline{\underline{\otimes}}X}(t) = \mathcal{I}_A^Y(t)\mathcal{P}_A^X(t^{-1})$ *if* $\mathrm{pd}_A X < \infty$.

(b) $\mathcal{P}_A^{\underline{\underline{\mathrm{Hom}}}(Y,X)}(t) = \mathcal{I}_A^Y(t)\mathcal{I}_A^X(t^{-1})$ *if* $\mathrm{id}_A X < \infty$.

Proof. See [F_1; Theorems 4.1 and 4.2]. □

2.6 COROLLARY. (a) $\mathcal{I}_A^X(t) = \mathcal{I}_A^A(t)\mathcal{P}_A^X(t^{-1})$ *if* $\mathrm{pd}_A X < \infty$.

(b) $\mathcal{P}_A^X(t) = \mathcal{I}_A^A(t)\mathcal{I}_A^X(t^{-1})$ *if* $\mathrm{id}_A X < \infty$. □

2.7 COROLLARY. (a) $\mathrm{depth}_A A = \mathrm{depth}_A X + \mathrm{pd}_A X$ *if* pd X $< \infty$.

(b) $\mathrm{depth}_A A = \mathrm{id}_A X - s(X)$ *if* id X $< \infty$.

(c) $s(\underline{\underline{\mathrm{Hom}}}(Y,X)) = \mathrm{depth}_A A - \mathrm{depth}_A Y + s(X)$ *if* $\mathrm{id}_A X < \infty$. □

3. *Flat base change*

In this section B is a flat A-algebra. Furthermore, A is supposed to be local, and the fibre ring $C = B/\underline{m}B$ is supposed to be non-trivial.

3.1 THEOREM. $\mathscr{I}_B^{X\otimes_A B}(t) = \mathscr{I}_A^X(t)\mathscr{I}_C^C(t)$.

Proof. See [FT]. $\square$

3.2 COROLLARY. $\mathrm{id}_B(X\otimes_A B) = \mathrm{id}_A X + \mathrm{id}_C C$ $(s \le \infty)$. $\square$

Recall that for a local ring the following are equivalent:

(1) D is Gorenstein;

(2) $\mathrm{id}_D D < \infty$;

(3) D is Cohen-Macaulay and $\mu_D^n(D) = 0$ when $n \ne \dim D$;

(4) $\mathscr{I}_D^D(t) = t^n$ when $n = \dim D$.

See [B], but see also (6.4). Recall also that C is Gorenstein if and only if all the localizations of C at prime ideals are Gorenstein.

For $\underline{p} \in \operatorname{Spec} A$ write $k(\underline{p}) = A_{\underline{p}}/\underline{p}_{\underline{p}}$ and $C(\underline{p}) = B\otimes_A k(\underline{p})$, the fibre ring at $\underline{p}$. Recall that the prime ideals of $C(\underline{p})$ are of the form $\underline{q}C(\underline{p})$ where $\underline{q} \in \operatorname{Spec} B$ has $\underline{q} \cap A = \underline{p}$.

3.3 COROLLARY. *If* $\mathrm{id}_A X < \infty$, *and if* $C = C(\underline{m})$ *is Gorenstein, then* $C(\underline{p})$ *is Gorenstein for all* $\underline{p} \in \operatorname{Supp} X$.

Proof. For $\underline{q} \in \operatorname{Spec} B$ with $\underline{q} \cap A = \underline{p}$ write $D = C(\underline{p})_{\underline{q}C(\underline{p})}$. We have

$$\begin{aligned}\mathrm{id}_D D + \mathrm{id}_{A_{\underline{p}}} X_{\underline{p}} &= \mathrm{id}_{B_{\underline{q}}}(X_{\underline{p}}\otimes_{A_{\underline{p}}} B_{\underline{q}}) \\ &= \mathrm{id}_{B_{\underline{q}}}(X\otimes_A B)_{\underline{q}} \\ &\le \mathrm{id}_B(X\otimes_A B) \\ &= \mathrm{id}_A X + \mathrm{id}_C C < \infty. \quad \square\end{aligned}$$

3.4 *Examples.* (a) Take $B = A[[T_1,\dots,T_n]]$, the ring of formal power series in n variables. Then $C = k[[T_1,\dots,T_n]]$ is Gorenstein.

(b) Take $B = \hat{A}$, the $\underline{m}$-adic completion of A. Then $C = k$.

4. *Dualizing complexes*

The ring A is local in this section.

4.1 *Definition.* A *dualizing complex* is a *bounded* complex of *injective* modules such that $H^{\ell}(D)$ is *f. g.* for all ℓ, and such that the canonical morphism $A \to \mathrm{Hom}(D,D)$ is a quasi-isomorphism. This is the usual definition. In this article we shall however also want a dualizing complex D to be minimal and to satisfy $i(D) = 0$. (These restrictions cause no loss of generality.) We write

$$X^{\dagger} = \mathrm{Hom}(X,D).$$

The following facts about dualizing complexes can be found in [Ha].

4.2 THEOREM. *If* D *is a dualizing complex then*

(a) D *is unique up to isomorphism,*

(b) $X \approx X^{\dagger\dagger}$,

(c) $\mathscr{I}_A^D(t) = t^{\dim A}$,

(d) $D^{\ell} \simeq \coprod_{\dim(A/\underline{p})=d-\ell} E(A/\underline{p})$ *when* $d = \dim A$,

(e) $\mathrm{Hom}_A(B,D)[t]$ *is a dualizing complex over* B *when* B *is a module finite* A*-algebra and* $t = \dim A - \dim B$, *and*

(f) $D_{\underline{p}}[n]$ *is a dualizing complex over the local ring* $A_{\underline{p}}$, *where* $\underline{p} \in \mathrm{Spec}\, A$ *and*

$$n = \dim A - \mathrm{ht}\,\underline{p} - \dim(A/\underline{p}).$$

(Here $[n]$ means translation n degrees to the left, so that $D_{\underline{p}}[n]^{\ell} = D_{\underline{p}}^{n+\ell}$.) □

4.3 COROLLARY. (a) $\mathrm{pd}_A X^{\dagger} = \mathrm{id}_A X - \dim A$ $(\le \infty)$.

(b) $\mathrm{id}_A X^{\dagger} = \mathrm{pd}_A X + \dim A$ $(\le \infty)$.

(c) $s(X^{\dagger}) = \dim A - \mathrm{depth}_A X$.

(d) $\mathrm{id}_A D = \dim A = \mathrm{depth}_A D_A$.

Proof. See (2.6). □

4.4 *Remark.* The ring A is Gorenstein if and only if A possesses a dualizing complex equivalent to A. In particular, if A is a homomorphic image of a Gorenstein ring, then A has a dualizing complex by (4.2)(e). Thus, "most rings possess a dualizing complex", and certainly a complete local ring does. Also A possesses a dualizing complex if A is essentially of finite type over a field or $\mathbb{Z}$ (or any other Gorenstein ring).

4.5 *Definition and Remark.*

$$\dim Z = \sup_{\underline{p}\in \operatorname{Spec} A} (\dim(A/\underline{p}) + s(X_{\underline{p}})) = \sup_{\ell\in\mathbb{Z}}(\dim H^{\ell}(X) + \ell);$$

see $[F_3]$.

The next result is due to Peskine, Szpiro, Roberts and Hochster, and it is called the New Intersection Theorem.

4.6 NEW INTERSECTION THEOREM (Peskine, Szpiro, Roberts, Hochster). $\dim A \le \operatorname{pd}_A X + \dim X$ *if* X *is non-trivial and* A *contains a field (as a subring).*

Proof. See $[PS_2]$, $[R_1]$, [Ho], or $[F_3]$. □

4.7 COROLLARY. *If* X *is non-trivial and* $\operatorname{id}_A X < \infty$, *then*

$$\dim A - \operatorname{depth} A \le s(X) - i(X),$$

provided A *contains a field.*

Proof. Pass to the completion (so that A has a dualizing complex). From $[F_3;\ (3.14.d)]$ and (4.2)(b) it follows that $i(X) = d - \dim X^{\dagger}$, and the desired assertion follows from (4.6), (4.3)(a) and (2.7)(b). □

(4.7) states that among non-trivial X with $\operatorname{id}_A X$ finite and all $H^{\ell}(X)$ f. g. the dualizing complex D has minimal amplitude (where the amplitude of a complex Y is $s(Y) - i(Y)$).

(4.7) also gives an affirmative answer to the following question of Bass (in the case where A contains a field): if there exists a non-zero f. g. A-module of finite injective dimension, is A then Cohen-Macaulay? See [B], but also $[PS_1]$.

5. *Localization*

The ring A is local in this section.

5.1 THEOREM. $\mathcal{I}_{A_{\underline{p}}}^{X_{\underline{p}}}(t)\,t^{v} \preccurlyeq \mathcal{I}_{A}^{X}(t)$, *where* $\underline{p} \in \operatorname{Spec} A$ *and* $v = \dim A/\underline{p}$.

Here $\preccurlyeq$ is the degreewise partial ordering of the formal

Laurent series.

A consequence of (5.1) is that $\mu_A^{\ell+v}(X) > 0$ if $\mu_A^{\ell}(\underline{p},X) > 0$. This is however well known and easy.

Proof. We divide the argument into two cases.

Case 1, *when* A *has a dualizing complex* D.

$$\mathscr{I}_A^X(t) = t^d \mathscr{P}_A^{X^\dagger}(t) \succeq t^d \mathscr{P}_{A_{\underline{p}}}^{(X^\dagger)_{\underline{p}}}(t) = t^v \mathscr{I}_{A_{\underline{p}}}^{X_{\underline{p}}}(t),$$

where d = dim A, by (2.5)(b), (4.2)(e),(f) and (2.5)(b) again.

Case 2: *the general case.* Pick $\underline{q}$ in $V_{\hat{A}}(\hat{\underline{p}})$ with $\dim \hat{A}/\underline{q} = v$. Then $C = \hat{A}_{\underline{q}} \otimes_{A_{\underline{p}}} k(\underline{p})$ is Artinian, so that in particular $\mu_C^0(C) \geq 1$, and hence

$$\mathscr{I}_A^X(t) = \mathscr{I}_{\hat{A}}^{\hat{X}}(t) \succeq t^v \mathscr{I}_{\hat{A}_{\underline{q}}}^{\hat{X}_{\underline{q}}}(t) = t^v \mathscr{I}_{A_{\underline{p}}}^{X_{\underline{p}}}(t) \mathscr{I}_C^C(t) \succeq t^v \mathscr{I}_{A_{\underline{p}}}^{X_{\underline{p}}}(t)$$

by (3.1) and case 1. □

6. *Vanishing of* $\mu^i(X)$

The ring A is local in this section.

6.1 LEMMA. *Assume that* A *has a dualizing complex. If* $\mu^\ell(X) = 0$ *then there exist complexes* U *and* V *such that* $X \approx U \oplus V$ *and* $\mathrm{id}_A U < \ell < \mathrm{depth}_A V$.

Proof (after Roberts [R_1]). Let L be a minimal free resolution of $X^\dagger$. Then $L^{d-\ell} = 0$ by (2.5)(b) and (1.6). That is, $L = L_- \oplus L_+$, where

$$L_- = 0 \to L^{d-\ell+1} \to L^{d-\ell+2} \to \ldots,$$

$$L_+ = \ldots \to L^{d-\ell-2} \to L^{d-\ell-1} \to 0.$$

Hence $X \approx L_-^\dagger \oplus L_+^\dagger$,

$$\mathrm{id}_A L_-^\dagger = \mathrm{pd}_A L_- + d \leq \ell - 1$$

and

$$\mathrm{depth}_A L_+^\dagger = d - s(L_+) \geq \ell + 1,$$

by (4.2)(b) and (4.3)(a),(c). □

6.2 THEOREM. $\mu_A^\ell(M) > 0$ *for* $\mathrm{depth}_A M \leq \ell \leq \mathrm{id}_A M$ ($\leq \infty$) *when* M *is a f. g. module.*

Proof. Assume that $\mu^{\ell}(M) = 0$, and that A is complete. Then $M \approx U \oplus V$ with U and V as in (6.1), and so $M \simeq H^0(U) \oplus H^0(V)$, $H^0(U) \approx U$ and $H^0(V) \approx V$. Furthermore,

$$\mathrm{id}_A H^0(U) = \mathrm{id}_A U < \ell < \mathrm{depth}_A V = \mathrm{depth}_A H^0(V),$$

and hence $s(\underline{\underline{\mathrm{Hom}}}(H^0(V), H^0(U))) < 0$ by (2.6)(c); thus either $H^0(V) = 0$ or $H^0(U) = 0$ and we are done. □

6.3 *Remark.* Often we have even $\mu^{\ell}(M) \geq 2$ when $\mathrm{depth}_A M < \ell < \mathrm{id}_A M$ ($\leq \infty$). This certainly holds if A is an integral domain possessing a dualizing complex: if L is the minimal free resolution of $M^{\dagger}$ then $L^{d-\ell} \simeq A$, so that L is of the form

$$\cdots \longrightarrow L^{d-\ell-1} \xrightarrow{\ \phi\ } A \xrightarrow{\ \psi\ } L^{d-\ell+1} \longrightarrow \cdots$$

where either ϕ or ψ is zero, and the proof can be completed as in (6.2) and (6.3). This argument can be found in [R_2;p.66], and it gives a very easy proof of the following result in the case of a domain with a dualizing complex.

6.4 THEOREM. *If* $\mu_A^{\dim A}(A) = 1$ *then* A *is Cohen-Macaulay, and therefore Gorenstein.*

That this holds (in general) was conjectured in [F_2] (where also some partial answers were given). Roberts states [R_2;p.67] that it holds in general, and his proof will appear in [R_3]. Roberts' proof actually shows that $\mu^{\ell}(M) \geq 2$ if $\mathrm{depth}_A M < \dim M = \ell$ (when M is a f. g. module), and the next result follows (see [F_2; (3.1)]).

6.5 THEOREM. *If* $\mu_A^{\dim M}(M) = 1$, *then* M *and* $B = A/\mathrm{ann}\, M$ *are Cohen-Macaulay, and* M *is a dualizing* B*-module (that is,* B *has a dualizing complex equivalent to* M*).* □

Let $d = \dim A$. In [F_2] it is proved that $\mu_A^{\ell}(M) \geq 2$ if $\mathrm{depth}_A M < \ell < d - 1$, and that $\mu_A^{d-1}(M) \geq 2$ if $\mathrm{depth}_A M < d - 1$ and either A contains a field or $d - \mathrm{depth}_A M \leq 3$ or $\dim \hat{A}/\underline{p} \geq d - 1$ for all $\underline{p} \in \mathrm{Ass}_{\hat{A}} \hat{A}$.

If A is Gorenstein, $a_1, \ldots, a_s$ is an A-regular sequence, and $M = A/(a_1, \ldots, a_s)$, then $\mu^{\ell}(M) = 1$ for $\ell = d - s = \mathrm{depth}_A M$ and for $\ell = d$.

If $A = \mathbb{Z}/(4)$ and $M = \mathbb{Z}/(2)$ then $\mu_A^{\ell}(M) = 1$ for all $\ell \geq 0$.

6.6 *Conjecture.* If $n = \dim X = \dim H^i(X) + i$, $i = i(X)$, and $\mu^n(X) = 1$, then there exists an integer t and an ideal $\underline{a}$ such that $X[t]$ is the dualizing complex over $A/\underline{a}$.

It is possible to prove that this conjecture is equivalent to the conjecture that the New Intersection Conjecture (4.6) holds for all local rings A.

7. *The small support*

The cohomology modules of the complex Z are not necessarily f. g. and the ring A is not necessarily local (but still commutative and Noetherian).

The usual support of Z is the set

$$\operatorname{Supp} Z = \bigcup_{\ell \in \mathbb{Z}} \operatorname{Supp} H^\ell(Z),$$

that is, the set of $\underline{p} \in \operatorname{Spec} A$ with $Z_{\underline{p}}$ non-trivial.

The *small support* supp Z consists of the prime ideals $\underline{p}$ with $\mu_A^\ell(\underline{p}, Z) > 0$ for some ℓ.

7.1 THEOREM. (a) $\operatorname{supp} Z \neq \emptyset$ *if* Z *is non-trivial.*

(b) $\operatorname{supp} Z \subseteq \operatorname{Supp} Z$ *with equality if* $H^\ell(Z)$ *is f. g. for all* ℓ.

(c) $\operatorname{supp}(Z \underline{\underline{\otimes}} W) = \operatorname{supp} Z \cap \operatorname{supp} W$.

Proof. See $[F_3]$. □

8. *Infinitely generated modules*

The ring is again supposed to be local, but the module M need not be f. g.

8.1 THEOREM. (a) *If* $\ell \geq \dim M$ *and* $\mu^\ell(M) = 0$, *then* $\mu^m(M) = 0$ *for* $m \geq \ell$.

(b) *If* $m > \operatorname{depth} A$ *and* $\mu^m(M) > 0$, *then* $\mu^\ell(M) > 0$ *for* $\ell \geq m$.

Proof. (a) Let I be the minimal injective resolution of M. Let $\Gamma_{\underline{m}}$ denote the section functor with support in $\underline{m}$ (that is, the 0-th local cohomology functor). It is known that $s(\Gamma_{\underline{m}}(I)) \leq \dim M \leq \ell$ and that $\Gamma_{\underline{m}}(I^\ell) = E(k)^{\mu^\ell(M)} = 0$: see [G]. This shows that the complex

$$J = 0 \to \Gamma_{\underline{m}}(I^{\ell+1}) \to \Gamma_{\underline{m}}(I^{\ell+2}) \to \cdots$$

is trivial; hence

$$H^m(\mathrm{Hom}(k,\Gamma_{\underline{m}}(I)) = H^m(\mathrm{Hom}(k,J)) = 0$$

for $m > \ell$. Directly from the definition of $\Gamma_{\underline{m}}$ it follows that $\mathrm{Hom}(k,I) \simeq \mathrm{Hom}(k,\Gamma_{\underline{m}}(I))$, and (a) has been established.

(b) We shall actually show that if Z is a bounded complex, $m > \mathrm{depth}\, A + s(Z)$, and $\mu^m(Z) > 0$, then $\mu^\ell(Z) > 0$ for $\ell \geq m$. We divide the argument into two cases.

Case 1, *when* depth A = 0. Choose $x \in \mathrm{ann}\,\underline{m} - 0$ and an integer $t > 0$ such that $x \in \underline{m}^t$. Then $(x) \simeq k$ and $(x) \cap \underline{m}^t = 0$. Write $\underline{a} = \underline{m}^t + (x)$. Since $\mathrm{Ext}^\ell(A,Z) = 0$ for $\ell > s(X)$ it follows from the long exact sequence of Ext modules induced by the short exact sequence $0 \to \underline{a} \to A \to A/\underline{a} \to 0$ that

$$\mathrm{Ext}^{\ell+1}(A/\underline{a},Z) \simeq \mathrm{Ext}^\ell(\underline{a},Z) \simeq \mathrm{Ext}^\ell(k,Z) \oplus \mathrm{Ext}^\ell(\underline{m}^t,Z);$$

thus the desired assertion follows by induction on ℓ, since the module $A/\underline{a}$ is of finite length.

Case 2, *when* depth A > 0. Let I be the minimal injective resolution of Z, and choose a maximal A-regular sequence $a_1,\ldots,a_t \in \underline{m}$, t = depth A. Write $B = A/(a_1,\ldots,a_t)$. Then $J = \mathrm{Hom}(B,I)$ is a minimal B-injective resolution of a bounded complex of B-modules, since $s(J) \leq \mathrm{pd}_A B + s(Z) = \mathrm{depth}\, A + s(Z)$. Also $\mu^\ell_A(Z) = \mu^\ell_B(J)$, and so the result follows from case 1. □

Acknowledgement

The author has been supported, in part, by the Danish Natural Science Research Council.

References

[B] H. Bass, "On the ubiquity of Gorenstein rings", Math. Z., 82 (1963), 8-28.

[F1] H.-B. Foxby, "Isomorphisms between complexes with applications to the homological theory of modules", Math. Scand., 40 (1977), 5-19.

[F2] H.-B. Foxby, "On the μ^i in a minimal injective resolution, II", Math. Scand., 41 (1977), 19-44.

[F3] H.-B. Foxby, "Bounded complexes of flat modules", J. Pure Appl. Algebra, 14 (1979), 149-172.

[G] A. Grothendieck (notes by R. Hartshorne), Local cohomology, Lecture Notes in Mathematics 41, (Springer, Berlin, Heidelberg, New York, 1967).

[Ha] R. Hartshorne, Residues and duality, Lecture Notes in Mathematics 20, (Springer, Berlin, Heidelberg, New York, 1966).

[Ho] M. Hochster, "Big Cohen-Macaulay modules and algebras and embeddability in rings of Witt vectors", Proceedings of the conference on commutative algebra, Queen's University, Kingston, Ontario, 1975 (Queen's University Papers on Pure and Applied Mathematics No. 42, 1975), pp. 106-195.

[PS_1] C. Peskine and L. Szpiro, "Dimension projective finie et cohomologie locale", Publications Mathématiques 42 (Institut des Hautes Études Scientifiques, Paris, 1973), pp. 47-119.

[PS_2] C. Peskine and L. Szpiro, "Syzygies et multiplicités", C. R. Acad. Sci. Paris Sér. A, 278 (1974), 1421-1424.

[R_1] P. Roberts, "Two applications of dualizing complexes over local rings", Ann. Sci. École Norm. Sup. (4), 9 (1976), 103-106.

[R_2] P. Roberts, Homological invariants of modules over commutative rings, Séminaire Math. Sup. 72 (Les Presses de l'Université de Montréal, Montréal, 1980).

[R_3] P. Roberts, "Rings of type 1 are Gorenstein", Bull. London Math. Soc., to appear.

Department of Mathematics,
University of Oklahoma,
Norman, Oklahoma 73019,
U.S.A.

and (from 1982)

Matematisk Institut,
Københavns Universitet,
Universitetsparken 5,
DK 2100 København Ø, Denmark.

THE LOCAL HOMOLOGICAL CONJECTURES

MELVIN HOCHSTER

Acknowledgments

The author wishes to thank once more the organizers of the Durham Symposium, Professor D.G.Northcott and Dr.R.Y.Sharp, for putting together an excellent conference.

Notes taken by Dr.Sharp were a great aid to the author in preparing this article.

The author was supported in part by a grant from the National Science Foundation, U.S.A.

1. *Rigidity, multiplicities, and other conjectures*

What follows is a modified version of a series of three lectures presented during the Durham Symposium on Commutative Algebra, which was held during the period July 15-25, 1981. Some proofs given during those lectures but which are to appear elsewhere have been omitted, while others which were sketched or not given in the original lectures have been provided in greater detail here.

I would like first to recall two conjectures which I feel have played a pivotal role in the development of local homological commutative algebra, and which have inspired much work and many results in several directions, not all of which are "homological".

In the sequel all rings are commutative, associative, with identity, all modules are unital, and "local ring" means Noetherian ring R with a unique maximal ideal $\underline{m}$.

The first is as follows.

(1.1) RIGIDITY CONJECTURE. *Let* R *be a local ring, let* M,N *be finitely generated* R-*modules, and suppose that* $\mathrm{pd}_R M$ *is finite. Suppose that* $\mathrm{Tor}_i^R(M,N) = 0$. *Then* $\mathrm{Tor}_j^R(M,N) = 0$ *for all* $j \geq i$.

The reader is referred to [A_1], [A_2], [L], [H_6] and [PS_1].

The conjecture is open even when pd M = 2; if pd M = 2, Peskine and Szpiro [PS_1] have done the case where grade Ann M > 0, but the general case remains open. It is true when $M = R/(x_1,\ldots,x_d)$, where $x_1,\ldots,x_d$ is an R-sequence, and when R is regular (see [L]). But the proofs in the regular case involve the trick of "reduction to the diagonal" (see [S]) and thus reduction to the case where $M = R/(x_1,\ldots,x_d)$ as above. Of course, for the most general case, some other ideas are necessary: one studies all the truncated Euler characteristics

$$\chi_i^R(M,N) = \sum_{j=i}^{\infty} (-1)^{j-i} \ell(\mathrm{Tor}_i^R(M,N));$$

here ℓ denotes length, and χ_i is defined if $\ell(\mathrm{Tor}_j^R(M,N))$ is finite for $j \geq i$ and vanishes for all sufficiently large j. (This last condition is automatic if pd $M < \infty$; for example, this is the case if R is regular.) The idea is to get as much information as possible about the $\chi_i^R(M,N)$ when $M = R/(x_1,\ldots,x_d)$ and then use reduction to the diagonal and some spectral sequence arguments to carry through. It is shown in [L] that, if R is unramified and regular, then $\chi_i^R(M,N) \geq 0$ (on the assumption that $\ell(\mathrm{Tor}_i^R(M,N))$ is finite, which implies the same for $\ell(\mathrm{Tor}_j^R(M,N))$, $j \geq i$), and it is shown that if $i \geq 2$ then $\chi_i^R(M,N) = 0$ implies that $\mathrm{Tor}_j^R(M,N) = 0$ if $j \geq i$. This is crucial in the proof of rigidity when R is ramified and regular. Recently the author noticed that Lichtenbaum's ideas suffice to prove that if $\chi_1^R(M,N) = 0$ then $\mathrm{Tor}_1^R(M,N) = 0$ as well when R is unramified and regular: see [H_6].

It is clear that the rigidity conjecture is closely akin to the following.

(1.2) SERRE'S MULTIPLICITIES CONJECTURE. *Let* R *be an arbitrary regular local ring and let* M,N *be nonzero modules such that* $\ell(M \otimes_R N)$ *is finite. Then we have the following:*

(0) dim M + dim N ≤ dim R;

(1) *if* dim M + dim N < dim R, *then* $\chi_0^R(M,N) = 0$;

(2) *if* dim M + dim N = dim R, *then* $\chi_0^R(M,N) > 0$.

[Henceforth, we shall usually write $e^R(M,N)$ or $e(M,N)$ for $\chi_0^R(M,N)$.]

In fact, (0) is no mere conjecture: Serre proves this in [S], as well as establishing all three parts when $\hat{R}$ is a formal power series ring over a DVR (discrete rank one valuation ring).

The conjecture is also known if $\dim R \le 4$ [H_1], or if the residual characteristic p kills both modules [MB]. Also the "vanishing part", statement (1), has been proved for $\dim R \le 5$ in [D_1] (where it is also shown for $\dim R \le 5$ that if M is perfect, p kills M, and $\dim M + \dim N = \dim R$, then $e(M,N) > 0$); see also [D_2], [D_3], [D_4].

This conjecture of Serre is part of the motivation for studying small C.-M. ("C.-M." means "Cohen-Macaulay") modules. To understand this, let R be a ramified regular local ring, and assume that $e(M,N) = 0$ whenever $\dim M + \dim N < \dim R$, that is assume the "vanishing part" of Serre's conjecture. To complete the proof of the conjecture, it suffices to show, for primes P,Q with P + Q primary to $\underline{m}_R = \underline{m}$ and $\dim(R/P) + \dim(R/Q) = \dim R$, that $e(R/P,R/Q) > 0$. *If* one knows that R/P has a C.-M. module M with $\dim M = \dim R/P$ and also that R/Q has a C.-M. module N with $\dim N = \dim R/Q$, then one can show that $e(R/P,R/Q) > 0$. The point is that under these conditions

$$\mathrm{Tor}_i^R(M,N) = 0, \quad i \ge 1,$$

so that $e(M,N) = \ell(M \otimes N) > 0$. But M has a prime filtration involving a positive number a of copies of R/P and other modules of lower dimension, and N has such a filtration involving $b > 0$ copies of R/Q and other modules of lower dimension; from the bi-additivity of e and the vanishing assumption we easily obtain that $e(M,N) = (ab)e(R/P,R/Q)$, whence $e(R/P,R/Q) > 0$.

Unfortunately, small (that is finitely generated) C.-M. modules are not known to exist even in dimension 3, even in the equicharacteristic case.

One can make Serre's conjecture into a more general conjecture by dropping the hypothesis that R be regular and assuming instead that $\mathrm{pd}_R M < \infty$. (Peskine and Szpiro have established this conjecture in a certain graded case: see [PS_2].) It is an open question even for hypersurfaces R whether, if M has finite length and finite projective dimension and dim N < dim R, it must then be true that $e(M,N) = 0$.

In this connection we raise the following.

(1.3) *Question*. Suppose that R is C.-M. Let G be the Grothendieck group of modules of finite length and finite projective dimension over R. (Thus, if $0 \to M' \to M \to M'' \to 0$ is exact then $[M] = [M'] + [M'']$, where $[M]$ denotes the class of M.) Let $d = \dim R$. *Is* G *generated by the classes* $[R/(x_1,\ldots,x_d)]$, *where* $x_1,\ldots,x_d$ *is an* R-*sequence?*

Question (1.3) has an affirmative answer if dim R = 1: this is essentially a result of MacRae [M]. It is also true in dimension 2, at least after enlargement of the field: see [H_1].

It turns out that if (1.3) has an affirmative answer, even for hypersurfaces, then the vanishing part of Serre's conjecture would follow for all ramified regular local rings. It would be enough if the classes $[R/(x_1,\ldots,x_d)]$ generated $\mathbb{Q} \otimes_{\mathbb{Z}} G$, where $\mathbb{Q}$ is the rational numbers.

It may be worth mentioning that A.Weil [W] has challenged the generality in which Grothendieck has developed algebraic geometry because one cannot establish a notion of multiplicity and prove that it has the "right" properties in such generality. Serre's idea gives what is probably the right notion, and whether one accepts Weil's point of view or not, it certainly provides additional motivation for proving that Serre's notion really does behave properly.

Before leaving the subject of multiplicities I want to discuss some ideas connected with lifting.

Let R be a complete local ring and let x be a non-zerodivisor. Let M be a finitely generated module over R/xR with $\mathrm{pd}_{R/xR} M < \infty$. The lifting question is as follows: *does there exist a finitely generated* R-*module* N *such that*

(1) x *is not a zerodivisor on* N, *and*

(2) $N/xN \cong M$.

This is true when R and A = R/xR are both complete unramified regular local rings, for then $R \cong A[[x]]$ and one can take

$N = M \otimes_A A[[x]]$.

It is an interesting question for if it were true when R is an unramified regular local ring and $A = R/xR$ is ramified and regular, it would establish Serre's conjecture: if M′ is a lifting of M from A to R it is easy to see that $e_A(M,N) = e_R(M',N)$ while

$\dim A - \dim M - \dim N = \dim R - \dim M' - \dim N$.

Part of the motivation of Buchsbaum and Eisenbud in their study of the structure of free resolutions is the problem of lifting such resolutions (which then automatically lift the module in our sense): see [BE_1] and [BE_2].

Peskine and Szpiro [PS_1] gave a counterexample to lifting in the case where pd M is assumed finite but R is not necessarily regular. In [H_4] a counterexample to lifting is given when R is regular and unramified, $A = R/xR$ is ramified and regular, and M is a cyclic A-module.

However, there are still weakenings of the lifting question which might have affirmative answers and which would give information about multiplicities.

For example, suppose that R is regular and $A = R/xR$ is a hypersurface. Let M be an A-module of finite length such that $\mathrm{pd}_A M < \infty$. We can then ask whether there exists a finitely generated R-module N such that

(1) $N/xN = M$, and

(2) x is *not nilpotent* on N.

An affirmative answer would yield a proof of the vanishing part of Serre's conjecture for ramified regular local rings.

While this question looks much weaker than lifting, since we are only asking that x not be nilpotent on N instead of that it be a non-zerodivisor, it is not easy to use the hypothesis that $\mathrm{pd}_A M < \infty$ (and one does need to use it). Moreover, it's not as weak as it sounds.

In fact, part of the usefulness of this kind of "weak" lifting is that if M is a *minimal* module of finite length and finite projective dimension (that is M has no non-zero proper submodule with these characteristics, or, equivalently, no non-trivial proper quotient with these characteristics), then if M has a weak lifting, it has a lifting.

The proof is simple. Let N be a weak lifting and let $N_0 = H^0_{(x)}(N) = \bigcup_t \mathrm{Ann}_N x^t$.

Then $N_0 \neq N$ and $\bar{N} = N/N_0 \neq 0$. Applying $- \otimes A$ to $0 \to N_0 \to N \to \bar{N} \to 0$ we get $N_0/xN_0 \to M \twoheadrightarrow \bar{N}/x\bar{N}$. Now $\bar{N}$ is an R-module, so that $\mathrm{pd}_R \bar{N} < \infty$, and x is not a zerodivisor on N. Hence, $\mathrm{pd}_A \bar{N}/x\bar{N} < \infty$. Since $\bar{N}/x\bar{N} \neq 0$ and since M is minimal, we must have $M \cong \bar{N}/x\bar{N}$, and $\bar{N}$ is a lifting, as required.

Now, it is clear that every module of finite length and finite projective dimension has a filtration in which all the factors are minimal modules of finite length and finite projective dimension, so that the minimal such modules generate the group G discussed in Question (1.3). For a liftable such M, $e_A(M,N) = 0$ when $\dim N < \dim A$. It follows that if we assume weak lifting for a local hypersurface A, then $e_A(M,N) = 0$ whenever M has finite length and finite projective dimension and $\dim N < \dim A$. This in turn would be enough to prove the vanishing of multiplicities in the case of ramified regular rings. Unfortunately, weak lifting appears intractable.

Notice that in the case where R and A are regular there is only one minimal module, namely $K = A/\underline{m}_A$, and K *is* liftable.

I would like next to discuss some related homological conjectures, eventually bringing the discussion to the subject of big C.-M. modules and the direct summand conjecture. But I would first like to discuss some history.

M.Auslander had noticed that the rigidity conjecture for modules of finite projective dimension implies the zerodivisor conjecture (that if R is local, $M \neq 0$ is a finitely generated R-module with $\mathrm{pd}_R M < \infty$, and $x \in R$ is a zerodivisor in R, then x is a zerdivisor on M): see [A_1], [A_2]. H.Bass [B] had noticed that a C.-M. local ring R always had a non-zero finitely generated module of finite injective dimension and asked whether a local ring possessing such a module must be C.-M.

A monumental breakthrough in the field occurred with the joint thesis of Peskine and Szpiro [PS_1]. They introduced the intersection conjecture, which asserts that if R is local, M,N are finitely generated non-zero modules, and $\ell(M \otimes N) < \infty$, then dim N $\le \mathrm{pd}_R M$. (Of course, one may as well assume that $\mathrm{pd}_R M$ is finite here.) They proved the implications

rigidity
⇓
intersection
⇙ ⇘
zerodivisor Bass'

(the words "conjecture" or "question" have been omitted) as well as discussing many other questions and implications and *they proved the intersection conjecture in characteristic* $p > 0$ *and in the most important cases in characteristic* 0. They showed the usefulness of the Frobenius endomorphism in this context, and in reducing in characteristic 0 they utilized Artin approximation to reduce to the case of algebras finitely generated over a field and from there to characteristic $p > 0$: this technique has repeatedly proved to be very fruitful.

They also observed that if, say, every complete local ring R of dimension d has a finitely generated module of depth d (a small C.-M. module), then the intersection conjecture follows.

The intersection conjecture is implied by and nearly equivalent to the following.

(1.4) CONJECTURE. *If* R *is Noetherian,* M *is a finitely generated* R-*module,* $I = \mathrm{Ann}\, M$, $R \to S$ *is a homomorphism into a Noetherian ring* S *and* Q *is a minimal prime of* IS, *then* ht $Q \le \mathrm{pd}_R M$.

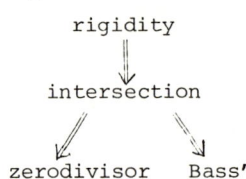

See [H_1]. Consider the example in which $R = \mathbb{Z}[x]$, $M = R/xR$, and S is an arbitrary Noetherian ring. To give a homomorphism $R \to S$ is the same as to specify an element s in S to serve as the image of x, and then IS = sS. Hence, in this case, (1.4) simply becomes the assertion that a minimal prime of a principal ideal in a Noetherian ring has height at most one, Krull's principal ideal theorem, which, geometrically, asserts that if a hypersurface meets a variety then the codimension of the intersection with that variety is at most one. This may help to explain the name "intersection conjecture". At the same time, we see that if (1.4) is true, then it is a generalized principal ideal theorem, and, in fact, all of the homological conjectures we consider now, suitably presented, can be viewed as generalized principal ideal theorems. This is often not apparent.

The author has conjectured that if R is a local ring with system of parameters (s.o.p.) $x_1,\ldots,x_d$ then there exists a module M, not necessarily finitely generated, such that

(1) $(x_1,\ldots,x_d)M \neq M$, and

(2) $x_1,\ldots,x_d$ is an M-sequence.

Such a module is called a big C.-M. module for R. These were shown to exist in the equicharacteristic case in [H_3]. At about the same time Paul Roberts [R_1] and Peskine and Szpiro [PS_2] proved a generalization of the intersection conjecture (although that may not be immediately clear from its statement) in characteristic $p > 0$ and, hence, by the "metatheorem" in [H_3], for all equicharacteristic local rings. This "new intersection theorem", as it came to be called, in its simplest form asserts that if R is local and

$$F. = 0 \to F_d \to \ldots \to F_0 \to 0$$

is a complex of finitely generated free R-modules such that $H_i(F.)$ has finite length for all i and $H_i(F.) \neq 0$ for some i, then $\dim R \leq d$.

The arguments given in the various proofs of the new intersection theorem actually prove a slightly stronger result, which we shall examine in the next section, and which, unfortunately, we shall refer to as the "new new intersection theorem". (It is a theorem in the equicharacteristic case: in the general case, it is a conjecture.) This result is of particular interest because it suffices to give a proof of the recent Evans-Griffith syzygy theorem (they use big C.-M. modules in their paper): see [EG].

It had been observed already in [H_1] that the existence of big C.-M. modules implies the intersection conjecture and hence the zerodivisor conjecture and an affirmative answer to Bass' question. It is not hard to use the same idea to prove the new or new new intersection theorem. (It is worth mentioning that while all three of the proofs just discussed rest ultimately on reduction to characteristic p, Paul Roberts [R_2] has proved the new intersection theorem in the analytic case by analytic techniques (including the Grauert-Riemenschneider vanishing theorem).)

Thus, the existence of big C.-M. modules seems to play a central role in the study of these conjectures. Now it is known that if R is regular local, S is local and module-finite over R, and S has a big C.-M. module, then R is a direct summand of S as an R-module (see [H_2]). This, of course, is conjectured to be true for arbitrary S: this is the "direct summand conjecture". The author has been able to show that the direct summand conjecture itself is enough to imply the new new intersection conjecture: we'll discuss this in the next section. See [H_7] for more details.

Thus, most of the applications of the existence of big C.-M. modules would be obtainable if we could prove the direct summand conjecture, which is perhaps the least "homological" of all the homological conjectures.

A diagram of implications is given in the accompanying Figure 1.

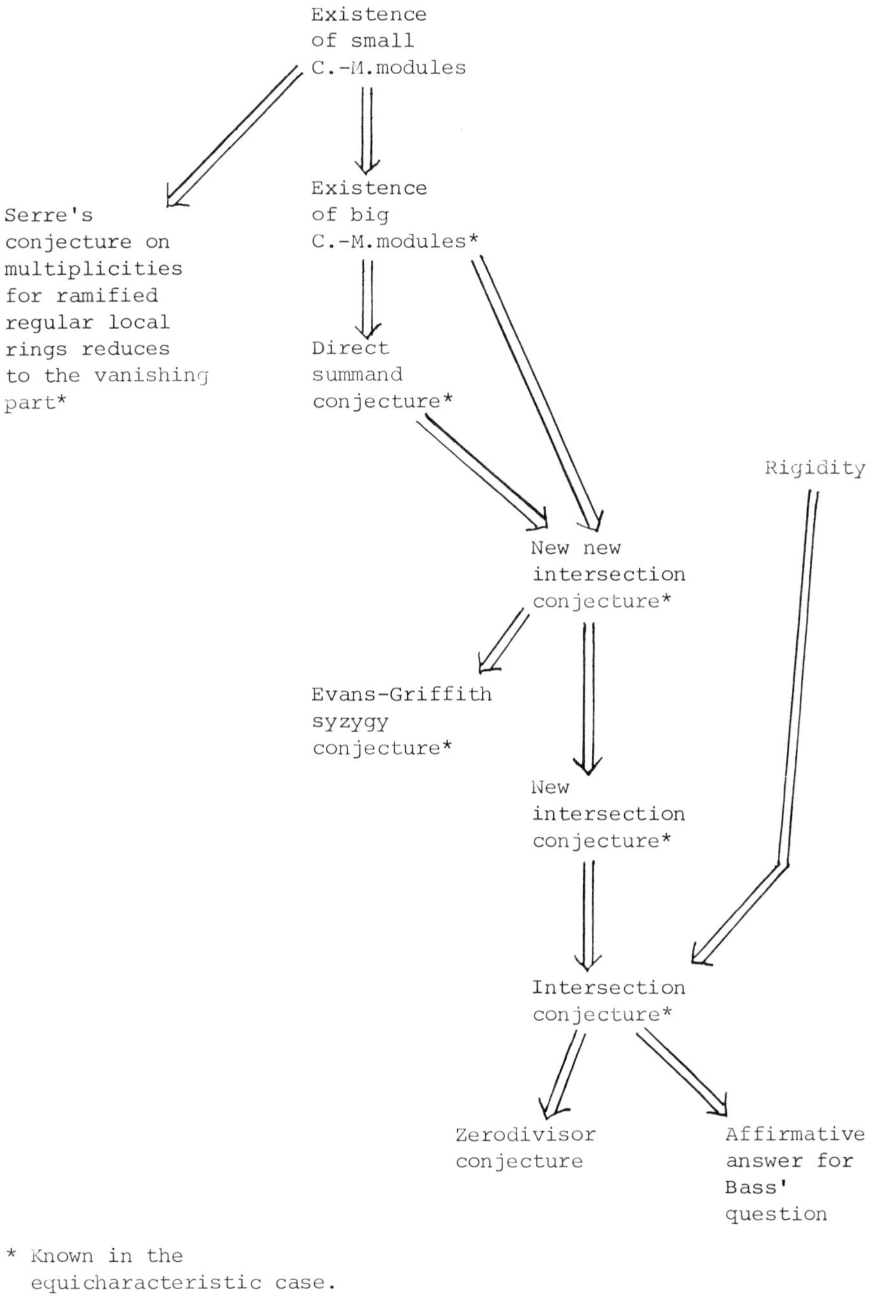
Existence of small C.-M.modules
Existence of big C.-M.modules*
Serre's conjecture on multiplicities for ramified regular local rings reduces to the vanishing part*
Direct summand conjecture*
Rigidity
New new intersection conjecture*
Evans-Griffith syzygy conjecture*
New intersection conjecture*
Intersection conjecture*
Zerodivisor conjecture
Affirmative answer for Bass' question
* Known in the equicharacteristic case.

Figure 1

2. *The direct summand conjecture and the new new intersection theorem*

The question of the existence of small C.-M. modules for complete local rings R immediately reduces to the case where R is a domain and then by Cohen's structure theorems R is module-finite over a complete regular local ring A. If M is a finitely generated R-module, M is a small (maximal) C.-M. module for R (that is depth M = dim R) if and only if M is non-zero and free as an A-module, say $M \cong A^r$. The action of R on M then gives a map $R \to \mathrm{End}_A(A^r) \cong \mathcal{M}_r(A)$ (r size matrices over A), which, it turns out, must be an embedding, so that R is embedded as a subring of r size matrices over A ($A \subset R$ is then identified with the scalar matrices in $\mathcal{M}_r(A)$). Clearly, if there exists such an embedding, then there exists a retraction $R \twoheadrightarrow A$ as A-modules, that is A is a direct summand of R as an A-module. In fact, $\mathcal{M}_r(A)$ retracts to A: simply map $(a_{ij}) \mapsto a_{11}$ (or a_{tt} for any fixed t).

This is one way of seeing how the existence of small C.-M. modules implies the direct summand conjecture. Big C.-M. modules also suffice: this is shown directly in [H_2], but one can also use the following beautiful result of Phil Griffith [G] (which we shall not prove here).

(2.1) THEOREM (P.Griffith). *Let* A *be a complete regular local ring and let* R *be a domain module-finite over* A. *If* R *has a big C.-M. module, then it has one which is countably generated and* A-*free.* □

[It is tempting to try to get from this result to the existence of small C.-M. modules: I believe this was part of Griffith's motivation. No one has succeeded.]

In any case, the existence of big C.-M. modules would imply the following.

(2.2) CONJECTURE. *Let* R *be a regular Noetherian ring and let* S *be a module-finite extension of* R. *Then* R *is a direct summand of* S *as an* R-*module.*

We make some elementary remarks about this problem. Given homomorphisms $R \to S \to T$, if $R \to T$ splits then $R \to S$ splits. Hence, in Conjecture (2.2) we are free to kill a minimal prime of S which meets R trivially and so assume that S is a domain. Now $R \to S$ splits if and only if $\mathrm{Hom}_R(S,R) \to \mathrm{Hom}_R(R,R)$ is onto. This proves that the problem is local and that we may assume that R is a regular local ring. Moreover, this argument also shows that we can replace R by a faithfully flat local extension (which is still regular), such as, for example, $\hat{R}$ (although we may have to kill another minimal prime to get the new S to be a domain again). By this trick we may assume that R has an algebraically closed residue field.

We next observe that when R contains the field of rationals the direct summand conjecture is trivially true - even if we only assume that R is normal instead of regular. For if L, L' are the fraction fields of R, S respectively and $[L' : L] = d$ then

$$(1/d)\mathrm{Tr}_{L'/L} : S \to R$$

gives the required retraction.

The case of characteristic $p > 0$ is handled using the Frobenius endomorphism F. We may assume that $R = K[[X_1,\ldots,X_n]]$, where K is an algebraically closed field of characteristic $p > 0$. First choose an R-linear map $\phi : S \to R$ such that $\phi(1) \neq 0$. (Because S is a torsion-free R-module it can be embedded in a free R-module.) Then choose e so large that $\phi(1) \notin \underline{m}^{p^e}$, where $\underline{m} = (X_1,\ldots,X_n)R$. Let $q = p^e$. Let $A = K[[X_1^q,\ldots,X_n^q]] \subset R$. Since K is perfect, $A = F^e(R)$. R is a free A-module and since $\phi(1) \notin (X_1^q,\ldots,X_n^q)R = \underline{m}_A R$, $\phi(1)$ is part of an A-free basis for R. It follows that there is an A-linear map $\psi : R \to A$ such that $\psi(\phi(1)) = 1$. Thus $\psi \circ \phi$ is an A-linear retraction of S to A, and its restriction to $F^e(S)$ is an A-linear retraction of $F^e(S)$ to $A = F^e(R)$. But we have a commutative diagram

$$\begin{array}{ccc}
S & \xrightarrow[\cong]{F^e} & F^e(S) \\
\uparrow & & \uparrow \\
R & \xrightarrow[\cong]{F^e} & F^e(R) = A,
\end{array}$$

so that the inclusion $R \hookrightarrow S$ is isomorphic to the inclusion $A \hookrightarrow F^e(S)$. It follows that there is an R-linear retraction $S \to R$.

This proves the direct summand conjecture in characteristic $p > 0$.

Our next objective in this section is to prove the new new intersection conjecture from the direct summand conjecture in characteristic $p > 0$. Of course, this proves it in the absolute sense: we don't need to say "conjecture". Later, we indicate the changes necessary to get the arguments to work in the mixed characteristic case, where we do not know whether the direct summand conjecture holds, but the implication

direct summand $\Longrightarrow$ new new intersection

remains valid.

A key point in the arguments to follow is the construction of a "funny" kind of Koszul complex, to which the usual one maps, but which is acyclic under very mild hypotheses. This construction can be made very generally in characteristic $p > 0$ and under somewhat greater restrictions in the mixed characteristic case: we discuss these later.

First recall that if $x_1, \ldots, x_n$ are non-zerodivisors in R, the usual Koszul complex $K_{\cdot}(x_1, \ldots, x_n; R)$ (or, briefly, $K_{\cdot}(\underline{x}; R)$), which is

$$\bigotimes_{i=1}^{n} (0 \longrightarrow R \xrightarrow{x_i} R \longrightarrow 0),$$

may be identified with

$$\bigotimes_{i=1}^{n} (0 \longrightarrow x_i R \hookrightarrow R \longrightarrow 0).$$

(Here, when $n = 1$, $K_0(x_1, R)$ is identified with R and $K_1(x_1; R)$ with $x_1 R$.)

Now let R be a ring of characteristic $p > 0$ and let

$$R^{\infty} = \varinjlim (R \xrightarrow{F} R \xrightarrow{F} R \xrightarrow{F} \ldots \longrightarrow R \xrightarrow{F} \ldots)$$

where every map is the Frobenius. R^{∞} may be thought of as R_{red} with all p^e-th roots adjoined. F is an automorphism of R^{∞} and for $y \in R^{\infty}$ we may write $F^{-1}(y) = y^{1/p}$.

Let $x_1, \ldots, x_n$ be non-zerodivisors in R. For simplicity assume that R is reduced, so that $R \subset R^{\infty}$. Then $x_1, \ldots, x_n$ are non-zero-divisors in R^{∞}. Let

$$(x^{\infty}) = \bigcup_{e=1}^{\infty} x^{1/p^e} R^{\infty}$$

when x is a non-zerodivisor in R^{∞}. Then (x^{∞}) is a *flat* ideal in R^{∞}.

Now let $K_{\cdot}^{\infty}(x_1, \ldots, x_n; R^{\infty})$ (or $K_{\cdot}^{\infty}(\underline{x}; R^{\infty})$ or simply $K_{\cdot}^{\infty}$) denote

$$\bigotimes_{i=1}^{n} (0 \longrightarrow (x_i^{\infty}) \hookrightarrow R^{\infty} \longrightarrow 0).$$

(When n = 1, $K_0^{\infty} = R^{\infty}$ and $K_1^{\infty} = (x_1^{\infty})$.) Then $K_{\cdot}^{\infty}$ is a flat complex. Note that

$$K_n^{\infty} = (x_1^{\infty}) \otimes \ldots \otimes (x_n^{\infty}) \cong (x_1^{\infty}) \ldots (x_n^{\infty}) = ((x_1 \ldots x_n)^{\infty}).$$

(2.3) THEOREM. *If* $x_1, \ldots, x_n$ *are non-zerodivisors in the ring* R *of characteristic* $p > 0$, *then* $K_{\cdot}^{\infty}(x_1, \ldots, x_n; R^{\infty})$ *is acyclic.*

We shall sketch a proof of this theorem. We use induction on n. In the inductive step we assume the acyclicity of $K_{\cdot}^{\infty}(x_1, \ldots, x_{n-1}; R^{\infty})$, which makes it a flat resolution of R^{∞}/J, where $J = (x_1^{\infty}) + \ldots + (x_{n-1}^{\infty})$. Let $I = (x_n^{\infty})$, so that $0 \to I \to R^{\infty} \to R^{\infty}/I \to 0$ is a flat resolution of R^{∞}/I. We want to show that if we tensor these flat resolutions (without their augmentations) and take the total complex, then we get an acyclic complex. But this simply says that

$$\operatorname{Tor}_i^{R^{\infty}}(R^{\infty}/I, R^{\infty}/J) = 0, \quad i \geq 1.$$

Since I has a flat resolution of length one, it suffices to show this when i = 1. But that particular Tor is simply $I \cap J/IJ$ and it will suffice to show that $IJ = I \cap J$. The crucial point is that I,J are stable under F^{-1}. Suppose that $u \in I \cap J$. Then $u^{1/p} \in I \cap J$, and $u = u^{1/p}(u^{1/p})^{p-1}$. We can think of $u^{1/p}$ as in I and $(u^{1/p})^{p-1}$ as in J. Thus, $u \in IJ$, as required. □

Note that the usual Koszul complex maps to our "funny" one: by tensoring the diagrams

$$\begin{array}{ccccccc}
0 & \longrightarrow & (x_i^\infty) & \hookrightarrow & R^\infty & \longrightarrow & 0 \\
 & & \uparrow\!\!\!\!\!\!\!\!\!\; & & \uparrow & & \\
 & & \cup & & \cup & & \\
0 & \longrightarrow & x_i R & \hookrightarrow & R & \longrightarrow & 0 \\
 & & \uparrow x_i & & \uparrow \mathrm{id} & & \\
0 & \longrightarrow & R & \xrightarrow{x_i} & R & \longrightarrow & 0,
\end{array}$$

we get a map $K.(\underline{x};R) \to K.^\infty(\underline{x};R^\infty)$.

The reader is referred to $[H_7]$ for more information about $K.^\infty$.

Notice that from (2.3) we have at once the following.

(2.4) COROLLARY. *The flat dimension of* $R^\infty/((x_1^\infty)+\ldots+(x_n^\infty))$ *is at most* n. □

Let $A = K[[x_1,\ldots,x_n]]$, where K is an algebraically closed field of characteristic $p > 0$. Let T be the integral closure of A in an algebraic closure of its fraction field. Then $T = T^\infty$, so that the theory described above applies, and $(x_1^\infty)+\ldots+(x_n^\infty)$ turns out to be the maximal ideal of T: call it $\underline{m}_T$. Now $K \cong T/\underline{m}_T$ (since K was algebraically closed). Thus, K has finite flat dimension over T. If T were Noetherian, this would imply that T is regular. Of course, here, T is a huge ring which is nothing like regular.

We now want to show how to use the acyclicity of $K.^\infty$ to prove the new (or even new new) intersection theorem. The theorem asserts that if R is local and $0 \to F_s \to \ldots \to F_0 \to 0$ is a finite complex of finitely generated free modules with non-zero finite length homology then $\dim R \le s$. It is easy to reduce to the case where $H_0(F.) \ne 0$. The "new" version relaxes the hypothesis slightly in this case: it is assumed that $H_i(F.)$ has finite length, $i \ge 1$, and that there is a minimal generator of $H_0(F.)$ which is killed by a power of $\underline{m} = \underline{m}_R$. In either version it is easy to reduce to the case where R is a complete local domain and, for simplicity, we henceforth assume this.

The slightly improved new version turns out to be just what is needed to prove the Evans-Griffith syzygy theorem: see [EG].

Assume now that the theorem is false and that $s < \dim R = n$.

Let $M = H_0(F)$. Let $x_1, \ldots, x_n$ be a system of parameters for R. We write $K.(\underline{x}^t;R)$ for $K_0(x_1^t, \ldots, x_n^t;R)$. To give the proof one first shows that for sufficiently large t there is a map of complexes $K.(\underline{x}^t;R) \to F.$, of the form

$$\begin{array}{ccccccccccccccc}
0 & \longrightarrow & 0 & \longrightarrow \cdots \longrightarrow & F_s & \longrightarrow \cdots \longrightarrow & F_1 & \xrightarrow{d_1} & F_0 & \xrightarrow{d_0} & M & \longrightarrow & 0 \\
 & & \uparrow & & \uparrow & & \uparrow & & \uparrow{\scriptstyle \phi_0} & & \uparrow & & \\
0 & \longrightarrow & R & \longrightarrow \cdots \longrightarrow & R^{\binom{n}{s}} & \longrightarrow \cdots \longrightarrow & R^{\binom{n}{1}} & \longrightarrow & R & \longrightarrow & R/(x_1^t, \ldots, x_n^t) & \longrightarrow & 0,
\end{array}$$

where ϕ_0 is chosen so that $d_0\phi_0(1)$ is the minimal generator v of M which, by hypothesis, is killed by a power of $\underline{m}$. Now $f = \phi_0(1)$ will be a free generator of F_0. One then constructs ϕ_1, etc., recursively, enlarging t if necessary, making repeated use of the standard map of Koszul complexes $K.(\underline{x}^{t+t'};R) \longrightarrow K.(\underline{x}^t;R)$.

In the inductive step one has a diagram

$$\begin{array}{ccc}
F_{i+1} & \xrightarrow{d_{i+1}} & F_i \xrightarrow{d_i} \\
 & & \uparrow{\scriptstyle \phi_i} \\
\vdots\uparrow{\scriptstyle \phi'_{i+1}} & & \longrightarrow K_i(\underline{x}^t;R) \longrightarrow \\
 & & \uparrow \\
K_{i+1}(\underline{x}^{t+t'};R) & \longrightarrow & K_i(\underline{x}^{t+t'};R)
\end{array}$$

and one wants to show that one can fill in the arrow ϕ'_{i+1} for sufficiently large t'. This would be trivial if we knew that $\operatorname{Im} d_{i+1} = \operatorname{Ker} d_i$, but all we know is that $\underline{m}^N \operatorname{Ker} d_i \subset \operatorname{Im} d_{i+1}$ for large N. However, for large t' the image of $K_{i+1}(\underline{x}^{t+t'};R)$ in F_i will be contained in $\underline{m}^{t'} F_i \cap \operatorname{Ker} d_i$ and so, by the Artin-Rees lemma, in $\operatorname{Im} d_{i+1}$; this permits the construction of ϕ'_{i+1}. For more details, the reader is referred to [H_7].

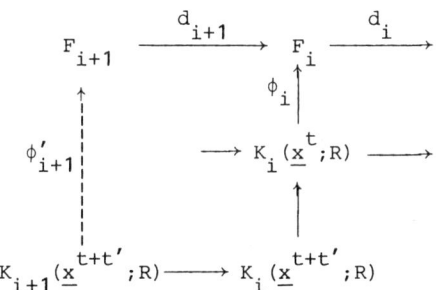

But now since F. is free and $K.^\infty$ is acyclic we can construct a map of complexes $F. \longrightarrow K.^\infty$ in such a way that the free generator $f = \phi_0(1)$ of F_0 discussed earlier maps to the element 1 in R^∞. Composition yields a map of complexes $K.(\underline{x}^t;R) \longrightarrow K.^\infty(\underline{x};R^\infty)$ of which the degree 0 piece is the inclusion map $R \hookrightarrow R^\infty$ while the degree n piece is 0 because it factors through $F_n = 0$ (we're assuming that $n > s$). On the other hand there is a "standard" map between these complexes constructed earlier, which behaves in the same way in degree 0 but in degree n maps the free generator 1 of $R = K_n(\underline{x}^t;R)$ to $x_1^t \dots x_n^t$ in $((x_1 \dots x_n)^{\infty})$. Since $K.(\underline{x}^t;R)$ is free while $K.^{\infty}$ is acyclic, these two maps are homotopic, so that their difference $i : R \hookrightarrow ((x_1 \dots x_n)^{\infty})$ factors via a map h (see diagram)

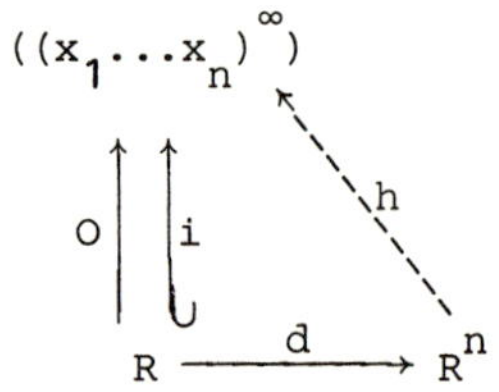

through d, whose matrix has entries $\pm x_1^t, \dots, \pm x_n^t$. It follows that

$$x_1^t \dots x_n^t \in (x_1^t, \dots, x_n^t)((x_1 \dots x_n)^{\infty}),$$

so that for some sufficiently large integer e there exist elements $v_1, \dots, v_n \in R^{\infty}$ such that

$$x_1^t \dots x_n^t = \sum_i x_i^t v_i (x_1 \dots x_n)^{1/p^e}.$$

Let $N = tp^e$ and $y_i = x_i^{1/p^e}$. R contains a regular ring A with $y_1, \dots, y_n$ as a system of parameters. Let $B = A[v_1, \dots, v_n]$.

The above equation can be rewritten as

$$(y_1 \dots y_n)^N = \sum_i v_i y_i^N (y_1 \dots y_n)$$

or as

$$(y_1 \dots y_n)^{N-1} = \sum_i v_i y_i^N.$$

In the case of characteristic $p > 0$ we can now apply a retraction $\theta : B \to A$ and get the equation

$$(y_1 \dots y_n)^{N-1} = \sum_i b_i y_i^N \qquad (b_i = \theta(v_i))$$

holding in the regular ring A, where it is easily seen to be impossible.

We next discuss how to get the idea of this proof to work in mixed characteristic. It turns out that we can construct an acyclic complex to play the part of $K_\cdot^\infty$: the difficulty is that we cannot prove the direct summand conjecture.

Specifically, let R be a complete local domain of mixed characteristic $p > 0$, let $x_1 = p, x_2, \ldots, x_n$ be a system of parameters and let R^∞ be a domain integral over R such that (1) if $s \in R^\infty$ then $s^{1/p} \in R^\infty$ and (2) R^∞ is integrally closed.

For example we can take R^∞ to be the integral closure of R in an algebraic closure of its fraction field. We can still define flat ideals $(x_i^\infty) = \bigcup_e (x_i^{1/p^e})R^\infty$ and as before let

$$K_\cdot^\infty = \bigotimes_i (0 \longrightarrow (x_i^\infty) \hookrightarrow R^\infty \longrightarrow 0).$$

(2.5) THEOREM. *Under the hypotheses above, the complex* $K_\cdot^\infty$ *is acyclic.*

For details we refer the reader to [H_7]. The proof is similar to that of the earlier result: one uses induction on the number of x_i. There are a couple of differences, however. When $n = 2$ one uses the fact that R^∞ is integrally closed, so that x_i^{1/p^e}, x_j^{1/p^f} $(i \neq j)$ is an R-sequence for all e and f, to establish acyclicity. In the inductive step the key point is still to show that

$$((x_1^\infty) + \ldots + (x_{n-1}^\infty)) \cap (x_n^\infty) = ((x_1^\infty) + \ldots + (x_{n-1}^\infty))(x_n^\infty), \; n \geq 3.$$

The idea is to work modulo (x_1^∞): the earlier result for charactistic $p > 0$ then yields that

$$((x_1^\infty) + \ldots + (x_{n-1}^\infty)) \cap (x_n^\infty) \subset (x_1^\infty) + [(x_2^\infty) + \ldots + (x_{n-1}^\infty)](x_n^\infty).$$

If z is an element of the left hand side, we have

$$z = x_1^{1/p^e} v + y,$$

and then $z - y = x_1^{1/p^e} v \in (x_n^\infty)$. By an R-sequence argument, it follows that $v \in (x_n^\infty)$, and the desired result holds. □

The reader has probably noticed that instead of the direct summand conjecture, one could use the "fact" that if $x_1, \ldots, x_n$ is a system of parameters of a local ring R, then

$$x_1^t \ldots x_n^t \notin (x_1^{t+1}, \ldots, x_n^{t+1})R.$$

This assertion, not surprisingly, turns out to be equivalent to the direct summand conjecture. In fact, a number of equivalent statements are given in Proposition (2.10) below. Before giving that result, we shall discuss briefly the canonical element conjecture studied in [H_7].

Let R be a local ring of dimension n with maximal ideal $\underline{m}$. We use $H^i_{\underline{m}}(M)$ to denote the i-th local cohomology module of the R-module M with support in $\underline{m}$: one definition is

$$H^i_{\underline{m}}(M) = \varinjlim_t \operatorname{Ext}^i(R/\underline{m}^t, M).$$

A finitely generated R-module Ω is called a *canonical* module for R if $\operatorname{Hom}_R(\Omega, E(K)) \cong H^n_{\underline{m}}(R)$, where E(K) is the injective hull of the residue class field $K = R/\underline{m}$. The module Ω is determined up to non-unique isomorphism, if it exists. (If R is a homomorphic image of a Gorenstein ring and, in particular, if R is complete, then such a module Ω always exists.)

Notice that for every module M we have a natural map $\operatorname{Ext}^n_R(K,M) \to H^n_{\underline{m}}(M)$ (using the definition above), which we denote by θ_M. If we choose a projective resolution of K over R we get an exact sequence

$$0 \to \operatorname{syz}^n K \to P_{n-1} \to \cdots \to P_1 \to P_0 \to K \to 0$$

which, under the Yoneda definition of Ext, represents an element ε of $\operatorname{Ext}^n_R(K, \operatorname{syz}^n K)$. We refer to $\eta_R = \theta_{\operatorname{syz}^n K}(\varepsilon)$ as the *canonical element* in $H^n_{\underline{m}}(\operatorname{syz}^n K)$. Given a different choice of resolution there is a (non-unique) map between the resolutions which induces a map from the original module $\operatorname{syz}^n K$ to the new n-th module of syzygies $(\operatorname{syz}^n K)'$. This in turn induces a map

$$H^n_{\underline{m}}(\operatorname{syz}^n K) \longrightarrow H^n_{\underline{m}}((\operatorname{syz}^n K)')$$

which turns out to take the canonical element in $H^n_{\underline{m}}(\operatorname{syz}^n K)$ to the one in $H^n_{\underline{m}}((\operatorname{syz}^n K)')$ independently of the choices made: moreover, when restricted to the cyclic modules generated by the two canonical elements, this map is an isomorphism. Thus, the canonical element is well defined and unique in a certain sense. In particular, whether $\eta_R \neq 0$ does not depend on the choices made. The canonical element conjecture asserts that for every local ring R, $\eta_R \neq 0$. We now quote without proof some results from [H_7] which show how this conjecture relates to some of the others.

(2.7) PROPOSITION. *The following conditions on a local ring* R *are equivalent:*

(1) $\theta_M \neq 0$ *for some module* M;

(2) $\theta_{syz^n K} \neq 0$;

(3) $\eta_R \neq 0$.

If R *has a canonical module* Ω *then the following fourth condition is also equivalent to the above three:*

(4) $\theta_\Omega \neq 0$. □

(2.8) PROPOSITION. *If a local ring* R *has a big Cohen-Macaulay module* M, *then* $\theta_M \neq 0$ *and, hence,* $\eta_R \neq 0$. □

(2.9) PROPOSITION. *If* R *is local,* $\eta_R \neq 0$, *and* R *is module-finite over a regular local ring* A, *then* A *is a direct summand of* R. □

Thus, the canonical element conjecture implies the direct summand conjecture. But the converse is also true. Before stating the result which contains this fact, we make the following notational convention: if A is a complete local domain, then T_A denotes the integral closure of A in an algebraic closure of the fraction field of A.

(2.10) PROPOSITION. *The following statements are equivalent:*

(1) *the direct summand conjecture holds for all regular local rings* A;

(2) *the direct summand conjecture holds for complete unramified regular local rings with algebraically closed residue class fields;*

(3) *if* A *is a complete unramified regular local ring, then* A *is a direct summand of* T_A;

(4) *if* A *is a complete unramified regular local ring, then* $\mathrm{Hom}_A(T_A, A) \neq 0$;

(5) *if* A *is a complete unramified regular local ring with*

maximal ideal $\underline{m}$, *then* $H^n_{\underline{m}}(T_A) \neq 0$;

(6) *for every local ring* R, *we have* $\eta_R \neq 0$;

(7) *for every complete local domain* R, *we have* $\eta_R \neq 0$;

(8) *if* $x_1, \ldots, x_n$ *is a system of parameters of a local ring* R *then there do not exist integers* $b > a \geq 0$ *and elements* $y_1, \ldots, y_n \in R$ *such that*

$$(x_1 \ldots x_n)^a = \sum_{i=1}^{n} y_i x_i^b. \ \square$$

Several remarks should be made. All of the statements are known in the equicharacteristic case. Thus, we might as well consider only the mixed characteristic situation. We could have fixed the residual characteristic p and/or the dimension n of the rings A and R discussed: the statements are equivalent for fixed p,n. In (8) it would suffice to prove the impossibility of the case where $a = t$, $b = t+1$. Moreover, it would suffice to do the case where $x_1 = p$. Many of the implications are trivial or easy ((1) ⇒ (2), (3) ⇒ (2), (3) ⇒ (4) ⇔ (5), (6) ⇔ (7)) but some ((2) ⇒ (1), (4) ⇒ (1)) are more subtle. The reader is referred to [H_7].

We should note that one needs infinitely many cases of the direct summand conjecture to prove that $\eta_R \neq 0$ for one local ring R.

One reason for studying η_R is that it behaves functorially under various kinds of change of rings.

The statement (8) seems to be the most down-to-earth form of these conjectures. For many years the author has been pointing out that even the case where $n = 3$, $a = 2$, $b = 3$ is open. The author has now eliminated this possibility, at least when $x_1 = p$. In fact he has shown [H_6] that for $n \geq 3$, if

$$(x_1 \ldots x_n)^a = \sum_{i=1}^{n} y_i x_i^b,$$

then $a/b > 2/n$. This is rather weak for large n, considering that what one really wants to show is that $a/b \geq 1$.

The question of whether the equation above can hold when $n = 3$, $a = 3$, $b = 4$ remains open, so far as the author knows.

We pointed out above that, in connection with Proposition (2.10), we might as well consider only the mixed characteristic situation. The following corollary, again quoted without proof from [H_7], is concerned with that situation.

(2.11) COROLLARY [H_7; Corollary (5.4)]. *Let* R *be an n-dimensional local ring which is a homomorphic image of a* q*-dimensional Gorenstein local ring* S; *say* R = S/I. *Assume that* R *is of mixed characteristic* p > 0, *and that* p *is not a zerodivisor in* R.

If p *is also not a zerodivisor on* $\Omega'_R = \mathrm{Ext}_S^{q-n+1}(R,S)$, *then* $\eta_R \neq 0$. □

It should be mentioned that the Matlis dual of Ω'_R is $H_{\underline{m}}^{n-1}(R)$, so that p is not a zerodivisor on Ω'_R if and only if $H_{\underline{m}}^{n-1}(R)$ is p-divisible. However, the latter condition is not always satisfied.

References

[A_1] M.Auslander, "Modules over unramified regular local rings", Illinois J. Math., 5 (1961), 631-645.

[A_2] M.Auslander, "Modules over unramified regular local rings", Proceedings of the International Congress of Mathematicians, 15-22 August 1962 (Institute Mittag-Leffler, Djursholm, 1963), pp.230-233.

[B] H.Bass, "On the ubiquity of Gorenstein rings", Math. Z., 82 (1963), 8-28.

[BE_1] D.A.Buchsbaum and D.Eisenbud, "Lifting modules and a theorem on finite free resolutions", Ring Theory (Academic Press, New York, 1972), pp.63-74.

[BE_2] D.A.Buchsbaum and D.Eisenbud, "Some structure theorems for finite free resolutions", Adv. in Math., 12 (1974), 84-139.

[D_1] S.P.Dutta, Thesis, University of Michigan, Ann Arbor, 1981.

[D_2] S.P.Dutta, "Weak linking and multiplicities", preprint, University of Pennsylvania, 1981.

[D_3] S.P.Dutta, "Generalized intersection multiplicities of modules", Trans. Amer. Math. Soc., to appear.

[D_4] S.P.Dutta, "Frobenius and multiplicities", preprint, University of Pennsylvania, 1981.

[EG] E.G.Evans and P.Griffith, "The syzygy problem", Ann. of Math. (2), 114 (1981), 323-333.

[G] P.Griffith, "A representation theorem for complete local rings", J. Pure Appl. Algebra, 7 (1976), 303-315.

[H_1] M.Hochster, "Cohen-Macaulay modules", Conference on commutative algebra, Lecture Notes in Mathematics 311 (eds. J.W. Brewer and E.A.Rutter, Springer, Berlin, Heidelberg, New York, 1973), pp.120-152.

[H_2] M.Hochster, "Contracted ideals from integral extensions of regular rings", Nagoya Math. J., 51 (1973), 25-43.

[H_3] M.Hochster, Topics in the homological theory of modules over commutative rings, C.B.M.S. Regional Conference Series in Mathematics 24 (American Mathematical Society, Providence, 1975).

[H_4] M.Hochster, "An obstruction to lifting cyclic modules", Pacific J. Math., 61 (1975), 457-463.

[H_5] M.Hochster, "Associated graded rings derived from integrally closed ideals and the local homological conjectures", Colloq. d'algèbre, Université de Rennes I, 1980 (Université de Rennes I, 1981), pp.1-27.

[H_6] M.Hochster, "Euler characteristics over unramified regular local rings", Illinois J. Math., to appear.

[H_7] M.Hochster, "Canonical elements in local cohomology modules and the direct summand conjecture", preprint, University of Michigan, Ann Arbor, 1982.

[HM] M.Hochster and J.McLaughlin, "Quadratic extensions of regular local rings", Illinois J.Math., to appear.

[L] S.Lichtenbaum, "On the vanishing of Tor in regular local rings", Illinois J. Math., 10 (1966), 220-226.

[M] R.E.MacRae, "On an application of the Fitting invariants", J. Algebra, 2 (1965), 153-169.

[MB] M.-P.Malliavin-Brameret, "Une remarque sur les anneaux locaux réguliers", Séminaire Dubreil-Pisot, 24ème année, 1970-71, exposé 13.

[PS_1] C.Peskine and L.Szpiro, "Dimension projective finie et cohomologie locale", Publications Mathématiques 42 (Institut des Hautes Études Scientifiques, Paris, 1973), pp.47-119.

[PS_2] C.Peskine and L.Szpiro, "Syzygies et multiplicités", C. R. Acad. Sci. Paris Sér.A, 278 (1974), 1421-1424.

[R_1] P.Roberts, "Two applications of dualizing complexes over local rings", Ann. Sci. École Norm. Sup. (4), 9 (1976), 103-106.

[R_2] P.Roberts, "Cohen-Macaulay complexes and an analytic proof of the new intersection conjecture", J. Algebra, 66 (1980), 220-225.

[S] J.-P.Serre, Algèbre locale: multiplicités, Lecture Notes in Mathematics 11 (Springer, Berlin, Heidelberg, New York, 1965).

[W] A.Weil, Foundations of algebraic geometry, American Mathematical Society Colloquium Publications 29 (American Mathematical Society, Providence, 1962).

Department of Mathematics,
University of Michigan,
Ann Arbor,
Michigan 48109, U.S.A.

THE RANK OF A MODULE

G. HORROCKS

Two simple invariants of a module are its rank and the codimension of the set of primes at which it fails to be locally free. In general these invariants are unrelated as can be seen by taking direct sums of ideals. However in the geometric context of extending a locally free sheaf given over an open set to its closure the modules that arise are reflexive and this naïve example fails. Restricting to regular rings and localizing at the non-free set of primes leads to the following problem: for a regular local ring A determine the ranks of those non-free reflexive A-modules which are locally free except at the maximal ideal.

Denote by $\underline{m}$ the maximal ideal of A, by k its residue field $A/\underline{m}$, and by X the spectrum of A punctured at $\underline{m}$. Call the modules in question X-bundles. The possible ranks of non-free X-bundles have been determined only when the dimension d of A is at most 5, and for these dimensions there are indecomposable X-bundles of all ranks. In the first section I review briefly some of the methods used for constructing X-bundles and in the second describe an approach to the problem of finding restrictions on the ranks especially in terms of their cohomology. Here the Syzygy Theorem of Evans and Griffith has interesting consequences [1,3].

1. *Constructions*

Reflexive modules are free for $d \le 2$. Assume that $d > 2$. Any syzygy with level at least two is reflexive (level 0 is the module itself), and those arising from artinian modules are X-bundles. Moreover local duality [5] shows that indecomposable artinian modules give indecomposable X-bundles. Second syzygies of cyclic artinian modules provide easy examples of indecomposable X-bundles of all ranks greater than d-2.

The p-th syzygies $T^{(d-p)}$ $(2 \leq p < d)$ of k serve as building-blocks for all X-bundles in the sense that any X-bundle can be obtained as the complement of a free direct summand of a bundle with a filtration whose quotients are the syzygies T. The size of the free direct summand can be found from the following.

LEMMA A. *Assume that* E,F *are* X-*bundles without free direct summands and that* $0 \to E \to M \to F \to 0$ *is exact. Choose minimal free resolutions* $P \to E^*$, $R \to F^*$ *for the duals of* E,F *and let* $0 \to E' \to M' \to F' \to 0$ *be the cokernel of the given exact sequence when it is embedded in* $P_o^* \oplus R_o^*$. *The rank of the free direct summand of* M *is equal to the rank of the connecting homomorphism*

$\mathrm{Tor}^1(F',k) \to E' \otimes k$.

With this criterion it is easy to find the ranks of the free direct summands of extensions where $E = T^p$, $F = T^q$. Except when d is even, $p = d-2$ and $q = 1$ the resulting non-free complements have rank at least $d-1$. In the exceptional case the module $\mathrm{Ext}^1(T^1, T^{d-2})$ is the second exterior power $\wedge^2(\underline{m}/\underline{m}^2)^*$ and provided the extension is a non-singular element the resulting bundle is a null-correlation bundle of rank d-2 [9]. (These calculations have been made independently by Moore [8] from a different viewpoint. He has also considered three-fold extensions.)

One bundle arising from the foregoing construction which may have special interest corresponds to $d = 7$, $p = 4$, $q = 2$. The module $\mathrm{Ext}^1(T^2, T^4)$ is $\wedge^3(\underline{m}/\underline{m}^2)^*$. The stabilizer of a general element of this space is the exceptional group G_2 and for a local ring A with a coefficient field the resulting X-bundle lifts from a vector bundle on $\mathbb{P}^6$ with a G_2-action, rank 9, and chern polynomial $1 + 3h^2 - 28h^6$ (where h is a generator of cohomology).

A non-free X-bundle of rank d-2 for any d has been constructed by Vetter [12] as the kernel of a mapping $(2d-3)A \to dA$ for which he writes down a quite explicit matrix. Tango [10] finds vector bundles of the same rank on $\mathbb{P}^{d-1}$ by factoring out sufficiently many general sections of the second exterior power of the tangent bundle. The dual of Vetter's bundle comes from a special choice of these sections. To construct these examples take a base $\xi_1,\ldots,\xi_d$ for a free module F of rank d, choose generators $a_1,\ldots,a_d$ for $\underline{m}$ and put $\xi = \Sigma a_i \xi_i$. The cokernel of

$$\xi : F \to \wedge^2 F$$

is a bundle G of rank (d-1)(d-2)/2. An element σ of $\wedge^2 F$ determines a rank 1 sub-bundle of G provided that $\sigma \notin F \wedge \xi$ mod p for any $p \in X$. Suppose that $\wedge^2 F$ has a direct summand V of rank N whose non-zero elements all satisfy this condition. The rank of G/ImV is (d-1)(d-2)/2 - N. In the geometric case $\wedge^2 F$ is the product of X and a vector space $\wedge^2(k^d)$, and it is sufficient to choose an N-dimensional subspace of $\wedge^2(k^d)$ containing no non-zero elements $\alpha \wedge \beta$ ($\alpha,\beta \in k^d$). When k is infinite this choice is possible provided N is less than the codimension of the grassmanian of lines in $\mathbb{P}^{d-1}$ in its ambient space. In this way a bundle of rank d-2 is obtained.

Vetter's construction is valid for any k (and any Cohen-Macaulay ring A). It is equivalent to choosing V to be the sub-module of $\wedge^2 F$ spanned by

$$\xi_i \wedge \xi_{j+1} + \xi_j \wedge \xi_{i+1} \quad (i+1 < j < d,\ 1 \le i < d-2).$$

In [9] these examples are shown to be indecomposable (the argument easily adapts to an arbitrary regular local ring). The problem remaining is to find whether there are non-free bundles with ranks r in the range [2,d-3].

The only known examples occur for (r,d) = (2,5), (3,6) (decomposable in characteristic 2), (2,6) in characteristic 2 [7,6,11]. They are subquotients $\mathrm{Ker}\beta/\mathrm{Im}\,\alpha$ of

$$F \xrightarrow{\alpha} E \xrightarrow{\beta} F',$$

where α,β are locally split, F, F' are free. When A has a coefficient field they lift from projective space and E is a homogeneous vector bundle. The homomorphisms α,β can be calculated explicitly by means of the Borel-Weil-Bott Theorem [6] and the mapping $(\alpha,\beta) \to \beta\alpha$ can be found by representation theory.

When E is a bundle on $\mathbb{P}^{d-1}$ generated by its sections α can be chosen (in general position) so that the rank of E/F is equal to the degree of the chern polynomial of E. The bundles of rank d-2 described earlier arise in this way [10]. The Borel-Weil-Bott

Theorem is an efficient method for finding those bundles associated with the tangent or null-correlation bundles that are generated by sections. I do not know whether any of these have chern polynomials with degree less than d-2.

2. *Restrictions*

Known restrictions on the rank of an X-bundle E involve the cohomology modules

$$H^i(E) = \mathrm{Ext}^i(E^*,A) \quad (i = 1,\dots,d-2).$$

(They are artinian and isomorphic to the cohomology modules $H^{i+1}_{\underline{m}}(E)$ [5].) When the first i of these modules vanish E is an (i + 2)-th syzygy and, at least for A with a coefficient field, either $\mathrm{rk}(E) \geq i+2$ or E is free. Because of duality this is also true if the last i vanish. These two consequences of the Syzygy Theorem are given in [1,3]. The two results that follow arise from attempts to find restrictions on the ranks in terms of invariants with support at the maximal ideal.

Let $\Xi(E)$ be the annihilator ideal of the canonical mapping $E^* \otimes E \to \mathrm{Hom}(E,E)$. It is $\underline{m}$-primary and its elements are the homotheties of E that factor through free modules. The annihilator ideals $\underline{u}_i$ of the cohomology modules give bounds

$$\underline{u}_1 \cap \underline{u}_2 \cap \dots \cap \underline{u}_{d-2} \supseteq \Xi(E) \supseteq \underline{u}_1\underline{u}_2\cdots\underline{u}_{d-2}.$$

Let N be the multiplicity of $\Xi(E)$, that is the minimum of the lengths $\ell(A/(y))$ over all systems of parameters y in $\Xi(E)$.

PROPOSITION B. $N.\mathrm{rk}(E) \geq \sum_{i=1}^{d-2} \binom{d-1}{i} \ell(H_i(E))$.

The proof is by the sub-additivity of cohomology and the vanishing of the morphisms of cohomology induced by elements of $\Xi(E)$. Equality is attained when E comes from a Koszul complex, but for syzygies of powers of $\underline{m}$ it trivializes.

The second result is a lower bound for rk(E) given by the ranks of mappings between syzygies of systems of parameters. Let $x_1,\dots,x_n$ belong to $\underline{m}$ and $T^p(x)$ be the module of (d-p-1)-th Koszul syzygies of x. When x is a base for $\underline{m}$ the module is just T^p. The last non-vanishing cohomology module is the (d-p-1)-th, isomorphic to $A/(x)$.

THEOREM C. *Let* p *be the least integer in* [1,d-2] *such that* $H^{d-p-1}(E) \neq 0$. *Choose any* $a_1,\ldots,a_n$ *in* $\Xi(E)$. *Then any*

$$\phi: A/(a_1,\ldots,a_n) \to H^{d-p-1}(E)$$

can be lifted to a homomorphism of the Koszul complex of $a_1,\ldots,a_n$ *into the dual of a projective resolution of* E^*.

Theorem C is deduced from a property of complexes that are modules over exterior algebras. Let E be the free exterior algebra over A (which may be any commutative ring) on generators $\xi_1,\ldots,\xi_n$. Take any complex K which is a graded E-module with a differential d given by multiplication by a fixed element $\Sigma x_i\xi_i$. Then we have the following.

LEMMA D. *Let* L *be a complex, with differential of degree* 1, *on which* $x_1,\ldots,x_n$ *induce homotheties chain homotopic to zero. Then any morphism of complexes* $K^i \to L^i$ $(i > p)$ *extends to* $K \to L$.

The extension is given by

$$f^p = \sum_{t=0}^{n-1} (-1)^{c(t)} \sum_{\alpha_0 < \ldots < \alpha_t} \sigma_{\alpha_0} \ldots \sigma_{\alpha_t} f^{p+1+t} \xi_{\alpha_0} \ldots \xi_{\alpha_t},$$

where $c(t) = tr + r + t(t-1)/2$, *and* α_i *is a chain homotopy inducing* x_i.

To obtain the connection between rk(E) and ranks of mappings between Koszul syzygies take n=d and a to be a system of parameters in Theorem C. Since a annihilates $H^{d-p-1}(E)$, ϕ can be chosen so that $\phi(1)$ belongs to a minimal generating set. Choose $\psi: H^{d-p-1}(E) \to k$ so that $\psi(\phi(1)) \neq 0$. It is covered by a mapping of the dual of the projective resolution of E into the Koszul complex of k. So there exists $g: E \to T^p$ inducing ψ on cohomology. By Theorem C there exists also $f: T^p(a) \to E$ inducing ϕ. So

$$gf: T^p(a) \to T^p$$

factors through E and determines a non-zero homomorphism $A/(a) \to k$ of cohomology. Define $\mu(x,p)$ to be the minimum of the ranks of homomorphisms $T^p(x) \to T^p$ non-zero on cohomology. Put $\mu(p) = \inf \mu(x,p)$ over all systems of parameters x. Thus

$$\mathrm{rk}(E) \geq \mu(a,p) \geq \mu(p). \qquad \text{(E)}$$

The functions $\mu(a,p)$, $\mu(p)$ have not been determined. I do not know whether this route leads to the consequences of the Syzygy Theorem described earlier. For that $\mu(p)$ would need to be at least $p+1$. This is known only for $p=d-2$ where it is a special case of Eisenbud and Evans [2] and for $p=1$ by unique factorization in regular local rings.

References

1. W. Bruns, E.G. Evans and P. Griffith, "Syzygies, ideals of height two and vector bundles", J.Algebra, 67 (1980), 143-162.

2. D. Eisenbud and E.G. Evans, "A generalized principal ideal theorem", Nagoya Math.J., 62 (1976), 41-53.

3. E.G. Evans and P. Griffith, "Syzygies of small rank", Bull. Amer. Math.Soc., 4 (1981), 329-330.

4. E.G. Evans and P. Griffith, "Order ideals of minimal generators", preprint, University of Illinois at Urbana-Champaign, 1981.

5. R. Hartshorne, Local cohomology, Lecture Notes in Mathematics 41 (Springer, Berlin, 1967).

6. G. Horrocks, "Examples of rank three vector bundles on five dimensional projective space", J.London Math.Soc. (2), 18 (1978), 15-27.

7. G. Horrocks and D. Mumford, "A rank 2 vector bundle on $\mathbb{P}^4$ with 15,000 symmetries", Topology, 12 (1973), 63-81.

8. R.R. Moore, "Extensions and cohomology of bundles on $\mathbb{P}^n$", D. Phil. Thesis, Oxford University, 1980.

9. C. Okonek, M. Schneider and H. Spindler, Vector bundles on complex projective space (Birkhäuser, Boston, 1980).

10. H. Tango, "An example of indecomposable vector bundle of rank n-1 on $\mathbb{P}^n$", J. Math. Kyoto Univ., 16 (1976), 137-141.

11. H. Tango, "On morphisms from projective space $\mathbb{P}^n$ to the Grassman variety Gr(n,d)", J.Math. Kyoto Univ., 16 (1976), 201-207.

12. U. Vetter, "Zu einem Satz von G. Trautmann über den Rang gewisser kohärenter analytische Moduln", Arch. Math. (Basel), 24 (1973), 158-161.

School of Mathematics,
University of Newcastle upon Tyne,
Newcastle upon Tyne NE1 7RU, U.K.

MODULES OF GENERALIZED FRACTIONS AND BALANCED BIG COHEN-MACAULAY MODULES

R.Y.SHARP AND H.ZAKERI

1. *Introduction*

It is well known that, for a Gorenstein ring A, the total ring of fractions of A provides the injective envelope of A. One of the motivations behind the work which led to our construction of modules of generalized fractions (which was outlined in a lecture at the Symposium, is reviewed in §2 below, and is described in detail in [9]) was a desire to find a similarly satisfactory description of the terms $E^i(A)$ for $i > 0$ in the minimal injective resolution for the Gorenstein ring A.

At the end of this paper it is shown that generalized fractions do provide such descriptions: whenever R is a commutative ring (with identity), M is an R-module, n is a positive integer and U is what is called a triangular subset of R^n, a module $U^{-n}M$ of generalized fractions may be constructed; in the case of a Gorenstein ring A, the set U_n of all poor A-sequences of length n forms a triangular subset of A^n (we say that a sequence $a_1,\ldots,a_n$ of elements of A forms a *poor* A-*sequence* if

$$((Aa_1 +\ldots+ Aa_{i-1}) : a_i) = (Aa_1 +\ldots+ Aa_{i-1})$$

for all $i = 1,\ldots,n$), and the module of generalized fractions $U_n^{-n}A$ turns out to be isomorphic to $E^{n-1}(A)$. This is just the sort of description that was hoped for, because it should be noted that $U_1^{-1}A$ is just the total ring of fractions of A.

However the main purpose of this paper is to establish connections between modules of generalized fractions and the concept of regular sequence in commutative algebra, and the above result is really just a consequence. The main result of the paper is the Exactness Theorem (Theorem 3.3) which is proved in §3; this leads to characterizations (given in §4), in terms of modules of generalized fractions, of balanced big Cohen-Macaulay modules [8] over a local (commutative, Noetherian) ring, and of modules which are big Cohen-Macaulay modules with respect to a specified system of parameters for such a local ring; §5 contains the above-mentioned result about the minimal injective resolution for a Gorenstein ring.

2. *Review of the construction of modules of generalized fractions*

In this section, we shall recall briefly the main ingredients in the construction of modules of generalized fractions: for more details, the reader is referred to [9].

Throughout the paper, A will denote a commutative ring (with identity) and n will denote a positive integer. We use $D_n(A)$ to denote the set of all $n \times n$ lower triangular matrices with entries in A; for $\underset{\sim}{H} \in D_n(A)$, the determinant of $\underset{\sim}{H}$ is denoted by $|\underset{\sim}{H}|$; and we use ^{T} to denote matrix transpose.

A *triangular subset* of A^n is a non-empty subset U of A^n such that

(i) whenever $(u_1,\ldots,u_n) \in U$, then $(u_1^{\alpha_1},\ldots,u_n^{\alpha_n}) \in U$ for all choices of positive integers $\alpha_1,\ldots,\alpha_n$, and

(ii) whenever $(u_1,\ldots,u_n)$ and $(v_1,\ldots,v_n) \in U$, then there exist $(w_1,\ldots,w_n) \in U$ and lower triangular matrices $\underset{\sim}{H},\underset{\sim}{K} \in D_n(A)$ such that

$$\underset{\sim}{H}[u_1 \ \ldots \ u_n]^T = [w_1 \ \ldots \ w_n]^T = \underset{\sim}{K}[v_1 \ \ldots \ v_n]^T.$$

Given such a U and an A-module M, one can construct (see §2 of [9]) the module of generalized fractions $U^{-n}M$ of M with respect to U. A general element of $U^{-n}M$ has the form $\dfrac{m}{(u_1,\ldots,u_n)}$ (with $m \in M$ and $(u_1,\ldots,u_n) \in U$), where $\dfrac{m}{(u_1,\ldots,u_n)}$ denotes the equivalence class of $(m,(u_1,\ldots,u_n))$ under the equivalence relation $\hat{\sim}$ on $M \times U$ defined as follows: for $x,y \in M$ and $(u_1,\ldots,u_n),(v_1,\ldots,v_n) \in U$, we write $(x,(u_1,\ldots,u_n)) \ \hat{\sim} \ (y,(v_1,\ldots,v_n))$ precisely when there exist $(w_1,\ldots,w_n) \in U$ and lower triangular matrices $\underset{\sim}{H},\underset{\sim}{K} \in D_n(A)$ such that

$$\underset{\sim}{H}[u_1 \ \ldots \ u_n]^T = [w_1 \ \ldots \ w_n]^T = \underset{\sim}{K}[v_1 \ \ldots \ v_n]^T$$

and

$$|\underset{\sim}{H}|x - |\underset{\sim}{K}|y \in \left(\sum_{i=1}^{n-1} Aw_i\right)M.$$

The A-module operations for $U^{-n}M$ are described as follows. If $x, y \in M$ and $(u_1,\ldots,u_n), (v_1,\ldots,v_n) \in U$, then

$$\frac{x}{(u_1,\ldots,u_n)} + \frac{y}{(v_1,\ldots,v_n)} = \frac{|\underset{\sim}{H}|x + |\underset{\sim}{K}|y}{(w_1,\ldots,w_n)}$$

for any choice of $(w_1,\ldots,w_n) \in U$ and $\underset{\sim}{H}, \underset{\sim}{K} \in D_n(A)$ such that $\underset{\sim}{H}\underset{\sim}{u} = \underset{\sim}{w} = \underset{\sim}{K}\underset{\sim}{v}$. (Here, and in the sequel, we use the notation of [9;§2] whereby, for a sequence $(s_1,\ldots,s_n) \in U$, the matrix $[s_1 \ldots s_n]^T$ is denoted by $\underset{\sim}{s}$.) Also, with the above notation, and for $a \in A$,

$$a\left(\frac{x}{(u_1,\ldots,u_n)}\right) = \frac{ax}{(u_1,\ldots,u_n)} .$$

The above concept is indeed a generalization of the familiar concept of ordinary module of fractions: see [9;3.1].

3. *Modules of generalized fractions and regular sequences: the Exactness Theorem*

Throughout this section, we shall assume that A is Noetherian and that M is an A-module (not necessarily finitely generated). We say that $a_1,\ldots,a_n \in A$ form a *poor M-sequence* if

$$((Aa_1 + \ldots + Aa_{i-1})M : a_i) = (Aa_1 + \ldots + Aa_{i-1})M$$

for all $i = 1,\ldots,n$, and that $a_1,\ldots,a_n$ form an *M-sequence* if they form a poor M-sequence and, in addition, $M \neq \left(\sum_{j=1}^{n} Aa_j\right)M$.

This section contains the statement and proof of the main theorem of the paper. Its formulation requires the introduction of some notation, which will be in force throughout the section.

3.1 *Notation.* We shall use $\mathbb{N}$ to denote the set of all positive integers. The symbol $\mathcal{U}$ will denote a family $(U_i)_{i\in\mathbb{N}}$ of sets such that

(i) U_i is a triangular subset of A^i for all $i \in \mathbb{N}$;

(ii) whenever $(u_1,\ldots,u_i) \in U_i$ with $1 < i \in \mathbb{N}$, then $(u_1,\ldots,u_{i-1}) \in U_{i-1}$;

(iii) whenever $(u_1,\ldots,u_i) \in U_i$ with $i \in \mathbb{N}$, then $(u_1,\ldots,u_i,1) \in U_{i+1}$;

(iv) $(1) \in U_1$.

It is straightforward to check that the above data allow us to construct a complex, as described in the following lemma.

3.2 LEMMA. *There are* A-*homomorphisms* $d^o : M \to U_1^{-1}M$ *and* $d^i : U_i^{-i}M \to U_{i+1}^{-i-1}M$ *for each* $i \in \mathbb{N}$ *such that* $d^o(m) = \frac{m}{(1)}$ *for all* $m \in M$ *and, for each* $i \in \mathbb{N}$,

$$d^i\left(\frac{m}{(u_1,\ldots,u_i)}\right) = \frac{m}{(u_1,\ldots,u_i,1)}$$

for all $m \in M$ *and* $(u_1,\ldots,u_i) \in U_i$. *Moreover, it follows from* [9;3.3(ii)] *that*

$$0 \xrightarrow{d^{-1}} M \xrightarrow{d^o} U_1^{-1}M \longrightarrow \ldots \longrightarrow U_i^{-i}M \xrightarrow{d^i} U_{i+1}^{-i-1}M \longrightarrow \ldots$$

is a complex of A-*modules and* A-*homomorphisms which we shall denote henceforth by* $C(\mathcal{U},M)$. □

We can now state the main result of this paper.

3.3 EXACTNESS THEOREM. *With the above notation, the complex* $C(\mathcal{U},M)$ *is exact if and only if, for all* $i \in \mathbb{N}$, *each member of* U_i *is a poor* M-*sequence*.

The proof of this result will be preceded by some preparatory lemmas and propositions.

3.4 *Notation: the diagrams* $D(s;t;\underset{\sim}{H})$. Let n be a positive integer and suppose that $s = (s_1,\ldots,s_{n+1})$, $t = (t_1,\ldots,t_{n+1})$ are two sequences in U_{n+1} for which there exists $\underset{\sim}{H} = [h_{ij}] \in D_{n+1}(A)$ such that $\underset{\sim}{H}\underset{\sim}{s} = \underset{\sim}{t}$. Note that, since $\underset{\sim}{H}$ is lower triangular, we have

$$\underset{\sim}{H}^* = \begin{bmatrix} h_{11} & 0 & \cdots & 0 \\ h_{21} & h_{22} & \cdots & 0 \\ \cdot & \cdot & \cdot & \cdot \\ \cdot & \cdot & \cdot & \cdot \\ \cdot & \cdot & \cdot & \cdot \\ h_{n1} & h_{n2} & \cdots & h_{nn} \end{bmatrix} \in D_n(A)$$

and $\underset{\sim}{H}^*[s_1 \ldots s_n]^T = [t_1 \ldots t_n]^T$. Set, for each $i = 1,\ldots,n+1$,

$$\underline{a}_i = As_1 + \ldots + As_i,$$

$$\underline{b}_i = At_1 + \ldots + At_i,$$

$$k_i = h_{11} \ldots h_{ii}.$$

We use $D(s;t;\underset{\sim}{H})$ to denote the diagram

$$\begin{array}{ccccccc}
M/\underline{a}_n M & \xrightarrow{\sigma_{n+1}} & M/\underline{a}_n M & \longrightarrow & M/\underline{a}_{n+1} M & \longrightarrow & 0 \\
\downarrow \gamma_n & & \downarrow \theta_{n+1} & & \downarrow \gamma_{n+1} & & \\
M/\underline{b}_n M & \xrightarrow{\tau_{n+1}} & M/\underline{b}_n M & \longrightarrow & M/\underline{b}_{n+1} M & \longrightarrow & 0 \\
\downarrow & & \downarrow & & \downarrow & & \\
M/(k_n,\underline{b}_n)M & \xrightarrow{\alpha_{n+1}} & M/(k_{n+1},\underline{b}_n)M & \longrightarrow & M/(k_{n+1},\underline{b}_{n+1})M & \longrightarrow & 0 \\
\downarrow & & \downarrow & & \downarrow & & \\
0 & & 0 & & 0 & &
\end{array}$$

of A-modules and A-homomorphisms in which each unnamed homomorphism is the obvious natural epimorphism, σ_{n+1} is multiplication by s_{n+1} and τ_{n+1} is multiplication by t_{n+1}, and the remaining homomorphisms are defined as follows.

By [9;2.2], $k_i \underline{a}_i M \subseteq \underline{b}_i M$ for $i = n, n+1$, and so we can define $\gamma_i : M/\underline{a}_i M \rightarrow M/\underline{b}_i M$ by $\gamma_i(\bar{m}) = k_i \tilde{m}$ for all $m \in M$. (The natural image of an element m of M in $M/\underline{a}_i M$ (respectively $M/\underline{b}_i M$) is denoted by $\bar{m}$ (respectively $\tilde{m}$).) Also, $\theta_{n+1} : M/\underline{a}_n M \rightarrow M/\underline{b}_n M$ is defined by $\theta_{n+1}(\bar{m}) = k_{n+1}\tilde{m} = |\underset{\sim}{H}|\tilde{m}$ for all $m \in M$.

Next, note that

$$t_{n+1}k_n = h_{11} \ldots h_{nn}\left(\sum_{i=1}^{n} h_{n+1i}s_i + h_{n+1n+1}s_{n+1}\right),$$

so that $t_{n+1}k_n \in Ak_{n+1} + \underline{b}_n$ by [9;2.2]. We thus see that

$$t_{n+1}(Ak_n + \underline{b}_n)M \subseteq (Ak_{n+1} + \underline{b}_n)M,$$

and this enables us to define $\alpha_{n+1} : M/(k_n,\underline{b}_n)M \rightarrow M/(k_{n+1},\underline{b}_n)M$ by $\alpha_{n+1}(\bar{\bar{m}}) = t_{n+1}\hat{m}$ for all $m \in M$ (and here the natural images of m in the appropriate modules are denoted by $\bar{\bar{m}}$ and $\hat{m}$).

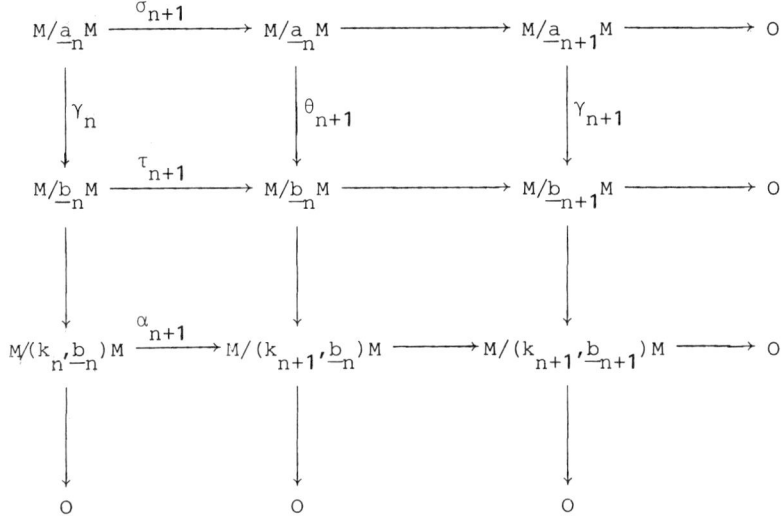

3.5 PROPOSITION. *With the above notation, the diagram* $D(s;t;\underset{\sim}{H})$ *is commutative and has exact rows and columns.*

Proof. It is clear that all the squares in the diagram apart from the upper left one commute; moreover, for $m \in M$,

$$t_{n+1}k_n m - k_{n+1}s_{n+1}m = k_n(t_{n+1} - h_{n+1\,n+1}s_{n+1})m = k_n\left(\sum_{i=1}^{n} h_{n+1\,i}s_i\right)m;$$

hence $t_{n+1}k_n m - k_{n+1}s_{n+1}m \in \underline{b}_n M$ by [9; 2.2], and so the upper left square commutes as well. It is easy to see that the rows and columns of $D(s;t;\underset{\sim}{H})$ are exact. □

3.6 LEMMA. *With the above notation,* $\underline{b}_{n+1}(\ker \alpha_{n+1}) = 0$.

Proof. It is clear that $\underline{b}_n(\ker \alpha_{n+1}) = 0$. If $m \in M$ is such that $\bar{\bar{m}} \in \ker \alpha_{n+1}$, then $t_{n+1}m \in (Ak_{n+1} + \underline{b}_n)M \subseteq (Ak_n + \underline{b}_n)M$, and so the claim follows. □

This last lemma will be used, in the proof of the Exactness Theorem, in conjunction with the following remark, the idea for which comes from Peskine's and Szpiro's "Lemme d'Acyclicité" [5;1.8].

3.7 *Remark.* If $\underline{b}$ is an ideal of A and $f : X \to Y$ is a homomorphism of A-modules for which $\underline{b}\ker f = 0$ but $\underline{b}$ contains a non-zerodivisor on X, then f is a monomorphism.

In the course of the proof of the Exactness Theorem, we shall need to use some local cohomology theory. For $i \geq 0$, we use $H^i_{\underline{a}}$ to denote the i-th right derived functor of the local cohomology functor $L_{\underline{a}}$ with respect to an ideal $\underline{a}$ of A [7;2.1]. In particular, we shall make use of the following.

3.8 *Observation.* Let X be an A-module, and let $\underline{p} \in \mathrm{Spec}(A)$. Then $\underline{p} \in \mathrm{Ass}(X)$ if and only if $H^0_{\underline{p}A_{\underline{p}}}(X_{\underline{p}}) \neq 0$.

3.9 *Notation.* Let n be a positive integer. For $s = (s_1,\ldots,s_{n+1})$ and $t = (t_1,\ldots,t_{n+1}) \in U_{n+1}$ and $\underset{\sim}{H} \in D_{n+1}(A)$ such that $\underset{\sim}{H}\underset{\sim}{s} = \underset{\sim}{t}$, we shall be interested not only in the diagram $D(s;t;\underset{\sim}{H})$ but also in its extension

$$\begin{array}{ccccccccc}
 & & 0 & & 0 & & 0 & & \\
 & & \downarrow & & \downarrow & & \downarrow & & \\
0 & \longrightarrow & M/\underline{a}_n M & \xrightarrow{\sigma_{n+1}} & M/\underline{a}_n M & \longrightarrow & M/\underline{a}_{n+1} M & \longrightarrow & 0 \\
 & & \downarrow{\scriptstyle \gamma_n} & & \downarrow{\scriptstyle \theta_{n+1}} & & \downarrow{\scriptstyle \gamma_{n+1}} & & \\
0 & \longrightarrow & M/\underline{b}_n M & \xrightarrow{\tau_{n+1}} & M/\underline{b}_n M & \longrightarrow & M/\underline{b}_{n+1} M & \longrightarrow & 0 \\
 & & \downarrow & & \downarrow & & \downarrow & & \\
0 & \longrightarrow & M/(k_n,\underline{b}_n)M & \xrightarrow{\alpha_{n+1}} & M/(k_{n+1},\underline{b}_n)M & \longrightarrow & M/(k_{n+1},\underline{b}_{n+1})M & \longrightarrow & 0 \\
 & & \downarrow & & \downarrow & & \downarrow & & \\
 & & 0 & & 0 & & 0 & & ,
\end{array}$$

which we shall denote by $E(s;t;\underset{\sim}{H})$. By 3.5, $E(s;t;\underset{\sim}{H})$ is commutative, but the exactness or otherwise of each of its rows and columns depends on whether or not the appropriate named homomorphism is injective.

The proof of one implication in the Exactness Theorem uses induction, and the next two results provide a basis for that induction.

3.10 LEMMA. (i) *If* $C(\mathcal{U},M)$ *is exact at* M, *then every member of* U_1 *is a poor* M-*sequence.*

(ii) *If* $C(\mathcal{U},M)$ *is exact at* M *and* $U_1^{-1}M$, *then every member of* U_2 *is a poor* M-*sequence.*

Proof. Part (i) is a routine exercise, and so we only deal with (ii). Suppose that $(u_1,u_2) \in U_2$ and $m \in M$ are such that $u_2m \in u_1M$. Then $d^1\left(\frac{m}{(u_1)}\right) = \frac{u_2m}{(u_1,u_2)} = 0$ by [9;3.3]. Thus, since $C(\mathcal{U},M)$ is exact at $U_1^{-1}M$, $\frac{m}{(u_1)} = \frac{m'}{(1)}$ for some $m' \in M$. Hence $m \in u_1M$. This, in conjunction with (i), shows that (u_1,u_2) is a poor M-sequence. □

3.11 PROPOSITION. *Assume that* $C(\mathcal{U},M)$ *is exact at* M *and* $U_1^{-1}M$. *Then, for each choice of* $s = (s_1,s_2)$, $t = (t_1,t_2) \in U_2$ *and* $\underset{\sim}{H} = [h_{ij}] \in D_2(A)$ *for which* $\underset{\sim}{H}\underset{\sim}{s} = \underset{\sim}{t}$, *the diagram* $E(s;t;\underset{\sim}{H})$ *has exact rows and columns.*

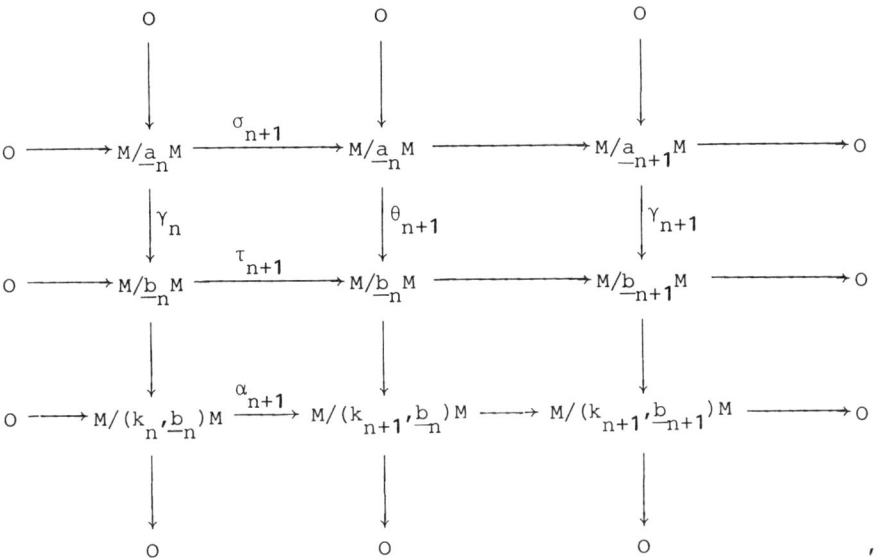

Proof. In the notation of 3.9 and 3.4, we have to show that $\gamma_1, \theta_2, \gamma_2, \sigma_2, \tau_2$ and α_2 are all monomorphisms. By 3.10, σ_2 and τ_2 are monomorphic. To see that γ_1 is a monomorphism, suppose $m \in M$ is such that $h_{11}m \in \underline{b}_1M = t_1M$. Thus $h_{11}m = t_1m'$ for some $m' \in M$, and so $t_1m = h_{11}s_1m = s_1t_1m'$. Hence, by 3.10, $m = s_1m' \in \underline{a}_1M$. Thus γ_1 is a monomorphism.

Next, we show that θ_2 is a monomorphism. Let $m \in M$ be such that $h_{11}h_{22}m \in \underline{b}_1M = t_1M$. Thus $h_{11}h_{22}s_2m \in t_1M$, and so $h_{11}(t_2-h_{21}s_1)m \in t_1M$. But $h_{11}s_1 = t_1$, and so $h_{11}t_2m \in t_1M$. Now (t_1,t_2) is a poor M-sequence, by 3.10, and so $h_{11}m \in t_1M$. Hence, since γ_1 is a monomorphism, $m \in \underline{a}_1M$. It follows that θ_2 is a monomorphism.

We now consider γ_2. Suppose that $m \in M$ is such that $h_{11}h_{22}m \in \underline{b}_2M = (At_1+At_2)M$. Thus there exists $m' \in M$ such that $h_{11}h_{22}m + t_2m' \in t_1M$. Therefore

$$s_1h_{11}h_{22}m + s_1t_2m' \in t_1M,$$

so that $t_2(s_1m') \in t_1M$. But (t_1,t_2) is a poor M-sequence by 3.10(ii), and so $s_1m' \in t_1M$; thus $s_1m' = h_{11}s_1m''$ for some $m'' \in M$. Hence, by 3.10(i), $m' = h_{11}m''$, and so

$$h_{11}h_{22}m + t_2h_{11}m'' \in t_1M.$$

Therefore $h_{11}h_{22}s_2m + s_2t_2h_{11}m'' \in t_1M$, so that

$$h_{11}(t_2-h_{21}s_1)m + s_2t_2h_{11}m'' \in t_1M.$$

Hence $t_2(h_{11}m + s_2h_{11}m'') \in t_1M$, so that $h_{11}m + s_2h_{11}m'' = s_1h_{11}m'''$ for some $m''' \in M$. But $s_1h_{11} = t_1$ is a non-zerodivisor on M, and so $m + s_2m'' = s_1m'''$; hence $m \in \underline{a}_2M$. It follows that γ_2 is a monomorphism.

It is now immediate from [3; Chapter 4, Theorem 5] that α_2 is a monomorphism. □

As mentioned earlier, the proof of one implication in the Exactness Theorem uses an inductive argument, and we isolate the inductive step in a separate theorem.

3.12 THEOREM. *Let* $n \in \mathbb{N}$ *with* $n > 1$. *Assume that* $C(\mathcal{U},M)$ *is exact at* $M, U_1^{-1}M, \ldots, U_n^{-n}M$, *that every element of* U_n *is a poor* M*-sequence, and that, for every* $i=1,\ldots,n-1$ *and for each choice of* $s',t' \in U_{i+1}$ *and* $\underset{\sim}{H}' \in D_{i+1}(A)$ *for which* $\underset{\sim}{H}'\underset{\sim}{s}' = \underset{\sim}{t}'$, *the diagram* $E(s';t';\underset{\sim}{H}')$ *has exact rows and columns.*

Then every element of U_{n+1} *is a poor* M-*sequence and, for each choice of* $s = (s_1,\ldots,s_{n+1})$, $t = (t_1,\ldots,t_{n+1}) \in U_{n+1}$ *and* $\underset{\sim}{H} = [h_{ij}] \in D_{n+1}(A)$ *for which* $\underset{\sim}{H}\underset{\sim}{s} = \underset{\sim}{t}$, *the diagram* $E(s;t;\underset{\sim}{H})$ *has exact rows and columns.*

Proof. We show first that each sequence $(u_1,\ldots,u_{n+1}) \in U_{n+1}$ must be a poor M-sequence. It follows from 3.1(ii) and the hypotheses that $(u_1,\ldots,u_n)$ is a poor M-sequence. So we suppose that $m \in M$ is such that $u_{n+1}m \in (Au_1 +\ldots+ Au_n)M$. Hence, by [9;3.3], we have

$$d^n\left(\frac{m}{(u_1,\ldots,u_n)}\right) = \frac{m}{(u_1,\ldots,u_n,1)} = \frac{u_{n+1}m}{(u_1,\ldots,u_n,u_{n+1})} = 0.$$

Hence, since $C(\mathcal{U},M)$ is exact at $U_n^{-n}M$, there exist $m' \in M$ and $(u'_1,\ldots,u'_{n-1}) \in U_{n-1}$ such that, in $U_n^{-n}M$,

$$\frac{m}{(u_1,\ldots,u_n)} = \frac{m'}{(u'_1,\ldots,u'_{n-1},1)}\,.$$

Thus there exist $(v_1,\ldots,v_n) \in U_n$ and $\underset{\sim}{L} = [\ell_{ij}]$, $\underset{\sim}{K} \in D_n(A)$ such that $\underset{\sim}{K}\underset{\sim}{u} = \underset{\sim}{v} = \underset{\sim}{L}[u'_1 \ \ldots \ u'_{n-1} \ \ 1]^T$ and

$$|\underset{\sim}{K}|m - |\underset{\sim}{L}|m' \in (Av_1 +\ldots+ Av_{n-1})M.$$

Now $v_n = \sum_{i=1}^{n-1} \ell_{ni}u'_i + \ell_{nn}$; hence

$$|\underset{\sim}{K}|m - \ell_{11}\ldots\ell_{n-1n-1}(v_n - \sum_{i=1}^{n-1} \ell_{ni}u'_i)m' \in (Av_1 +\ldots+ Av_{n-1})M.$$

We may now use [9;2.2] to see that $|\underset{\sim}{K}|m \in (Av_1 +\ldots+ Av_n)M$. Exactness of the right hand column in the diagram $E((u_1,\ldots,u_n);(v_1,\ldots,v_n);\underset{\sim}{K})$ now shows that $m \in (Au_1 +\ldots+ Au_n)M$.

It follows that every element of U_{n+1} must be a poor M-sequence.

We now show that, in the notation of the statement of the theorem, the diagram $E(s;t;\underset{\sim}{H})$ has exact rows and columns. We use the notation of 3.9. By hypothesis, γ_n is a monomorphism, and, by what we have just proved, σ_{n+1} and τ_{n+1} are monomorphisms. We show now that θ_{n+1} is monomorphic.

Let $m \in M$ be such that $k_{n+1}m \in \underline{b}_nM$. Then

$$k_n(s_{n+1}h_{n+1n+1})m \in \underline{b}_nM,$$

so that, in view of [9;2.2], $k_nt_{n+1}m \in \underline{b}_nM$. But we now know that $(t_1,\ldots,t_n,t_{n+1})$ is a poor M-sequence, and so $k_nm \in \underline{b}_nM$. Since

γ_n is monomorphic, it follows that $m \in \underline{a}_n M$. Hence θ_{n+1} is monomorphic.

To complete the proof it is now, in view of [3; Chapter 4, Theorem 5] sufficient to show that α_{n+1} is monomorphic; and, to do this, in view of 3.6, 3.7 and the fact that we now know that t_{n+1} is a non-zerodivisor on $M/\underline{b}_n M$, it is sufficient to show that

$$\mathrm{Ass}(M/(Ak_n+\underline{b}_n)M) \subseteq \mathrm{Ass}(M/\underline{b}_n M) \qquad (1)$$

(for this will ensure that t_{n+1} is a non-zerodivisor on $M/(Ak_n+\underline{b}_n)M$).

So suppose that $\underline{p} \in \mathrm{Ass}(M/(Ak_n+\underline{b}_n)M)$ (so that $Ak_n + \underline{b}_n \subseteq \underline{p}$). Assume that $\underline{p} \notin \mathrm{Ass}(M/\underline{b}_n M)$ and look for a contradiction. By 3.8, this means that $H^0_{\underline{p}A_{\underline{p}}}((M/\underline{b}_n M)_{\underline{p}}) = 0$. Assume, inductively, that j is an integer such that $0 \leq j < n$ and we have proved that $H^i_{\underline{p}A_{\underline{p}}}((M/\underline{b}_{n-j}M)_{\underline{p}}) = 0$ for all $i = 0,\ldots,j$. Consider the exact sequence

$$0 \longrightarrow M/\underline{b}_{n-j-1}M \xrightarrow{t_{n-j}} M/\underline{b}_{n-j-1}M \longrightarrow M/\underline{b}_{n-j}M \longrightarrow 0;$$

here, $\underline{b}_0 = 0$. Localization of this at $\underline{p}$ and use of the inductive hypothesis shows that, for each $i = 0,\ldots,j+1$, multiplication by $\frac{t_{n-j}}{1}$ provides a monomorphism of $H^i_{\underline{p}A_{\underline{p}}}((M/\underline{b}_{n-j-1}M)_{\underline{p}})$ into itself, and since each element of this module is annihilated by some power of $\underline{p}A_{\underline{p}}$ (which contains $\frac{t_{n-j}}{1}$) we therefore deduce that this module is zero. Thus, by induction, for each $j = 0,\ldots,n$, we have

$$H^i_{\underline{p}A_{\underline{p}}}((M/\underline{b}_{n-j}M)_{\underline{p}}) = 0 \text{ for all } i = 0,\ldots,j. \qquad (2)$$

In particular,

$$H^i_{\underline{p}A_{\underline{p}}}(M_{\underline{p}}) = 0 \text{ for all } i = 0,\ldots,n. \qquad (3)$$

We can now deduce from this, by use of the exact sequences

$$0 \longrightarrow M/\underline{a}_i M \xrightarrow{s_{i+1}} M/\underline{a}_i M \longrightarrow M/\underline{a}_{i+1}M \longrightarrow 0$$

$(0 \leq i \leq n)$ (where $\underline{a}_0 = 0$) that, for each $j = 0,\ldots,n$,

$$H^i_{\underline{p}A_{\underline{p}}}((M/\underline{a}_j M)_{\underline{p}}) = 0 \text{ for all } i = 0,\ldots,n-j. \qquad (4)$$

We can also deduce from (3), since $k_1 = h_{11}$ is a non-zerodivisor on M (because $t_1 = h_{11}s_1 \in U_1$) that

$$H^i_{\underline{p}A_{\underline{p}}}((M/k_1M)_{\underline{p}}) = 0 \text{ for all } i = 0,\dots,n-1,$$

that is

$$H^i_{\underline{p}A_{\underline{p}}}((M/(k_1,\underline{b}_1)M)_{\underline{p}}) = 0 \text{ for all } i = 0,\dots,n-1. \qquad (5)$$

Next, for each $j = 1,\dots,n-1$, we use the exact sequence

$$0 \longrightarrow M/\underline{a}_jM \longrightarrow M/\underline{b}_jM \longrightarrow M/(k_{j+1},\underline{b}_j)M \longrightarrow 0$$

from the centre column of the diagram

$$E((s_1,\dots,s_{j+1});(t_1,\dots,t_{j+1});\underset{\sim}{H}_{j+1})$$

(where $\underset{\sim}{H}_{j+1}$ is the $(j+1)\times(j+1)$ matrix formed from the first $j+1$ rows and the first $j+1$ columns of $\underset{\sim}{H}$) in conjunction with (2) and (4) above to see that

$$H^i_{\underline{p}A_{\underline{p}}}((M/(k_{j+1},\underline{b}_j)M)_{\underline{p}}) = 0 \text{ for all } i = 0,\dots,n-j-1. \qquad (6)$$

Now suppose that j is an integer such that $1 \le j < n$ and that we have already proved that

$$H^i_{\underline{p}A_{\underline{p}}}((M/(k_j,\underline{b}_j)M)_{\underline{p}}) = 0 \text{ for all } i = 0,\dots,n-j; \qquad (7)$$

this is the case when $j=1$, by (5). Then we use the exact sequence

$$0 \longrightarrow M/(k_j,\underline{b}_j)M \longrightarrow M/(k_{j+1},\underline{b}_j)M \longrightarrow M/(k_{j+1},\underline{b}_{j+1})M \longrightarrow 0$$

from the last row of the diagram $E((s_1,\dots,s_{j+1});(t_1,\dots,t_{j+1});\underset{\sim}{H}_{j+1})$ in conjunction with (7) and (6) to see that

$$H^i_{\underline{p}A_{\underline{p}}}((M/(k_{j+1},\underline{b}_{j+1})M)_{\underline{p}}) = 0 \text{ for all } i = 0,\dots,n-j-1.$$

It therefore follows by induction that $H^0_{\underline{p}A_{\underline{p}}}((M/(Ak_n+\underline{b}_n)M)_{\underline{p}}) = 0$, so that $\underline{p} \notin \mathrm{Ass}(M/(Ak_n+\underline{b}_n)M)$ by 3.8. This is a contradiction. Hence $\underline{p} \in \mathrm{Ass}(M/\underline{b}_nM)$, and (1) is proved.

This completes the proof of the theorem. □

We are now in a position to prove the Exactness Theorem.

Proof of the Exactness Theorem. (⇒) This can now be proved by an induction argument based on 3.10, 3.11 and 3.12.

(⇐) Assume that, for all $i \in \mathbb{N}$, each member of U_i is a poor M-sequence. It is easy to prove that $C(\mathcal{U},M)$ is exact at M and $U_1^{-1}M$, and so we suppose that $i > 1$ and that $m \in M$ and $(u_1,\dots,u_i) \in U_i$ are such that $\frac{m}{(u_1,\dots,u_i)} \in \ker d^i$. Thus $\frac{m}{(u_1,\dots,u_i,1)} = 0$ in

$U_{i+1}^{-i-1}M$. Hence there exist $(v_1,\ldots,v_{i+1}) \in U_{i+1}$ and $\underset{\sim}{H} = [h_{jk}]$ $\in D_{i+1}(A)$ such that $\underset{\sim}{H}[u_1 \ \ldots \ u_i \ \ 1]^T = [v_1 \ \ldots \ v_i \ \ v_{i+1}]^T$ and $|\underset{\sim}{H}|m \in (Av_1 +\ldots+ Av_i)M$. Thus, since

$$v_{i+1} = \sum_{j=1}^{i} h_{i+1j}u_j + h_{i+1i+1},$$

it follows that $h_{11}\ldots h_{ii}v_{i+1}m \in (Av_1 +\ldots+ Av_i)M$, in view of [9;2.2]. Since $(v_1,\ldots,v_i,v_{i+1})$ is a poor M-sequence, it follows that

$$h_{11}\ldots h_{ii}m = v_1m_1' +\ldots+ v_im_i'$$

for some $m_1',\ldots,m_i' \in M$. Now the matrix $\underset{\sim}{H}_i$ obtained by deleting the last row and last column from $\underset{\sim}{H}$ belongs to $D_i(A)$ and is such that $\underset{\sim}{H}_i[u_1 \ \ldots \ u_i]^T = [v_1 \ \ldots \ v_i]^T$. Hence, in $U_i^{-i}M$,

$$\frac{m}{(u_1,\ldots,u_i)} = \frac{|\underset{\sim}{H}_i|m}{(v_1,\ldots,v_i)} = \frac{h_{11}\ldots h_{ii}m}{(v_1,\ldots,v_i)} = \frac{v_1m_1' +\ldots+ v_im_i'}{(v_1,\ldots,v_i)}.$$

In view of [9;3.3], it thus follows that

$$\frac{m}{(u_1,\ldots,u_i)} = \frac{v_im_i'}{(v_1,\ldots,v_i)} = \frac{m_i'}{(v_1,\ldots,v_{i-1},1)} \in \operatorname{im} d^{i-1}.$$

Hence $C(\mathscr{U},M)$ is exact at $U_i^{-i}M$. □

It should be noted that the inductive method of proof of the implication (⇒) in the Exactness Theorem yields additional information, as detailed in the following corollary.

3.13 COROLLARY. *Suppose that* $C(\mathscr{U},M)$ *is exact, or, equivalently, that, for all* $i \in \mathbb{N}$ *each member of* U_i *is a poor* M*-sequence. Then, for all* $n \in \mathbb{N}$ *and* $s = (s_1,\ldots,s_{n+1}), t = (t_1,\ldots,t_{n+1}) \in U_{n+1}$ *and* $\underset{\sim}{H} = [h_{ij}] \in D_{n+1}(A)$ *for which* $\underset{\sim}{H}\underset{\sim}{s} = \underset{\sim}{t}$*, the following hold:*

(i) *the diagram* $E(s;t;\underset{\sim}{H})$ *has exact rows and columns;*

(ii) *for each* $i = 1,\ldots,n+1$*, the* A*-homomorphism*

$$\gamma_i : M/(As_1 +\ldots+ As_i)M \longrightarrow M/(At_1 +\ldots+ At_i)M$$

induced by multiplication by $h_{11}\ldots h_{ii}$ *is a monomorphism;*

(iii) *for each* $i = 1,\ldots,n$*, the* A*-homomorphism*

$$\theta_{i+1} : M/(As_1 +\ldots+ As_i)M \longrightarrow M/(At_1 +\ldots+ At_i)M$$

induced by multiplication by $h_{11}\ldots h_{ii}h_{i+1i+1}$ *is a monomorphism.* □

Let $m \in M$, $n \in \mathbb{N}$, and $(u_1,\ldots,u_n) \in U_n$. If $m \in \left(\sum_{i=1}^{n-1} Au_i\right)M$, then, by [9;3.3], we have $\frac{m}{(u_1,\ldots,u_n)} = 0$ in $U_n^{-n}M$. We can now see that the converse of this is true when $C(\mathcal{U},M)$ is exact.

3.14 COROLLARY. *Suppose that* $C(\mathcal{U},M)$ *is exact. Let* $m \in M$, $n \in \mathbb{N}$, *and* $(u_1,\ldots,u_n) \in U_n$. *Then* $\frac{m}{(u_1,\ldots,u_n)} = 0$ *in* $U_n^{-n}M$ *if and only if* $m \in \left(\sum_{i=1}^{n-1} Au_i\right)M$.

Proof. (⇒) Assume that $\frac{m}{(u_1,\ldots,u_n)} = 0$. Then there exist $(v_1,\ldots,v_n) \in U_n$ and $\underset{\sim}{H} \in D_n(A)$ such that $\underset{\sim}{H}\underset{\sim}{u} = \underset{\sim}{v}$ and $|\underset{\sim}{H}|m \in \left(\sum_{i=1}^{n-1} Av_i\right)M$. Hence, by (3.13)(iii), $m \in \left(\sum_{i=1}^{n-1} Au_i\right)M$.

(⇐) Use [9;3.3(ii)]. □

It may be thought that the Exactness Theorem and its Corollaries 3.13 and 3.14 will only be of use when one is working with a family $\mathcal{U} = (U_i)_{i\in\mathbb{N}}$ of sets as in 3.1. We give the following further corollary to show how these results may be applied to a more extensive range of situations.

3.15 COROLLARY. *Let* V *be a triangular subset of* A^n *and let* M *be an* A-*module such that each member of* V *is a poor* M-*sequence. Then, for* $m \in M$ *and* $(v_1,\ldots,v_n) \in V$, *it is the case that*

$$\frac{m}{(v_1,\ldots,v_n)} = 0 \textit{ in } V^{-n}M \textit{ if and only if } m \in \left(\sum_{i=1}^{n-1} Av_i\right)M.$$

Proof. Let U_n be the expansion of V [9;3.2]. For each $i=1,\ldots,n-1$, let U_i be the restriction [9;3.6] of U_n to A^i; for each $j \in \mathbb{N}$ with $j > n$, let

$$U_j = \{(u_1,\ldots,u_n,1,\ldots,1) \in A^j : (u_1,\ldots,u_n) \in U_n\},$$

a triangular subset of A^j. Then the family $\mathcal{U} = (U_i)_{i\in\mathbb{N}}$ satisfies the conditions 3.1(i)-(iv), and, by hypothesis, it is the case that, for all $i \in \mathbb{N}$, each member of U_i is a poor M-sequence. The result therefore follows from 3.14 in conjunction with the fact [9;3.2] that the natural A-homomorphism $\mu(M) : V^{-n}M \longrightarrow U_n^{-n}M$, for which

$$\mu(M)\left(\frac{m}{(v_1,\ldots,v_n)}\right) = \frac{m}{(v_1,\ldots,v_n)}$$

for all $m \in M$ and $(v_1,\ldots,v_n) \in V$, is an isomorphism. □

Before we move on to applications of the Exactness Theorem to big Cohen-Macaulay modules in Section 4, there is one further observation which should be made about the complex $C(\mathcal{U},M)$. This is contained in the second of the following two lemmas, the proofs of which are straightforward and left to the reader.

3.16 LEMMA. *Let* V *be a triangular subset of* A^n. *Then there is an* A-*isomorphism*

$$\psi : V^{-n}A \otimes_A M \longrightarrow V^{-n}M$$

which is such that, for $a \in A$, $m \in M$ *and* $(v_1,\ldots,v_n) \in V$, *we have*

$$\psi\left(\frac{a}{(v_1,\ldots,v_n)} \otimes m\right) = \frac{am}{(v_1,\ldots,v_n)} . \square$$

3.17 LEMMA. *There is an isomorphism of complexes of* A-*modules and* A-*homomorphisms*

$$\Phi = (\phi^i)_{i\geq 0} : C(\mathcal{U},A) \otimes_A M \longrightarrow C(\mathcal{U},M)$$

such that $\phi^0 : A \otimes_A M \longrightarrow M$ *is the natural isomorphism and, for each* $n \in \mathbb{N}$, $\phi^n : U_n^{-n}A \otimes_A M \longrightarrow U_n^{-n}M$ *is the isomorphism given by* 3.16.$\square$

4. *Applications to big Cohen-Macaulay modules*

Throughout this section, we shall assume that A is Noetherian and local, that $\underline{m}$ is the maximal ideal of A, that dim A = n (≥ 1), and that M is an A-module. It will be convenient to use the following abbreviations: 's.o.p.' will stand for 'system of parameters', while 's.s.o.p.' will stand for 'subset of a system of parameters'.

Let $x_1,\ldots,x_n$ be a s.o.p. for A. We say that M is a *big Cohen-Macaulay module with respect to* $x_1,\ldots,x_n$ if $x_1,\ldots,x_n$ is an M-sequence. The reader is referred to the work and writings of Hochster, such as [2], for details of the relationship between this concept and the various homological conjectures in commutative algebra. Furthermore, we say that M is a *balanced big Cohen-Macaulay* A-*module* if it is a big Cohen-Macaulay module with respect to every s.o.p. for A: see [8].

The Exactness Theorem of Section 3, used in conjunction with 3.17, allows us to characterize balanced big Cohen-Macaulay A-modules as follows.

4.1 THEOREM. (A *is local and has maximal ideal* $\underline{m}$ *and dimension* n (≥ 1).) *For each* $i \in \mathbb{N}$, *we set*

$$U_i = \{(s_1,\ldots,s_i) \in A^i : \textit{there exists } j \textit{ with } 0 \le j \le i \textit{ such that } s_1,\ldots,s_j \textit{ form an s.s.o.p. and } s_{j+1} = \ldots = s_i = 1\}.$$

Then the family $\mathcal{U} = (U_i)_{i \in \mathbb{N}}$ *satisfies the conditions of* 3.1(i)-(iv), *and we may form the complexes* $C(\mathcal{U},M)$ *and* $C(\mathcal{U},A)$. *In fact,* M *is a balanced big Cohen-Macaulay* A*-module if and only if* $C(\mathcal{U},A) \otimes_A M$ ($\cong C(\mathcal{U},M)$, *by* 3.17) *is exact and* $M \neq \underline{m}M$. □

We can also use the same results in conjunction with [9;3.9] to characterize those A-modules which are big Cohen-Macaulay modules with respect to a specified s.o.p. for A.

4.2 THEOREM. *Let* $x_1,\ldots,x_n$ *be a s.o.p. for the local ring* A *(which has maximal ideal* $\underline{m}$ *and dimension* n (≥ 1)). *For each* $i \in \mathbb{N}$, *let* $U(x)_i$ *be the expansion (see* [9;3.2]) *of the triangular subset*

$$\{(x_1^{\alpha_1},\ldots,x_i^{\alpha_i}) : \alpha_j \in \mathbb{N} \textit{ for all } j = 1,\ldots,i\}$$

of A^i, *where* x_r *is interpreted as* 1 *whenever* $r > n$.

Then the family $(U(x)_i)_{i\in\mathbb{N}}$, $= \mathcal{U}(x)$ *say, satisfies the conditions of* 3.1(i)-(iv) *and we may form the complexes* $C(\mathcal{U}(x),M)$ *and* $C(\mathcal{U}(x),A)$. *In fact,* M *is a big Cohen-Macaulay module with respect to* $x_1,\ldots,x_n$ *if and only if* $C(\mathcal{U}(x),A) \otimes_A M$ ($\cong C(\mathcal{U}(x),M)$, *by* 3.17) *is exact and* $M \neq \underline{m}M$. □

Of course, the ideas and results in this paper can be applied to Cohen-Macaulay (commutative Noetherian) rings. We end this section with one such application.

4.3 THEOREM. *Suppose that* A *is a Cohen-Macaulay local ring (having maximal ideal* $\underline{m}$ *and dimension* n (≥ 1)). *Let* $x_1,\ldots,x_n$ *be a s.o.p. for* A. *Then*

$$H^n_{\underline{m}}(A) \cong U(x)^{-n-1}_{n+1}A,$$

where

$$U(x)_{n+1} = \{(x_1^{\alpha_1},\ldots,x_n^{\alpha_n},1) : \textit{there exists } j \textit{ with } 0 \le j \le n \textit{ such that } \alpha_1,\ldots,\alpha_j \in \mathbb{N} \textit{ and } \alpha_{j+1} = \ldots = \alpha_n = 0\}.$$

Remark. This result has connections with the known fact that, if k is a field, $X_1,\ldots,X_n$ are independent indeterminates and

$B = k[[X_1,\ldots,X_n]]$, and we use $\underline{r}$ to denote the maximal ideal of the regular local ring B, then

$$H^n_{\underline{r}}(B) \cong (E_B(B/\underline{r}) \cong)\ k[X_1^{-1},\ldots,X_n^{-1}],$$

the B-module of "inverse polynomials": see [4; Theorem 3]. As the arguments used in [9; 3.11 and 3.2] can easily be modified to show that there is a B-isomorphism

$$k[X_1^{-1},\ldots,X_n^{-1}] \cong U(X)^{-n-1}_{n+1}B,$$

where we are using the notation of 4.3 and X refers to the s.o.p. $X_1,\ldots,X_n$ for B, it follows that 4.3 may be viewed as a generalization of the fact that $H^n_{\underline{r}}(B) \cong k[X_1^{-1},\ldots,X_n^{-1}]$.

Proof. We use the notation of 4.2: by that theorem, $C(\mathscr{U}(x),A)$ is exact. We show that, for each $i \in \mathbb{N}$, multiplication by x_i provides an automorphism of $U(x)^{-i}_i A$. Since, for $a \in A$ and $(x_1^{\alpha_1},\ldots,x_i^{\alpha_i}) \in U(x)_i$, we have

$$\frac{a}{(x_1^{\alpha_1},\ldots,x_i^{\alpha_i})} = \frac{x_1\ldots x_i a}{(x_1^{\alpha_1+1},\ldots,x_i^{\alpha_i+1})},$$

we just need to show that x_i is a non-zerodivisor on $U(x)^{-i}_i A$; this is immediate from 3.15 because each member of $U(x)_i$ is a poor A-sequence.

This done, we can now deduce that, for all $i = 1,\ldots,n$,

$$H^j_{\underline{m}}(U(x)^{-i}_i A) = 0 \text{ for all } j \geq 0.$$

The result can now be established by use of the n short exact sequences which can be obtained from the exact sequence $C(\mathscr{U}(x),A)$: bear in mind that $U(x)^{-i}_i A = 0$ for all $i > n+1$. □

5. *Applications to the minimal injective resolution of a Gorenstein ring*

Throughout this section, we shall assume that A is Noetherian; we shall only assume that A is Gorenstein when this is explicitly stated.

We begin with a general lemma.

5.1 LEMMA. *Let* U *be a triangular subset of* A^n, *let* m *be an element of the* A-*module* M *and let* $(u_1,\ldots,u_n) \in U$.

(i) *If* $\frac{u_n m}{(u_1,\ldots,u_n)} = 0$ *in* $U^{-n}M$, *then* $\frac{m}{(u_1,\ldots,u_n)} = 0$.

(ii) *If* $(u_1,\ldots,u_{n-1},1) \in U$ *also, then*

$$\left(0 : \frac{m}{(u_1,\ldots,u_{n-1},1)}\right) = \left(0 : \frac{m}{(u_1,\ldots,u_{n-1},u_n)}\right).$$

Proof. (i) There exist $(w_1,\ldots,w_n) \in U$ and $\underset{\sim}{H} = [h_{ij}] \in D_n(A)$ such that $\underset{\sim}{H}\underset{\sim}{u} = \underset{\sim}{w}$ and $|\underset{\sim}{H}|u_n m \in \left(\sum_{i=1}^{n-1} Aw_i\right)M$. Hence

$$\left(\prod_{i=1}^{n-1} h_{ii}\right)\left(w_n - \sum_{i=1}^{n-1} h_{ni}u_i\right)m \in \left(\sum_{i=1}^{n-1} Aw_i\right)M.$$

Therefore, by [9;2.2], $h_{11}\ldots h_{n-1n-1}w_n m \in \left(\sum_{i=1}^{n-1} Aw_i\right)M$; hence, by [9;3.3(ii)],

$$\frac{h_{11}\ldots h_{n-1n-1}w_n m}{(w_1,\ldots,w_{n-1},w_n^2)} = 0$$

in $U^{-n}M$. It follows from this that $\frac{h_{11}\ldots h_{n-1n-1}h_{nn}m}{(w_1,\ldots,w_{n-1},w_n)} = 0$, from which the desired conclusion follows since $\underset{\sim}{H}\underset{\sim}{u} = \underset{\sim}{w}$.

(ii) This is an easy consequence of (i). □

We now introduce some notation which will be in force throughout the section.

5.2 *Notation*. We adopt the convention whereby the ideal A of A has height ∞. For each $i \in \mathbb{N}$, we set

$$U_i = \{(u_1,\ldots,u_i) \in A^i : \mathrm{ht}(Au_1 + \ldots + Au_j) \geq j \text{ for all } j = 1,\ldots,i\}.$$

It is an elementary exercise to check that U_i is a triangular subset of A^i; in fact, the family $\mathcal{U} = (U_i)_{i\in\mathbb{N}}$ satisfies the conditions (i)-(iv) of 3.1, and so we may form the complex $C(\mathcal{U},A)$ as in 3.2.

For $(u_1,\ldots,u_n) \in U_n$, we shall denote by $P(u_1,\ldots,u_n)$ the (finite) set

$$\{\underline{p} \in \mathrm{Spec}(A) : \underline{p} \supseteq Au_1 + \ldots + Au_n \text{ and } \mathrm{ht}\,\underline{p} = n\}.$$

5.3 *Remark*. It should be noted that, in the special case in which A is Cohen-Macaulay, we have, for all $i \in \mathbb{N}$,

$$U_i = \{(u_1,\ldots,u_i) \in A^i : u_1,\ldots,u_i \text{ form a poor A-sequence}\},$$

so that, in view of the Exactness Theorem 3.3, the complex $C(\mathcal{U},A)$ is actually exact in this case.

5.4 *Remark*. It is immediate from [9;3.3(ii)] that, for each $i \in \mathbb{N}$,

$$\mathrm{Supp}(U_i^{-i}A) \subseteq \{\underline{p} \in \mathrm{Spec}(A) : \mathrm{ht}\,\underline{p} \geq i-1\}.$$

In fact, we can say rather more.

5.5 LEMMA. *For each* $i \in \mathbb{N}$,

$$\mathrm{Ass}(U_i^{-i}A) \subseteq \{\underline{p} \in \mathrm{Spec}(A) : \mathrm{ht}\,\underline{p} = i-1\}.$$

Proof. Let $\underline{p} \in \mathrm{Ass}(U_i^{-i}A)$, so that, by 5.4, $\mathrm{ht}\,\underline{p} \geq i-1$. Now $\underline{p} = (0 : \alpha)$ for some $\alpha = \dfrac{a}{(u_1,\ldots,u_i)} \in U_i^{-i}A$, where $a \in A$ and $(u_1,\ldots,u_i) \in U_i$. Suppose that $\mathrm{ht}\,\underline{p} \geq i$. Then, using the notation of 5.2, and interpreting $P(u_1,\ldots,u_{i-1})$ as the set $\{\underline{p} \in \mathrm{Spec}(A) : \mathrm{ht}\,\underline{p} = 0\}$ when $i = 1$, we have

$$\underline{p} \not\subseteq \bigcup_{\underline{q} \in P(u_1,\ldots,u_{i-1})} \underline{q},$$

so that we may select $v_i \in \underline{p} \setminus \left[\bigcup_{\underline{q} \in P(u_1,\ldots,u_{i-1})} \underline{q}\right]$. Then

$(u_1,\ldots,u_{i-1},v_i) \in U_i$, and, by (5.1)(ii),

$$\underline{p} = \left[0 : \frac{a}{(u_1,\ldots,u_{i-1},u_i)}\right] = \left[0 : \frac{a}{(u_1,\ldots,u_{i-1},1)}\right] \qquad (8)$$

$$= \left[0 : \frac{a}{(u_1,\ldots,u_{i-1},v_i)}\right].$$

But $v_i \in \underline{p}$, and so $\dfrac{a}{(u_1,\ldots,u_{i-1},1)} = \dfrac{v_i a}{(u_1,\ldots,u_{i-1},v_i)} = 0$, contradicting (8). The result follows. □

5.6 COROLLARY. *Let* $(u_1,\ldots,u_i) \in U_i$ *(where* $i \in \mathbb{N}$*) and let* $a \in A$. *Suppose that* $\dfrac{a}{(u_1,\ldots,u_i)} \neq 0$ *in* $U_i^{-i}A$. *Then each associated prime of* $\left[0 : \dfrac{a}{(u_1,\ldots,u_i)}\right]$ *has height* $i-1$. □

5.7 THEOREM. *Let* $i \in \mathbb{N}$. *Since, in view of* 5.6, *a given element of* $U_i^{-i}A$ *has annihilator which is contained in at most finitely many prime ideals of height* $i-1$, *there is an* A-*homomorphism*

$$\theta : U_i^{-i}A \longrightarrow \bigoplus_{\substack{\underline{p} \in \mathrm{Spec}(A) \\ \mathrm{ht}\underline{p}=i-1}} (U_i^{-i}A)_{\underline{p}}$$

which is such that, for $\alpha \in U_i^{-i}A$ *and* $\underline{p} \in \mathrm{Spec}(A)$ *of height* $i-1$, *the component of* $\theta(\alpha)$ *in* $(U_i^{-i}A)_{\underline{p}}$ *is* $\frac{\alpha}{1}$.

The map θ *is an isomorphism.*

Proof. Suppose that $\ker\theta \neq 0$, and let $\underline{p} \in \mathrm{Ass}(\ker\theta)$. Thus $\underline{p} = (0 : \alpha)$ for some $\alpha \in \ker\theta$. By 5.4 and the definition of θ, we have $\mathrm{ht}\,\underline{p} \geq i$; but 5.5 shows that $\mathrm{ht}\,\underline{p} = i-1$. This contradiction shows that θ is monomorphic.

Let $\underline{p}_1$ be a prime ideal of A of height $i-1$. Let

$$0 \neq \delta = (\delta_{\underline{p}})_{\mathrm{ht}\underline{p}=i-1} \in \bigoplus_{\substack{\underline{p}\in \mathrm{Spec}(A)\\ \mathrm{ht}\underline{p}=i-1}} (U_i^{-i}A)_{\underline{p}},$$

where

$$\delta_{\underline{p}} = \begin{cases} 0 & \text{if } \underline{p} \neq \underline{p}_1, \\ \dfrac{\frac{a}{(v_1,\ldots,v_i)}}{t} & \text{if } \underline{p} = \underline{p}_1; \end{cases}$$

here, $a \in A$, $(v_1,\ldots,v_i) \in U_i$ and $t \in A \setminus \underline{p}_1$. It is enough, in order to show that θ is surjective, to show that $\delta \in \mathrm{im}\,\theta$.

Set $\underline{a} = \left(0 : \frac{a}{(v_1,\ldots,v_i)}\right)$, a proper ideal of A. By 5.6, all the associated primes of $\underline{a}$ have height $i-1$; we thus see that $\underline{p}_1$ must be one of these. Let $\underline{a} = \bigcap_{j=1}^{k} \underline{q}_j$, with $r(\underline{q}_j) = \underline{p}_j$ for $j = 1,\ldots,k$, be the minimal primary decomposition for $\underline{a}$.

Since $Av_1 + \ldots + Av_{i-1} \subseteq \underline{a} \subseteq \underline{p}_1$ and $\mathrm{ht}\,\underline{p}_1 = i-1$, it follows that $v_i \notin \underline{p}_1$. Also, we may choose $s \in \left[\bigcap_{j=2}^{k} \underline{q}_j\right] \setminus \underline{p}_1$. Then

$$\delta_{\underline{p}_1} = \frac{\frac{sa}{(v_1,\ldots,v_{i-1},1)}}{stv_i}.$$

Also, use of 5.1 enables us to see that

$$\left(0 : \frac{sa}{(v_1,\ldots,v_{i-1},1)}\right) = \left(0 : \frac{sa}{(v_1,\ldots,v_{i-1},v_i)}\right) = (\underline{a} : s) = \underline{q}_1. \quad (9)$$

Next, with the notation of 5.2 (and the understanding that $P(v_1,\ldots,v_{i-1}) = \{\underline{p} \in \mathrm{Spec}(A) : \mathrm{ht}\,\underline{p} = 0\}$ when $i = 1$), we note that

$$Astv_i + \underline{q}_1 \not\subseteq \bigcup_{\underline{p}\in P(v_1,\ldots,v_{i-1})} \underline{p},$$

for if this were not the case it would follow that $\mathrm{Astv}_i + \underline{q}_1 \subseteq \underline{p}'$ for some prime ideal $\underline{p}'$ of height i-1, and the inclusions $\underline{q}_1 \subseteq \underline{p}'$ and $\mathrm{Astv}_i \subseteq \underline{p}'$ would lead, respectively, to the contradictory statements that $\underline{p}_1 = \underline{p}'$ and $\underline{p}_1 \neq \underline{p}'$! Thus we may choose

$$v_i' = bstv_i + c \in A \setminus \left[\bigcup_{\underline{p} \in P(v_1,\ldots,v_{i-1})} \underline{p} \right]$$

with $b \in A$ and $c \in \underline{q}_1$. Note that $(v_1,\ldots,v_{i-1},v_i') \in U_i$ and that, by 5.1 and (9),

$$\left(0 : \frac{sa}{(v_1,\ldots,v_{i-1},v_i')}\right) = \left(0 : \frac{sa}{(v_1,\ldots,v_{i-1},1)}\right) = \underline{q}_1. \tag{10}$$

Now, in $U_i^{-i}A$,

$$\frac{sa}{(v_1,\ldots,v_{i-1},1)} = \frac{v_i'sa}{(v_1,\ldots,v_{i-1},v_i')} = \frac{(bstv_i + c)sa}{(v_1,\ldots,v_{i-1},v_i')}$$
$$= stv_i\left(\frac{bsa}{(v_1,\ldots,v_{i-1},v_i')}\right)$$

in view of (10) and the fact that $c \in \underline{q}_1$. Hence

$$\delta_{\underline{p}_1} = \frac{\frac{bsa}{(v_1,\ldots,v_{i-1},v_i')}}{1}.$$

Moreover, it follows from (10) that, for all prime ideals $\underline{p}$ of height i-1 with $\underline{p} \neq \underline{p}_1$, we have $\dfrac{\frac{bsa}{(v_1,\ldots,v_{i-1},v_i')}}{1} = 0$ in $(U_i^{-i}A)_{\underline{p}}$.

Hence $\delta \in \operatorname{im} \theta$, as required. □

We are now in a position to prove the result about the minimal injective resolution for a Gorenstein ring that was promised in the Introduction.

5.8 THEOREM. *Let* A *be a Gorenstein ring, and note that, by* 5.3,

$$U_i = \{(u_1,\ldots,u_i) \in A^i : u_1,\ldots,u_i \textit{ form a poor } A\textit{-sequence}\}$$

for all $i \in \mathbb{N}$. *Also, by* 5.2, *the family* $\mathcal{U} = (U_i)_{i \in \mathbb{N}}$ *satisfies conditions* (i)-(iv) *of* 3.1; *in fact the complex* $C(\mathcal{U},A)$ *(see* 3.2*)*

$$0 \xrightarrow{d^{-1}} A \xrightarrow{d^0} U_1^{-1}A \longrightarrow \cdots \longrightarrow U_i^{-i}A \xrightarrow{d^i} U_{i+1}^{-i-1}A \longrightarrow \cdots$$

is exact, and actually provides the minimal injective resolution for A.

In particular, $E^i(A) \cong U_{i+1}^{-i-1}A$ *for all* $i \geq 0$. *(Here,* $E^i(A)$ *denotes the* i-*th term in the minimal injective resolution for* A.*)*

Proof. We remarked in 5.3 that $C(\mathcal{U},A)$ is exact in this situation. Since it is an immediate consequence of 5.1 that, for each $i \in \mathbb{N}$, the module $U_i^{-i}A$ is an essential extension of im d^{i-1}, it remains only to show that each $U_i^{-i}A$ is an injective A-module.

We achieve this by induction on i. Suppose, inductively, that $i \in \mathbb{N}$ and we have shown that $U_1^{-1}A,\ldots,U_{i-1}^{-i+1}A$ are all A-injective. (This is certainly so when i=1!) Let $\underline{p}$ be a prime ideal of height i-1, so that the $A_{\underline{p}}$-module $A_{\underline{p}}$ has injective dimension i-1 [1]. If we now localize $C(\mathcal{U},A)$ at $\underline{p}$ and use 5.4, we obtain an exact sequence

$$0 \longrightarrow A_{\underline{p}} \longrightarrow (U_1^{-1}A)_{\underline{p}} \longrightarrow \cdots \longrightarrow (U_{i-1}^{-i+1}A)_{\underline{p}} \longrightarrow (U_i^{-i}A)_{\underline{p}} \longrightarrow 0$$

of $A_{\underline{p}}$-modules and $A_{\underline{p}}$-homomorphisms. Since $(U_j^{-j}A)_{\underline{p}}$ is $A_{\underline{p}}$-injective for all j = 1,...,i-1, it follows that $(U_i^{-i}A)_{\underline{p}}$ is $A_{\underline{p}}$-injective (since $A_{\underline{p}}$ has injective dimension i-1 over itself). Hence $(U_i^{-i}A)_{\underline{p}}$ is A-injective. Therefore, by 5.7 and the fact that our ring A is Noetherian,

$$U_i^{-i}A \cong \bigoplus_{\substack{\underline{p} \in \mathrm{Spec}(A)\\ \mathrm{ht}\underline{p}=i-1}} (U_i^{-i}A)_{\underline{p}}$$

is also A-injective. The result now follows by induction. □

5.9 *Remarks.* Another natural description of the minimal injective resolution for the Gorenstein ring A is provided by the Cousin complex C(A) of A: see [6;(5.4)]. However, the new natural description provided above by $C(\mathcal{U},A)$ is perhaps simpler to explain and is clearly related to the fact that the total ring of fractions of A, when viewed as an A-module, provides the injective envelope of A.

Our work thus shows that, for a Gorenstein ring A, the Cousin complex C(A) is isomorphic to a complex of modules of generalized fractions. Other situations in which certain exact Cousin complexes turn out to be isomorphic to complexes of modules of generalized fractions are discussed in [10].

5.10 *Concluding remarks.* We feel that the results presented in this paper (which, incidentally, were almost all mentioned in a lecture at the Symposium) demonstrate that the concept of module of generalized fractions has significant interactions with several of the topics discussed at the Symposium; we believe that modules of generalized fractions are worthy of further investigation.

References

1. H. Bass, "On the ubiquity of Gorenstein rings", Math. Z., 82 (1963), 8-28.

2. M. Hochster, Topics in the homological theory of modules over commutative rings, Conference Board of the Mathematical Sciences Regional Conference Series in Mathematics 24 (American Mathematical Society, Providence, 1975).

3. D.G. Northcott, An introduction to homological algebra (Cambridge University Press, Cambridge, 1960).

4. D.G. Northcott, "Injective envelopes and inverse polynomials", J. London Math. Soc. (2), 8 (1974), 290-296.

5. C. Peskine and L. Szpiro, "Dimension projective finie et cohomologie locale", Publications Mathématiques 42 (Institut des Hautes Études Scientifiques, Paris, 1973), pp. 47-119.

6. R.Y. Sharp, "The Cousin complex for a module over a commutative Noetherian ring", Math. Z., 112 (1969), 340-356.

7. R.Y. Sharp, "Local cohomology theory in commutative algebra", Quart. J. Math. Oxford (2), 21 (1970), 425-434.

8. R.Y. Sharp, "Cohen-Macaulay properties for balanced big Cohen-Macaulay modules", Math. Proc. Cambridge. Philos. Soc., 90 (1981), 229-238.

9. R.Y. Sharp and H. Zakeri, "Modules of generalized fractions", Mathematika, 29 (1982), 32-41.

10. H. Zakeri, "Modules of generalized fractions and their application in commutative algebra", Ph.D. Thesis, University of Sheffield, 1982.

Department of Pure Mathematics,
University of Sheffield,
Hicks Building,
Sheffield S3 7RH, U.K.

SUR LA THÉORIE DES COMPLEXES PARFAITS

L.SZPIRO

Soit A un anneau local noethérien; pour notre propos, un *complexe parfait* sur A est un complexe fini

$$0 \to L_s \to L_{s-1} \to \ldots \to L_t \to 0$$

de A-modules libres de rang fini. On s'intéresse principalement aux propriétés "d'intersection" d'un complexe parfait L. avec un A-module de type fini N. A savoir, si X = Spec A et Y = Support(H(L.)), on veut des renseignements sur l'application

$$\chi : K.(X) \to K.(Y)$$

où K. est le foncteur "Groupe de Grothendieck" des faisceaux cohérents et

$$\chi(N) = \Sigma(-1)^i [H_i(L. \otimes N)].$$

Si A est un anneau local régulier non ramifié, la situation est controlée et décrite dans le cours de Serre [10]. Dans un cas plus général le seul résultat patent est le Théoreme d'Intersection dont nous parlons plus bas. Partant de ce théoreme et des conjectures qu'on aimerait démontrer, l'examen du cas gradué nous conduit à introduire une conjecture du type 'Riemann-Roch". Cette conjecture a été montrée par W.Fulton dans le cas des anneaux locaux géométriques.

1. *Le Théoreme d'Intersection comme théoreme d'annulation*

Définition. Soit L. un complexe parfait sur un anneau local noethérien A; on appelle *longueur* de L., et on note long(L.), l'entier définit par long(L.) $\leq$ s si L. est homotope à un complexe parfait F. de la forme

$$\ldots \to 0 \to 0 \to \ldots \to 0 \to F_{s+t} \to \ldots \to F_t \to 0 \to 0 \to \ldots .$$

Par exemple, un complexe parfait homotope a zéro est dit de longueur -1!

THÉOREME D'INTERSECTION. *Soit* A *un anneau local noethérien contenant un corps et* L. *un complexe parfait sur* A *tel que* long(H.(L.)) < ∞ . *Alors si* long(L.) > -1 *on a*

long(L.) ≥ dim A.

On connait trois démonstrations de ce théoreme.

(a) La démonstration qu'on trouve dans [6] et [7] consiste à montrer *l'annulation de certains groupes de cohomologie locale* $H^i_{\underline{m}}(A)$. Remarquons que le lecteur surpris que cette démonstration donne le cas equicaractéristique zéro et non seulement le cas formellement géométrique peurra faire l'exercice simple suivant: si k est un corps et $B = k[X^{\ell}_{ij}]/X^{\ell}\circ X^{\ell+1} = 0$, l'anneau universel pour les complexes parfaits L. dont on a fixé les rangs des L_i (cf. §3), l'hypothèse du Théoreme d'Intersection donne un homomorphisme fini $\hat{B} \to \hat{A}$; en approximant $\hat{A}$ sur $\hat{B}$ on se ramène au cas formellement géométrique.

(b) La démonstration de Hochster [4], en utilisant le *Lemme d'Acyclicité* [6], conclue à l'annulation de la cohomologie de L. après tensorisation par un gros module de Cohen-Macaulay [4].

(c) La démonstration de P.Roberts [8,9] montre *l'annulation de certains groupes de cohomologie du complexe dualisant.*

On voit que ce théoreme donne des renseignements de dimension par des "vanishing theorems" mais par contre aucun renseignement sur la caractéristique d'Euler-Poincaré

$\Sigma(-1)^i \mathrm{long}(H_i(L.))$.

Une situation analogue s'est développée au 19e siècle dans l'étude des zéros et des poles des fonctions méromorphes sur une surface de Riemann. On pense donc tout naturellement qu'un résultat de nature "Riemann-Roch" devrait clarifier la situation. Pour illustrer ce propos nous allons regarder le cas gradué à la place du cas local. Nous précisons d'abord les conjectures qu'on aimerait démontrer.

2. *Les conjectures*

A est un anneau local noethérien, et L. est un complexe parfait sur A.

Conjecture C1. Si L. est la résolution projective d'un A-module M et si N est un A-module de type fini tel que $\dim M \otimes N = 0$, alors $\dim N \leq \operatorname{grade} M$; de plus $\operatorname{grade} M + \dim M = \dim A$.

(Ici $\operatorname{grade} M = \inf\{i : \operatorname{Ext}^i_A(M,A) \neq 0\}$.)

Conjecture C2. Si L. a tous ses groupes de cohomologie de longueur finie et si $L.^{\vee} = \operatorname{Hom}(L.,A)$, on a

$$\chi(L.^{\vee}) = (-1)^d \chi(L.),$$

où $d = \dim A$. De plus, si caractéristique(A) = $p > 0$, si F est l'homomorphisme de Frobenius de A et si $L.^{(p)} = F^*L.$, alors on a

$$\chi(L.^{(p)}) = p^d \chi(L.).$$

Conjecture C3. Si L. a tous ses groupes de cohomologie de longueur finie et si N est un A-module de type fini tel que $\dim N < \dim A$, alors

$$\chi(L. \otimes N) = 0.$$

Conjecture C4. Le Théoreme de Riemann-Roch local (voir §4) est vrai.

3. *Le cas gradué*

Dans notre note [7] avec C.Peskine nous montrons les Conjectures C1 à C4 pour les complexes parfaits homogènes sur un anneau gradué. Nous rappelons d'abord le lemme qui joue le role de Théoreme de Riemann-Roch gradué et donnons les détails de la démonstration de la deuxième partie de C1, le reste coulant de source.

Soit A un anneau gradué, $A = \bigoplus_{n=0}^{\infty} A_n$, tel que A_0 *soit artinien* et A de type fini sur A_0. Nous notons A(n) le A-module gradué tel que $A(n)_k = A_{n+k}$. Un *complexe parfait homogène* sur A est un complexe parfait de la forme

$$\cdots \longrightarrow \bigoplus_{j=1}^{r_i} A(-n_{i\,j}) \xrightarrow{\ \phi_i\ } \bigoplus_{j=1}^{r_{i-1}} A(-n_{i-1\,j}) \to \cdots$$

où ϕ_i est un homomorphisme homogène de degré zéro.

Par exemple, si k est un corps, $\underline{r} = (r_0,\ldots,r_s)$ une suite d'entiers, $X^i_{\ell,m}$ $(i = 0,\ldots,s-1,\ \ell = 1,\ldots,r_i,\ m = 1,\ldots,r_{i+1})$ des variables (coefficients des matrices $\phi_i = (X^i_{\ell,m})_{\ell,m}$) et

$$A_{\underline{r}} = k[X^i_{\ell,m}]/(\phi_{i+1} \circ \phi_i)_{i=0,\ldots,s-1},$$

alors $A_{\underline{r}}$ est l'anneau universel pour les complexes parfaits L. tels que $\mathrm{rg}(L_i) = r_i$. Le complexe parfait $L_{\underline{r},\cdot}$, universel sur $A_{\underline{r}}$, est homogène quand on écrit

$$L_{\underline{r},i} = A_{\underline{r}}(-i) \oplus A_{\underline{r}}(-i) \oplus \ldots \oplus A_{\underline{r}}(-i) \quad (r_i \text{ fois}).$$

D'autre part, pour un A-module gradué de type fini M nous noterons $P_M(\nu)$ le *polynôme de Samuel de* M.

LEMME (Riemann-Roch gradué). *Avec les notations ci-dessus, soit* N *un* A-*module gradué de type fini. Alors on a*

$$\sum_{i=0}^{s} (-1)^i P_{H_i(L.\otimes N)}(\nu) = \sum (-1)^k \rho_k P_N^{(k)}(\nu),$$

où

$$\rho_k = \frac{1}{k!} \sum_{i=0}^{s} (-1)^i \sum_{j=1}^{r_i} n_{ij}^k$$

et $P^{(k)}(\nu)$ *est la k-ième dérivée du polynôme* $P(\nu)$.

La démonstration de ce lemme se trouve dans [7]. L'idée est d'écrire le terme de gauche comme une triple somme incluant les $P_{N(-n_{ij})}(\nu)$ et d'utiliser le développement de Taylor pour calculer $P_{N(-n_{ij})}(\nu)$ en terme des $P_N^{(k)}(\nu)$.

Indiquons une justification de l'appellation de ce lemme. Posons

$$O(L.) = \sum \rho_k \frac{\partial^k}{\partial \nu^k},$$

$H(A) = \mathbb{Q}[T]/T^{d+1}$ où $d = \dim A$, et $\tau(N) = P_N(\nu)$. Le lemme dit que

$$\tau \circ \chi(L. \otimes N) = O(L.) \cup \tau(N)$$

dans le $\mathbb{Q}$ espace vectoriel H(A) des polynômes de degré au plus d, où $\frac{\partial^k}{\partial T^k}$ opère de la façon habituelle qu'on note ici cup ($\cup$).

THÉOREME DU GRADE (cas gradué). *Soit* A *un anneau gradué du type décrit au début de ce paragraphe et soit* M *un* A-*module gradué de type fini et de dimension projective finie dont la résolution projective est le complexe parfait homogène* L. . *Soit* $g = \inf\{k : \rho_k \neq 0\}$. *Alors*

(i) *si* N *est un* A-*module gradué de type fini tel que* $\dim M \otimes N = 0$, *on a* $\dim N \leq g$;

(ii) $g = \mathrm{grade}\, M = \dim A - \dim M$.

Le (i) et $g = \dim A - \dim M$ sont clairs à partir du lemme ci-dessus.

Pour montrer que $g = \operatorname{grade} M$ il suffit de prouver que pour tout idéal $\underline{p}$ dans Support(M) on a $\dim A_{\underline{p}} \geq g$. En effet

$$\operatorname{grade} M = \inf_{\underline{p}\in \operatorname{Ass}(M)} \operatorname{prof} A_{\underline{p}},$$

et par le Théoreme d'Intersection $\operatorname{prof} A_{\underline{p}} = \dim A_{\underline{p}}$ pour tout $\underline{p}$ minimal de M.

Soient $\underline{q}_i$, $i = 1,\dots,t$, les idéaux premiers minimaux de M tels que $\dim A_{\underline{q}_i} < g$. Soit $c = \sup \dim A_{\underline{q}_i}$ et $n = \sup \dim A/\underline{q}_i$. Posons d'autre part $S = \bigcap_{i=1}^{t} (A-\underline{q}_i)$ et $\underline{a} = \ker(A \to S^{-1}A)$.

LEMME 1. $\dim A/\underline{a} \leq c+n$.

En effet, si x est diviseur de zéro dans $A/\underline{a}$ alors $x \notin S$, car $A/\underline{a} \hookrightarrow S^{-1}A$. Donc

$$\bigcup_{\underline{p} \text{ min.de } A/\underline{a}} \underline{p} \subseteq \bigcup \underline{q}_i,$$

d'où $\dim A/\underline{a} \leq \sup(\dim A_{\underline{q}_i} + \dim A/\underline{q}_i)$. □

LEMME 2. *Si* $\underline{m}$ *est un idéal premier minimal de* $M \otimes A/\underline{a}$ *et si* $\dim A/\underline{m} \geq n$, *alors* $\underline{m}$ *est l'un des* $\underline{q}_i$.

Il suffit de montrer que $\dim(A/\underline{a})_{\underline{m}} \geq g$ si $\underline{m} \not\supset \underline{q}_i$ pour tout $i = 1,\dots,t$. En effet on aurait dans ce cas

$$\dim A/\underline{a} \geq \dim A/\underline{m} + \dim(A/\underline{a})_{\underline{m}} \geq n+g > n+c,$$

ce qui contredit le Lemme 1. Or on sait que si $\underline{m} \not\supset \underline{q}_i$ pour tout $i = 1,\dots,t$ on a

$$\operatorname{prof} A_{\underline{m}} \geq \operatorname{prof} A_{\underline{p}} = \dim A_{\underline{p}} \geq g$$

pour $\underline{p}$ minimal de M contenu dans $\underline{m}$. Donc il nous suffit de montrer que $\operatorname{prof} A_{\underline{m}} = \dim(A/\underline{a})_{\underline{m}}$. On sait par le Théoreme d'Intersection que $\operatorname{prof} A_{\underline{m}} \geq \dim(A/\underline{a})_{\underline{m}}$. Soit donc $f_1,\dots,f_r$ une suite $A_{\underline{m}}$-régulière qui soit un système de paramètres de $(A/\underline{a})_{\underline{m}}$. Il nous suffit de montrer que

$$\operatorname{prof}(A/\underline{f})_{\underline{m}} = 0.$$

Comme $\underline{f}$ est une suite $A_{\underline{m}}$-régulière il suffit de prouver que $\operatorname{prof}(A/\underline{f}^t)_{\underline{m}} = 0$ pour un entier $t > 0$.

On sait qu'il existe x et y non dans $\underline{m}$ et s dans S tel que l'on ait $sx\underline{m}^n \subset y\underline{f}$ pour un entier positif n. Si $s \notin \underline{m}$ on a $\underline{m}^n A_{\underline{m}} \subset \underline{f}A_{\underline{m}}$ donc $\dim(A/\underline{f})_{\underline{m}} = 0$; donc $\operatorname{prof}(A/\underline{f})_{\underline{m}} = 0$. Si $s \in \underline{m}$ soit t l'entier définit par $s \in \underline{f}_{\underline{m}}^{t-1}$ et $s \notin \underline{f}_{\underline{m}}^{t}$; on a

$$s \notin \underline{f}^t_{\underline{m}} \text{ et } s\underline{m}^n x \subset y\underline{f}^t,$$

donc $\mathrm{prof}(A/\underline{f}^t)_{\underline{m}} = 0$. □

Pour finir de montrer le théoreme, remarquons que $(A/\underline{a})_{\underline{q}_i} = A_{\underline{q}_i}$ pour tout $i = 1,\ldots,t$, et que les idéaux premiers minimaux $\underline{p}$ de $M \otimes A/\underline{a}$ tels que $\dim A/\underline{p} \geq n$ sont parmi les $\underline{q}_i$ d'après le Lemme 2. Donc $\dim M \otimes A/\underline{a} = n$ et

$$\dim(\mathrm{Tor}_i^A(A/\underline{a},M)) < n$$

pour $i > 0$. On voit qu'on a ainsi montré que $\Sigma(-1)^i P_{H_i(L.\otimes A/\underline{a})}(\nu)$ est un polynôme de degré n. D'après le lemme dit de Riemann-Roch gradué on a donc

$$\dim A/\underline{a} \geq n+g > n+c,$$

ce qui contredit le Lemme 1. □

4. *Riemann-Roch*

Au vu du §3 et des travaux [1] on est tenté de faire la conjecture suivante.

Conjecture C4. Soit A un anneau local noethérien, L. un complexe parfait sur A, $X = \mathrm{Spec}\, A$ et $Y = \mathrm{Support}(H.(L.))$. Notons $K.(X)$ (resp. $K.(Y)$) le groupe de Grothendieck des modules de type fini sur X (resp. sur Y) et $H.(X)$ (resp. $H.(Y)$) le groupe de Chow de X (resp. de Y). Alors il existe un diagramme commutatif

$$\begin{array}{ccc} K.(X) & \xrightarrow{\ \tau\ } & H.(X)\otimes\mathbb{Q} \\ \downarrow{\scriptstyle \chi(L.\otimes\,\cdot)} & & \downarrow{\scriptstyle O(L.)\cup\,\cdot} \\ K.(Y) & \xrightarrow{\ \tau\ } & H.(Y)\otimes\mathbb{Q} \end{array}$$

où

(i) τ (homomorphisme de Todd) est fonctoriel, et la partie homogène de degré $d = \dim A$, $\tau_d(A)$, de $\tau(A)$ est égale à la classe de

$$\sum_{\dim A/\underline{p}=d} \mathrm{long}\, A_{\underline{p}}\,[\underline{p}]$$

dans $H_d(X)$, et

(ii) $O(L.)$ est un opérateur (par l'opération notée cup) $H.(X) \to H.(Y)$ qui dépend fonctorielement de A et L.

THÉOREME (W.Fulton). *Soit* A *un anneau essentiellement de type fini sur un corps. Alors la Conjecture* C4 *est vérifiée.*

Nous espérons que Fulton publiera bientôt sa démonstration.

Les constructions de la classe de Todd et de $O(L.)$ sont celles qu'on trouve dans [1] ou [5].

COROLLAIRE. *Soit* A *un anneau local de dimension* d *et de caractéristique* $p > 0$, *pour lequel on a établi le Théoreme de Riemann-Roch local. Soit* L. *un complexe parfait sur* A *dont l'homologie est de longueur finie; alors* $\frac{1}{p^{nd}}\chi(L.^{(p^n)})$ *tend vers une limite* $\bar{\chi}(L.)$ *qui est égale à* $O_d(L.) \cup \tau_d(A)$.

Cet énoncé est clair à partir du Théoreme de Riemann-Roch une fois qu'on a remarqué que $O_j(L.^{(p^n)}) = p^{nj}O_j(L.)$ pour tout j. □

Je ne sais pas si l'existence de la limite $\bar{\chi}(L.)$ est connue de Sankar P.Dutta (University of Pennsylvania). En tout cas il semble utiliser ce fait dans quelques cas particulier dans sa thèse.

Il est facile de vérifier que $\bar{\chi}(L.)$ vérifie les conjectures C2 et C3 (on sait que $O_j(L.^{\vee}) = (-1)^j O_j(L.)$). Il semble donc raisonnable de se demander si $\bar{\chi} = \chi$ ou encore $O_j(L.) = 0$ pour tout $j < d$ quand $H(L.)$ est de longueur finie.

Pour finir remarquons que dans la démonstration du Théoreme du Grade au §3, les Lemmes 1 et 2 n'utilisent que le Théoreme d'Intersection et sont donc vrais si A contient un corps. On peut donc se demander si, posant

$$g(L.) = \inf\{i : O_i(L.) \cup \tau_d(A) \neq 0\}$$

(où $d = \dim A$) on a $g(L.) \leq g(L.\otimes B)$ quand on a un morphisme d'anneaux $A \to B$.

Bibliographie

1. P.Baum, W.Fulton et R.Macpherson, "Riemann-Roch for singular varieties", Publications Mathématiques 45 (Institut des Hautes Études Scientifiques, Paris, 1975), pp.101-146.

2. E.G.Evans et P.Griffith, "The syzygy problem", Ann.of Math., 114 (1981), 323-333.

3. M.Flexor et L.Szpiro, "Un théoreme de structure locale pour les complexes parfaits", Algebraic geometry, Proceedings, Tromsø, Norway 1977, Lecture Notes in Mathematics 687 (Springer, Berlin, 1978), pp.236-244.

4. M.Hochster, Topics in the homological theory of modules over commutative rings, Conference Board of the Mathematical Sciences Regional Conference Series in Mathematics 24 (American Mathematical Society, Providence, 1975).

5. B.Iversen, "Local Chern classes", Ann.Sci.École Norm.Sup.(4), 9 (1976), 155-169.

6. C.Peskine et L.Szpiro, "Dimension projective finie et cohomologie locale", Publications Mathématiques 42 (Institut des Hautes Études Scientifiques, Paris, 1973), pp.47-119.

7. C.Peskine et L.Szpiro, "Syzygies et multiplicités", C.R.Acad. Sci.Paris Sér.A, 278 (1974), 1421-1424.

8. P.Roberts, "Two applications of dualizing complexes over local rings", Ann.Sci.École Norm.Sup.(4), 9 (1976), 103-106.

9. P.Roberts, "Cohen-Macaulay complexes and an analytic proof of the new intersection conjecture", J.Algebra, 66 (1980), 220-225.

10. J.-P.Serre, Algèbre locale : multiplicités, Lecture Notes in Mathematics 11 (Springer, Berlin, 1965).

École Normale Supérieure,
45, Rue d'Ulm,
75230 Paris Cedex 05, France.

PART II

DETERMINANTAL IDEALS, FINITE FREE RESOLUTIONS, AND RELATED TOPICS

SOME EXACT COMPLEXES AND FILTRATIONS RELATED TO CERTAIN SPECIAL YOUNG DIAGRAMS

KAAN AKIN AND DAVID A. BUCHSBAUM

1. *Introduction*

In this article we shall deal with a few straightforward applications and interpretations of the material covered in the lectures we gave at this Symposium (which in turn were based on the articles [2] and [3]). In those lectures, skew-Schur functors and complexes were described, and some of their fundamental properties were derived. Here, we shall use these properties to

(a) generalize the Koszul complex;

(b) generalize the complex of tensor products of "hooks" given in [4];

(c) sketch the proof of a characteristic-free version of the Pieri formulas.

In §2 we recall some of the basic definitions and properties of skew-Schur functors and complexes that we shall use.

In §3 we generalize the Koszul complex in the following sense. For a free R-module F, we have the complex

$$\cdots \to S_{q-1}F \otimes \Lambda^{p+1}F \to S_qF \otimes \Lambda^pF \to S_{q+1}F \otimes \Lambda^{p-1}F \to \cdots,$$

which may be regarded as one of the strands of the Koszul complex associated to the ideal $(X_1,\ldots,X_n)$ in $R[X_1,\ldots,X_n]$, where $n = \text{rank } F$. The image of the map $S_qF \otimes \Lambda^pF \to S_{q+1}F \otimes \Lambda^{p-1}F$ is denoted in [5] by $L^p_{q+1}F$, or $L_{(p,\underbrace{1,\ldots,1}_{q})}F$ in [3], and is the Schur functor whose shape is the "hook"

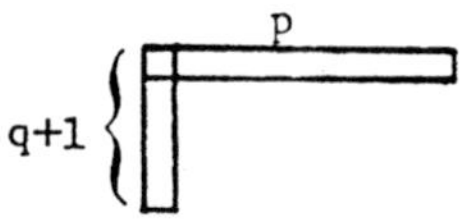

Thus, we may regard the Koszul complex as a "resolution" of $L^p_{q+1}F$:

$$0 \to S_{p+q-n}F \otimes \Lambda^n F \to \dots \to S_{q-1}F \otimes \Lambda^{p+1}F \to S_q F \otimes \Lambda^p F \to L^p_{q+1}F \to 0.$$

A natural question, then, is the following. If λ is any partition, and $S_q F \otimes L_\lambda F \to L_{(\lambda,\underbrace{1,\dots,1}_{q})}F$ is the natural surjection (to be defined in §2) which generalizes the surjection $S_q F \otimes \Lambda^p F \to L^p_{q+1}F$, is there some exact complex

$$\dots \to S_{q-2}F \otimes X_2 \to S_{q-1}F \otimes X_1 \to S_q F \otimes L_\lambda F \to L_{(\lambda,\underbrace{1,\dots,1}_{q})}F \to 0,$$

where X_i depends on λ in such a way that, when $L_\lambda F = \Lambda^p F$, we have $X_i = \Lambda^{p+i}F$? We answer this question affirmatively in §3.

In §4 we generalize the "hooks" $L^p_q F$ to "skew-hooks" $L^{p_1\dots p_k}_{q_1\dots q_k}F$ and essentially complete a project first proposed in [4;pp. 558-559]. In particular, we show that what we supposed to be true there is indeed very far from the truth, that is we do not come anywhere near to recovering all the irreducible representations of GL(F) in characteristic zero.

Finally, in §5 we sketch a proof of a characteristic-free version of the Pieri formulas. We show that if λ is any partition, then $L_\lambda F \otimes S_q F$ has a natural filtration (with respect to the action of GL(F)) whose associated graded module is $\Sigma\ L_\nu F$, where the sum is taken over all partitions ν containing λ such that $|\nu| = |\lambda| + q$ and the shape of ν/λ has at most one entry in each row ($|\nu|$ means the weight of ν (see §2)). The method of proof does not yield the corresponding result for $L_\lambda F \otimes \Lambda^q F$. We show, however, that by means of a basis theorem for Schur complexes (rather than for Schur functors), one does get the Pieri formula for $L_\lambda F \otimes \Lambda^q F$.

Throughout this article, R is a commutative ring and all modules F,G, etc., are finitely generated free R-modules.

2. *Recall of basic definitions and properties*

For detailed definitions of the terms and complete proofs of the propositions sketched in this section, the reader is referred to [3].

A *partition* is a (weakly) decreasing sequence $\lambda = (\lambda_1,\ldots,\lambda_q)$ of non-negative integers. The *length* of the partition λ is the number of positive terms in the sequence, and the *weight* of λ is the sum of the terms of the sequence.

Thus, the length of (4,4,3,2,0,0) is 4, and its weight is 13.

The partition μ is *contained* in the partition λ ($\mu \subseteq \lambda$) if $\mu_i \leq \lambda_i$ for all i.

The *diagram*, or *shape*, of λ, denoted by Δ_λ, is the set of ordered pairs of integers (i,j) with $1 \leq i \leq$ length of λ and $1 \leq j \leq \lambda_i$. We use the ordering of the pairs (i,j) that we use for matrices. That is, the i's increase downward and the j's increase from left to right. Thus, if $\lambda = (4,4,3,2,0,0)$, Δ_λ is usually depicted as

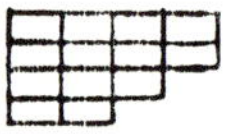 or  .

If μ and λ are partitions with $\mu \subseteq \lambda$, then the *skew-shape*, or *skew-diagram* of shape λ/μ, $\Delta_{\lambda/\mu}$, is $\Delta_\lambda - \Delta_\mu$. Thus, if $\lambda =$ (4,4,3,2,0,0) and $\mu = (2,1,0)$, then $\Delta_{\lambda/\mu}$ is

 .

Suppose now that $A = (a_{ij})$ is an s×t matrix (s rows and t columns) whose entries are either 0 or 1. For each $i = 1,\ldots,s$, let $a_i = \sum_{j=1}^{t} a_{ij}$ and for each $j = 1,\ldots,t$, let $b_j = \sum_{i=1}^{s} a_{ij}$. Then for a free R-module F we can define the map

$$d_A\colon \Lambda^{a_1}F \otimes \ldots \otimes \Lambda^{a_s}F \longrightarrow S_{b_1}F \otimes \ldots \otimes S_{b_t}F$$

as follows. We diagonalize $\Lambda^{a_i}F$ to $\Lambda^{a_{i1}}F \otimes \ldots \otimes \Lambda^{a_{it}}F$ and, taking the tensor product, we obtain the map

$$\delta\colon \Lambda^{a_1}F \otimes \ldots \otimes \Lambda^{a_s}F$$

$$\longrightarrow \Lambda^{a_{11}}F \otimes \ldots \otimes \Lambda^{a_{1t}}F \otimes \ldots \otimes \Lambda^{a_{s1}}F \otimes \ldots \otimes \Lambda^{a_{st}}F.$$

Permuting these terms, and using the fact that $\Lambda^{a_{ij}}F \approx S_{a_{ij}}F$, we obtain the isomorphism

$$\theta\colon\ \Lambda^{a_{11}}F \otimes \ldots \otimes \Lambda^{a_{1t}}F \otimes \ldots \otimes \Lambda^{a_{s1}}F \otimes \ldots \otimes \Lambda^{a_{st}}F$$

$$\approx S_{a_{11}}F \otimes \ldots \otimes S_{a_{s1}}F \otimes \ldots \otimes S_{a_{1t}}F \otimes \ldots \otimes S_{a_{st}}F.$$

Finally, by multiplication in the symmetric algebra S(F), we obtain the map

$$m\colon\ S_{a_{11}}F \otimes \ldots \otimes S_{a_{s1}}F \otimes \ldots \otimes S_{a_{1t}}F \otimes \ldots \otimes S_{a_{st}}F$$

$$\longrightarrow S_{b_1}F \otimes \ldots \otimes S_{b_t}F.$$

The map d_A is defined to be the composite map

$$d_A = m\theta\delta.$$

The map d_A is called the *Schur map associated to the matrix* A.

Now let μ and λ be partitions with $\mu \subseteq \lambda$. Let us write $\lambda = (\lambda_1,\ldots,\lambda_q)$ with $q = \text{length}(\lambda)$, and $\mu = (\mu_1,\ldots,\mu_q)$, and let $A_{\lambda/\mu}$ be the $q \times \lambda_1$ matrix whose entries a_{ij} are defined as follows: $a_{ij} = 1$ if $\mu_i + 1 \le j \le \lambda_i$, and $a_{ij} = 0$ otherwise. Clearly, $a_i = \lambda_i - \mu_i$ for $i = 1,\ldots,q$, and b_j is equal to the number of boxes in the jth column of $\Delta_{\lambda/\mu}$ for $j = 1,\ldots,\lambda_i$. The map $d_{A_{\lambda/\mu}}$ is called the *Schur map of shape* λ/μ, and the image of this map is denoted by $L_{\lambda/\mu}F$. (One usually writes $d_{\lambda/\mu}$ in place of $d_{A_{\lambda/\mu}}$.) The module $L_{\lambda/\mu}F$ is called the *skew-Schur functor of* F *of shape* λ/μ.

Notice that if $\mu = 0$, then $\Delta_{\lambda/\mu} = \Delta_\lambda$ and we may write $L_\lambda F$ and d_λ for $L_{\lambda/\mu}F$ and $d_{\lambda/\mu}$. Also notice that if λ is a partition, and if we denote by $\tilde{\lambda}_j$ the number of boxes in the jth column of Δ_λ, then the sequence $\tilde{\lambda} = (\tilde{\lambda}_1,\tilde{\lambda}_2,\ldots)$ is also a partition, called the *transpose* of the partition λ. Thus, letting $A_{\lambda/\mu}$ denote the matrix described above, we see that the integers b_j associated to $A_{\lambda/\mu}$ are simply $\tilde{\lambda}_j - \tilde{\mu}_j$, so that the map $d_{\lambda/\mu}$ is a map from $\Lambda^{\lambda_1-\mu_1}F \otimes \ldots \otimes \Lambda^{\lambda_q-\mu_q}F$ to $S_{\tilde{\lambda}_1-\tilde{\mu}_1}F \otimes \ldots \otimes S_{\tilde{\lambda}_t-\tilde{\mu}_t}$ where $t = \lambda_1$.

This is usually abbreviated to $d_{\lambda/\mu}\colon\ \Lambda_{\lambda/\mu}F \to S_{\tilde{\lambda}/\tilde{\mu}}F$.

For each $i = 1,\ldots,q$, let $p_i = \lambda_i - \mu_i$, $t_i = \mu_i - \mu_{i+1} + 1$, and let $\square_i'\colon \sum_{\alpha=t_i}^{p_{i+1}} \Lambda^{p_i+\alpha}F \otimes \Lambda^{p_{i+1}-\alpha}F \longrightarrow \Lambda^{p_i} \otimes \Lambda^{p_{i+1}}$ be the map, for $i = 1,\ldots,q-1$, which for each α is the composite

$$\Lambda^{p_i+\alpha}F \otimes \Lambda^{p_{i+1}-\alpha}F \xrightarrow{\beta_1} \Lambda^{p_1}F \otimes \Lambda^{\alpha}F \otimes \Lambda^{p_{i+1}-\alpha}F \xrightarrow{\beta_2} \Lambda^{p_i}F \otimes \Lambda^{p_{i+1}}F,$$

where β_1 diagonalizes $\Lambda^{p_i+\alpha}F$ to $\Lambda^{p_i}F \otimes \Lambda^{\alpha}F$ and β_2 multiplies $\Lambda^{\alpha}F \otimes \Lambda^{p_{i+1}-\alpha}F$ to $\Lambda^{p_{i+1}}F$. Thus, for $i = 1,\ldots,q-1$, we have maps $\square_i = 1 \otimes \ldots \otimes \square_i' \otimes \ldots \otimes 1$ where

$$\square_i\colon \sum_{\alpha=t_i}^{p_{i+1}} \Lambda^{p_1}F \otimes \ldots \otimes \Lambda^{p_{i-1}}F \otimes \Lambda^{p_i+\alpha}F \otimes \Lambda^{p_{i+1}-\alpha}F \otimes \ldots \otimes \Lambda^{p_q}F$$
$$\longrightarrow \Lambda^{p_1}F \otimes \ldots \otimes \Lambda^{p_q}F.$$

Now set $\square_{\lambda/\mu}$ to be the sum of the maps $\square_i$:

$$\square_{\lambda/\mu}\colon \sum_{i=1}^{q-1} \sum_{\alpha=t_i}^{p_{i+1}} \Lambda^{p_1}F \otimes \ldots \otimes \Lambda^{p_i+\alpha}F \otimes \Lambda^{p_{i+1}-\alpha}F \otimes \ldots \otimes \Lambda^{p_q}F$$
$$\longrightarrow \Lambda^{p_1}F \otimes \ldots \otimes \Lambda^{p_q}F = \Lambda_{\lambda/\mu}F.$$

Finally, denote by $\bar{L}_{\lambda/\mu}F$ the cokernel of $\square_{\lambda/\mu}$. The main theorem about the modules $L_{\lambda/\mu}F$ and $\bar{L}_{\lambda/\mu}F$ and the maps $d_{\lambda/\mu}$ and $\square_{\lambda/\mu}$ is the following.

THEOREM 2.1 [3]. *The kernel of* $d_{\lambda/\mu}$ *is the image of* $\square_{\lambda/\mu}$ *and thus* $\bar{L}_{\lambda/\mu}F$ *and* $L_{\lambda/\mu}F$ *are naturally isomorphic.* $L_{\lambda/\mu}F$ *is a direct summand of* $S_{\tilde{\lambda}/\tilde{\mu}}F$ *and is a universally free module.*

A basis for the module $L_{\lambda}F$ can be explicitly described in terms of a basis for the module F, but we shall discuss this after we define Schur complexes, which generalize Schur functors.

A Schur complex of shape λ/μ is a complex associated to a map $\phi\colon\ G \to F$ of free modules, and a pair of partitions $\mu \subseteq \lambda$. Given the map $\phi\colon\ G \to F$, we define complexes $\Lambda^k(\phi)$ and $S_k(\phi)$ as follows:

$$\Lambda^k(\phi):\quad 0 \to D_kG \to D_{k-1}G \otimes F \to \dots \to D_2G \otimes \Lambda^{k-2}F \to G \otimes \Lambda^{k-1}F \to \Lambda^kF \to 0;$$

$$S_k(\phi):\quad 0 \to \Lambda^kG \to F \otimes \Lambda^{k-1}G \to \dots \to S_{k-2}F \otimes \Lambda^2G \to S_{k-1}F \otimes G \to S_kF \to 0.$$

The module D_jG is the jth divided power module of G and the map $D_jG \otimes \Lambda^{k-j}F \to D_{j-1}G \otimes \Lambda^{k-j+1}F$ is the composite

$$D_jG \otimes \Lambda^{k-j}F \xrightarrow{\gamma_1} D_{j-1}G \otimes G \otimes \Lambda^{k-j}F \xrightarrow{\gamma_2} D_{j-1}G \otimes F \otimes \Lambda^{k-j}F \xrightarrow{\gamma_3} D_{j-1}G \otimes \Lambda^{k-j+1}F,$$

where γ_1 diagonalizes D_jG as indicated, γ_2 maps G to F by the map ϕ, and γ_3 multiplies $F \otimes \Lambda^{k-j}F$ to $\Lambda^{k-j+1}F$. The maps in $S_k(\phi)$ are essentially Koszul complex maps.

It is easy to show [3] that the complexes $\Lambda^k(\phi)$ diagonalize; that is, there are diagonal maps from $\Lambda^k(\phi)$ to $\Lambda^j(\phi) \otimes \Lambda^{k-j}(\phi)$ which are maps of complexes. Also, the complexes $\Lambda^k(\phi)$ and $S_k(\phi)$ multiply so that we have maps $S_k(\phi) \otimes S_\ell(\phi) \to S_{k+\ell}(\phi)$ (and similarly for $\Lambda(\phi)$). Thus if $\mu \subseteq \lambda$ are partitions, we may define

$$d_{\lambda/\mu}(\phi):\quad \Lambda^{\lambda_1-\mu_1}(\phi) \otimes \dots \otimes \Lambda^{\lambda_q-\mu_q}(\phi) \longrightarrow S_{\tilde\lambda_1-\tilde\mu_1}(\phi) \otimes \dots \otimes S_{\tilde\lambda_t-\tilde\mu_t}(\phi)$$

and

$$\square_{\lambda/\mu}(\phi):\quad \sum_{i=1}^{q-1} \sum_{\alpha=t_i}^{p_{i+1}} \Lambda^{p_1}(\phi) \otimes \dots \otimes \Lambda^{p_i+\alpha}(\phi) \otimes \Lambda^{p_{i+1}-\alpha}(\phi) \otimes \dots \otimes \Lambda^{p_q}(\phi) \longrightarrow \Lambda^{p_1}(\phi) \otimes \dots \otimes \Lambda^{p_q}(\phi)$$

in a manner analogous to the definitions of $d_{\lambda/\mu}$ and $\square_{\lambda/\mu}$.

We define $\mathbb{L}_{\lambda/\mu}(\phi)$ to be the image of $d_{\lambda/\mu}(\phi)$, and $\overline{\mathbb{L}}_{\lambda/\mu}(\phi)$ to be the cokernel of $\square_{\lambda/\mu}(\phi)$, and we call $\mathbb{L}_{\lambda/\mu}(\phi)$ the *Schur complex of* ϕ *of shape* λ/μ. Notice that when $G = 0$, $(\mathbb{L}_{\lambda/\mu}(\phi))_k = 0$ for $k > 0$, and $(\mathbb{L}_{\lambda/\mu}(\phi))_0 = L_{\lambda/\mu}F$.

THEOREM 2.2 [3]. $\mathrm{Im}(\square_{\lambda/\mu}(\phi)) = \mathrm{Ker}(d_{\lambda/\mu}(\phi))$ *and thus* $\overline{\mathbb{L}}_{\lambda/\mu}(\phi)$ *and* $\mathbb{L}_{\lambda/\mu}(\phi)$ *are naturally isomorphic. The modules* $(\mathbb{L}_{\lambda/\mu}(\phi))_k$ *are universally free and are direct summands of*

$$(S_{\tilde\lambda_1-\tilde\mu_1}(\phi) \otimes \dots \otimes S_{\tilde\lambda_t-\tilde\mu_t}(\phi))_k.$$

In order to describe a basis for the module $(\mathbb{L}_{\lambda/\mu}(\phi))_k$, we must introduce tableaux. If $\mu \subseteq \lambda$ are partitions, and S is a totally ordered set, we define a *tableau of shape* λ/μ *with values in* S to be a map from $\Delta_{\lambda/\mu}$ to S, and we denote the set of all such tableaux by $\mathrm{Tab}_{\lambda/\mu}(S)$. Pictorially, a tableau is simply a filling in of the diagram of λ/μ by elements of S. If G and F are free modules, let $\{y_1,\ldots,y_m\}$ and $\{x_1,\ldots,x_n\}$ be bases of G and F, and let $S = \{y_1,\ldots,y_m\} \cup \{x_1,\ldots,x_n\}$. We assume a total ordering on the y's and x's as given by their indexing, and we put a total order on S by choosing a fixed shuffle of the x's through the y's. (Usually one chooses the ordering on S by selecting the identity shuffle,in which case all the y's precede all the x's, or by selecting the shuffle which passes all the x's past all the y's, so that the x's precede the y's in this order.) If $T \in \mathrm{Tab}_{\lambda/\mu}(S)$, we say that T *is standard modulo* G if, in each row of $\Delta_{\lambda/\mu}$, the entries are weakly increasing but the inequality between adjacent x's is strict, and in each column of $\Delta_{\lambda/\mu}$ the entries are weakly increasing but the inequality between adjacent y's is strict.

To each tableau which is regular mod G, one can assign in a canonical (and obvious) way a basis element of $\Lambda_{\lambda/\mu}(\phi)$. In the proof of Theorem 2.2, one shows that $\{d_{\lambda/\mu}(\phi)(T) \mid T \text{ is standard mod } G\}$ is a basis for $\mathbb{L}_{\lambda/\mu}(\phi)$. In particular, the subset of these elements in which the number of y's is equal to k forms a basis for $(\mathbb{L}_{\lambda/\mu}(\phi))_k$. When one takes G to be O, one obtains a description of the basis for $L_{\lambda/\mu}F$ in terms of standard tableaux (which agrees with the classical description).

3. *The Koszul complex somewhat generalized*

In §1 we said that the image of the boundary map ∂: $S_qF \otimes \Lambda^pF \to S_{q+1}F \otimes \Lambda^{p-1}F$ in the Koszul complex was $L^p_{q+1}F$ or, in the notation of §2, $L_{(p,\underbrace{1,\ldots,1}_{q})}F$. Recall that the map ∂ is defined as the composite $S_qF \otimes \Lambda^pF \to S_qF \otimes F \otimes \Lambda^{p-1}F \to S_{q+1}F \otimes \Lambda^{p-1}F$. If we let $\lambda = (p,\underbrace{1,\ldots,1}_{q})$, it is easy to see that the map

$d_\lambda: \Lambda^p F \otimes \underbrace{F \otimes \cdots \otimes F}_{q} \to S_{q+1}F \otimes \underbrace{F \otimes \cdots \otimes F}_{p-1}$ is the composite

$$\Lambda^p F \otimes \underbrace{F \otimes \cdots \otimes F}_{q} \xrightarrow{\beta_1} \Lambda^p F \otimes S_q F \xrightarrow{\partial} \Lambda^{p-1}F \otimes S_{q+1}F$$

$$\xrightarrow{\beta_2} S_{q+1}F \otimes \underbrace{F \otimes \cdots \otimes F}_{p-1},$$

where β_1 multiplies q copies of F to $S_q F$ and β_2 (in addition to interchanging $S_{q+1}F$ and $\Lambda^{p-1}F$) diagonalizes $\Lambda^{p-1}F$ to $\underbrace{F \otimes \cdots \otimes F}_{p-1}$.

Since β_1 is a surjection and β_2 is an injection, we see immediately that $\operatorname{Im} \partial \approx L_\lambda F$. We also see that if we let λ' be the partition p then $L_{\lambda'}F = \Lambda^p F$ and we have the surjection $S_q F \otimes L_{\lambda'}F \to L_\lambda F$ where $\lambda = (\lambda', \underbrace{1,\ldots,1}_{q})$.

In general, for any partition λ, we may define the surjection $S_q F \otimes L_\lambda F \to L_{(\lambda,\underbrace{1,\ldots,1}_{q})}F$ as follows. If we let $\lambda = (\lambda_1,\ldots,\lambda_k)$ and $\tilde\lambda = (\tilde\lambda_1,\ldots,\tilde\lambda_t)$, then $(\lambda,\underbrace{1,\ldots,1}_{q})^\sim = (\tilde\lambda_1 + q, \tilde\lambda_2,\ldots,\tilde\lambda_t)$. Thus $S_q F \otimes L_\lambda F$ is the image of

$$1 \otimes d_\lambda: S_q F \otimes \Lambda^{\lambda_1}F \otimes \cdots \otimes \Lambda^{\lambda_k}F \to S_q F \otimes S_{\tilde\lambda_1} \otimes \cdots \otimes S_{\tilde\lambda_t}F$$

and $L_{(\lambda,\underbrace{1,\ldots,1}_{q})}F$ is the image of

$$d_{(\lambda,\underbrace{1,\ldots,1}_{q})}: \Lambda^{\lambda_1}F \otimes \cdots \otimes \Lambda^{\lambda_k}F \otimes \overbrace{F \otimes \cdots \otimes F}^{q}$$

$$\to S_{\tilde\lambda_1+q}F \otimes S_{\tilde\lambda_2}F \otimes \cdots \otimes S_{\tilde\lambda_t}F.$$

It is easy to see that $d_{(\lambda,\underbrace{1,\ldots,1}_{q})}$ is the composite

$$\Lambda^{\lambda_1}F \otimes \cdots \otimes \Lambda^{\lambda_k}F \otimes \overbrace{F \otimes \cdots \otimes F}^{q} \xrightarrow{u_1} \Lambda^{\lambda_1}F \otimes \cdots \otimes \Lambda^{\lambda_k}F \otimes S_q F$$

$$\xrightarrow{d_\lambda \otimes 1} S_{\tilde\lambda_1}F \otimes \cdots \otimes S_{\tilde\lambda_t}F \otimes S_q F \xrightarrow{u_2} S_{\tilde\lambda_1+q}F \otimes S_{\tilde\lambda_2}F \otimes \cdots \otimes S_{\tilde\lambda_t}F,$$

where u_1 multiplies $\overbrace{F \otimes \cdots \otimes F}^{q}$ onto $S_q F$ and u_2 multiplies $S_{\tilde\lambda_1}F \otimes S_q F$ onto $S_{\tilde\lambda_1+q}F$. Since u_1 is surjective, the image of $1 \otimes d_\lambda$ (or $d_\lambda \otimes 1$) clearly gets mapped surjectively onto the image of

$d_{(\lambda,\underbrace{1,\ldots 1}_{q})}$ and it is this surjection of $S_qF \otimes L_\lambda F$ onto $L_{(\lambda,\underbrace{1,\ldots,1}_{q})}F$ which generalizes the surjection of $S_qF \otimes \Lambda^p F$ onto $L_{(p,\underbrace{1,\ldots,1}_{q})}F$, and that we shall consider in this section.

To facilitate matters, we introduce some notation. If λ is a partition and ℓ is a non-negative integer, we shall denote by $\lambda + \ell$ the skew partition $(\lambda_1+\ell,\ldots,\lambda_k+\ell)/(\ell,\ldots,\ell)$, where k is the length of λ and $(\ell,\ldots,\ell)$ is of length k-1.

Now let ℓ be a positive integer, and let λ be a partition of length k. For every r, there is a canonical map

$$\partial:\quad S_rF \otimes L_{\lambda+\ell}F \to S_{r+1}F \otimes L_{\lambda+(\ell-1)}F$$

which is induced by the map

$$S_rF \otimes \Lambda^{\lambda_1}F \otimes \ldots \otimes \Lambda^{\lambda_{k-1}}F \otimes \Lambda^{\lambda_k+\ell}F$$
$$\to S_{r+1}F \otimes \Lambda^{\lambda_1}F \otimes \ldots \otimes \Lambda^{\lambda_{k-1}}F \otimes \Lambda^{\lambda_{k+\ell-1}}F$$

which, itself, is induced by the Koszul map

$$\partial:\quad S_rF \otimes \Lambda^{\lambda_k+\ell}F \to S_{r+1}F \otimes \Lambda^{\lambda_k+\ell-1}F.$$

This is easy to check, particularly when one realizes that the diagram of $\lambda+\ell$ is the diagram of λ augmented by ℓ boxes added to the *left* of the last row of λ:

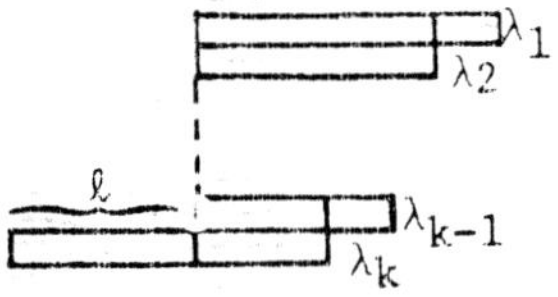

Thus, the Schur map $d_{\lambda+\ell}$ is the composition

$$\Lambda^{\lambda_1}F \otimes \ldots \otimes \Lambda^{\lambda_{k-1}}F \otimes \Lambda^{\lambda_k+\ell}F \to \Lambda^{\lambda_1}F \otimes \ldots \otimes \Lambda^{\lambda_{k-1}}F \otimes \Lambda^{\lambda_k}F \otimes \Lambda^{\ell}F$$

$$\xrightarrow{d_\lambda \otimes 1} S_{\tilde{\lambda}_1}F \otimes \ldots \otimes S_{\tilde{\lambda}_t}F \otimes \Lambda^{\ell}F \to \underbrace{F \otimes \ldots \otimes F}_{\ell} \otimes S_{\tilde{\lambda}_1}F \otimes \ldots \otimes S_{\tilde{\lambda}_t}F, \qquad (*)$$

where the first map diagonalizes $\Lambda^{\lambda_k+\ell}F$ to $\Lambda^{\lambda_k}F \otimes \Lambda^{\ell}F$ and the last one diagonalizes $\Lambda^{\ell}F$ to $\underbrace{F \otimes \ldots \otimes F}_{\ell}$.

THEOREM 3.1. *If* F *is a free* R-*module*, λ *is a partition of length* k, *and* q *is a positive integer, then*

$$\cdots \to S_{q-2}F \otimes L_{\lambda+2}F \to S_{q-1}F \otimes L_{\lambda+1}F \to S_qF \otimes L_\lambda F \qquad (**)$$
$$\to L_{(\lambda,\underbrace{1,\ldots,1}_{q})}F \to 0$$

is an exact sequence.

Proof. Notice that when k = 1, $L_\lambda F = \Lambda^{\lambda_1}F$ and $L_{\lambda+\ell}F = \Lambda^{\lambda_1+\ell}F$, so that (**) reduces to the Koszul complex in that case.

The proof proceeds by induction on q. In order to treat the case in which q = 1, as well as to handle the inductive step, we must first consider some additional canonical maps. For every positive integer ℓ, we have a surjection $\Lambda^\ell F \otimes L_\lambda F \xrightarrow{\alpha} L_{(\lambda,1)+\ell-1}F$ and an injection $L_{\lambda+\ell}F \xrightarrow{\beta} \Lambda^\ell F \otimes L_\lambda F$. The injection β is clear from the factorization (*) of the map $d_{\lambda+\ell}$, and we see that, on the generators, the map is induced from the diagonalization map

$$\Lambda^{\lambda_1}F \otimes \cdots \otimes \Lambda^{\lambda_{k-1}}F \otimes \Lambda^{\lambda_k+\ell}F \xrightarrow{\tilde{\beta}} \Lambda^{\lambda_1}F \otimes \cdots \otimes \Lambda^{\lambda_k}F \otimes \Lambda^\ell F.$$

The map α is the map induced from the identity map

$$\Lambda^\ell F \otimes \Lambda^{\lambda_1}F \otimes \cdots \otimes \Lambda^{\lambda_k}F \to \Lambda^{\lambda_1}F \otimes \cdots \otimes \Lambda^{\lambda_k}F \otimes \Lambda^\ell F.$$

From 2.1 we know that we may regard $\Lambda^\ell F \otimes L_\lambda F$ as the cokernel of $1 \otimes \Box_\lambda$, and $L_{(\lambda,1)+\ell-1}$ as the cokernel of $\Box_{(\lambda,1)+\ell-1}$, and we observe that $\Box_{(\lambda,1)+\ell-1} = 1 \otimes \Box_\lambda + \tilde{\beta}$. We therefore have proved the following.

LEMMA 3.2. *For every partition* λ, *and every positive integer* ℓ, *the sequence*

$$0 \to L_{\lambda+\ell}F \xrightarrow{\beta} \Lambda^\ell F \otimes L_\lambda F \xrightarrow{\alpha} L_{(\lambda,1)+\ell-1}F \to 0$$

is exact. □

Lemma 3.2 proves Theorem 3.1 for the case in which q = 1, for the special case of 3.2 in which $\ell = 1$ gives us (**) when q = 1. Assuming now that the Theorem is true for q and setting $\mu = (\lambda,1)$, we consider the map of complexes

$$\begin{array}{ccccccccc}
\cdots \to S_{q-2}F\otimes\Lambda^3F\otimes L_\lambda F & \xrightarrow{\partial\otimes 1} & S_{q-1}F\otimes\Lambda^2F\otimes L_\lambda F & \xrightarrow{\partial\otimes 1} & S_qF\otimes F\otimes L_\lambda F & \xrightarrow{\partial\otimes 1} & S_{q+1}F \otimes L_\lambda F & \to & 0 \\
\downarrow 1\otimes\alpha & & \downarrow 1\otimes\alpha & & \downarrow 1\otimes\alpha & & \downarrow \partial & & \\
\cdots \to S_{q-2}F \otimes L_{\mu+2}F & \xrightarrow{\partial} & S_{q-1}F \otimes L_{\mu+1}F & \xrightarrow{\partial} & S_qF \otimes L_\mu F & \xrightarrow{\partial} & L_{(\mu,\underbrace{1,\ldots,1}_{q})}F & \to & 0.
\end{array}$$

The reader can check that this is indeed a commutative diagram, with the horizontal maps $\partial\otimes 1$ those of the ordinary Koszul complex tensored with the identity, and the horizontal maps ∂ those of the complex (**) with λ replaced by μ. The kernels of the maps $1\otimes\alpha$ are, by Lemma 3.2, the modules $S_{q-r}F\otimes L_{\lambda+r+1}F$ for $r \geq 0$. The induction hypothesis on q, and the acyclicity of the Koszul complex, tell us that the two complexes above are exact, and a simple homological argument completes the proof that (**) is exact for q+1. □

4. *Skew hooks*

As we have already said, the modules $L_{(p,\underbrace{1,\ldots,1}_{q-1})}F$ were denoted by L^p_qF in [5]. These modules, or shapes, are called *hooks*, as the diagram of the partition (p,1,...,1) looks like

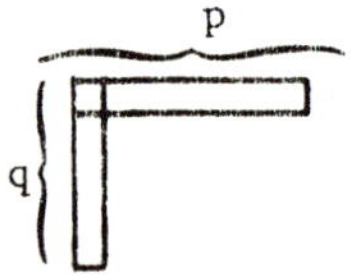

.

Prompted by the discussion in [4; pp.558-559], we were led to consider skew shapes of the form

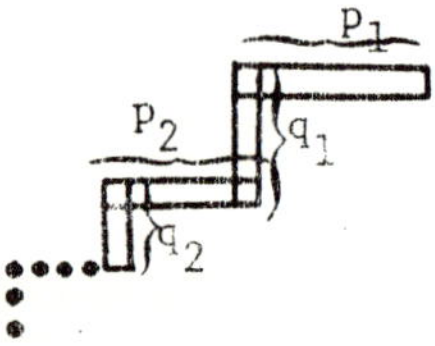

,

which we naturally call *skew-hooks*, and the corresponding skew-Schur functors, which we again call skew-hooks. More technically, we make the following definition.

Definition 4.1. Let $p_1,\ldots,p_k,q_1,\ldots,q_k$ be positive integers, and set $a_j = \sum_{i=j}^{k} p_i - (k-j)$. Consider the partitions

$$\lambda = (a_1,\overbrace{a_2,\ldots,a_2}^{q_1-1},\overbrace{a_3,\ldots,a_3}^{q_2-1},\ldots,\overbrace{a_k,\ldots,a_k}^{q_{k-1}-1},\overbrace{1,\ldots,1}^{q_k-1})$$

and

$$\mu = (\overbrace{a_2-1,\ldots,a_2-1}^{q_1-1},\overbrace{a_3-1,\ldots,a_3-1}^{q_2-1},\ldots,\overbrace{a_k-1,\ldots,a_k-1}^{q_{k-1}-1}).$$

The skew-shape λ/μ is called a *skew-hook of type* $(p_1,\ldots,p_k;\ q_1,\ldots,q_k)$, and the corresponding skew-Schur functor will be denoted by $L_{q_1\cdots q_k}^{p_1\cdots p_k}F$.

For $k = 1$, we see easily that we have our original hooks, and that for $k > 1$ the modules $L_{q_1\cdots q_k}^{p_1\cdots p_k}F$ are generally not irreducible, even in characteristic zero.

A few elementary facts about these skew-hooks can be seen quite easily, and we state them in the following proposition.

PROPOSITION 4.2. *Let* $p_1,\ldots,p_k,\ q_1,\ldots,q_k$ *be positive integers.*

(i) *If* $q_i = 1$ *for some* $i = 1,\ldots,k-1$, *then*

$$L_{q_1\cdots q_k}^{p_1\cdots p_k}F = L_{q_1\cdots \hat{q}_i\cdots q_k}^{p_1\cdots p_i+p_{i+1}-1\ p_{i+2}\cdots p_k}F.$$

(ii) *If* $p_i = 1$ *for some* $i = 2,\ldots,k$, *then*

$$L_{q_1\cdots q_k}^{p_1\cdots p_k}F = L_{q_1\cdots q_{i-1}+q_i-1\ q_{i+1}\cdots q_k}^{p_1\cdots \hat{p}_i\cdots p_k}F.$$

(iii) *If* $p_1 = 1 = q_k$, *then* $L_{q_1\cdots q_k}^{p_1\cdots p_k}F \approx L_{q_{k-1}q_{k-2}\cdots q_1}^{p_k p_{k-1}\cdots p_2}F.$ □

Because of 4.2, we may as well assume, when studying $L_{q_1\cdots q_k}^{p_1\cdots p_k}F$, that $q_i > 1$ for $i = 1,\ldots,k-1$, that $p_i > 1$ for $i = 2,\ldots,k$, and that $p_1q_k > 1$.

In [4] we defined a map from $L_{q_1-1}^{p_1}F \otimes L_{q_2}^{p_2}F$ to $L_{q_1}^{p_1}F \otimes L_{q_2}^{p_2-1}F$, which is the action of the trace element $C_F \in F \otimes F^*$ considered as an element of $SF \otimes \Lambda F^*$ (C_F is the element of $F \otimes F^*$ which corresponds to the identity map of $F \to F$ under the natural isomorphism $\mathrm{Hom}(F,F) \approx F\otimes F^*$). In fact, $SF \otimes \Lambda F$ is an $SF \otimes \Lambda F^*$ module and the action of C_F on $SF \otimes \Lambda F$ is precisely the Koszul complex boundary map (since $C_F^2 = 0$, one automatically obtains a complex). Since L_q^pF is the image (or kernel) of the action of C_F on a suitable homogeneous strand of SF ⊗ ΛF, $\Sigma L_q^p F$ is an SF ⊗ ΛF-module so that we have the action of C_F on $L_{q_1-1}^{p_1} F \otimes L_{q_2}^{p_2} F$, considering the SF action on the first factor, and the ΛF* action on the second. It is easy to check that $L_{q_1 q_2}^{p_1 p_2} F$ is the image of this action in $L_{q_1}^{p_1} F \otimes L_{q_2}^{p_2-1} F$ and since the complex

$$0 \to L_{q_2}^{p_1+p_2+q_1-2} F \xrightarrow{\delta} L_1^{p_1} F \otimes L_{q_2}^{p_2+q_1-2} F \to \cdots \to L_{q_1-1}^{p_1} F \otimes L_{q_2}^{p_2} F$$

$$\xrightarrow{C_F} L_{q_1}^{p_1} F \otimes L_{q_2}^{p_2-1} F \xrightarrow{C_F} L_{q_1+1}^{p_1} F \otimes L_{q_2}^{p_2-2} F \to \cdots \quad (*)$$

$$\to L_{p_2+q_1-2}^{p_1} F \otimes L_{q_2}^{1} F \xrightarrow{\partial} L_{p_2+q_1+q_2-2}^{p_1} F \to 0$$

is exact, we see that $L_{q_1 q_2}^{p_1 p_2} F$ is also the kernel of the map $L_{q_1}^{p_1} F \otimes L_{q_2}^{p_2-1} F \xrightarrow{C_F} L_{q_1+1}^{p_1} F \otimes L_{q_2}^{p_2-2} F$. Moreover, we see that $\Sigma L_{q_1 q_2}^{p_1 p_2} F$ is also an SF ⊗ ΛF*-module. Thus, iterating the above procedure, we can define $L_{q_1 q_2-1}^{p_1 p_2} F \otimes L_{q_3}^{p_3} F$ as an SF ⊗ ΛF*-module, operate by C_F, and get a map to $L_{q_1 q_2}^{p_1 p_2} F \otimes L_{q_3}^{p_3-1} F$. The image of this action is easily seen to be $L_{q_1 q_2 q_3}^{p_1 p_2 p_3} F$. Thus, by iteration, we have a well-defined action of C_F on $L_{q_1 \cdots q_\ell - 1}^{p_1 \cdots p_\ell} F \otimes L_{q_{\ell+1} \cdots q_k}^{p_{\ell+1}-1 \cdots p_k} F$ to $L_{q_1 \cdots q_\ell}^{p_1 \cdots p_\ell} F \otimes L_{q_{\ell+1} \cdots q_k}^{p_{\ell+1}-1 \cdots p_k} F$ whose image is $L_{q_1 \cdots q_k}^{p_1 \cdots p_k} F$. A proof similar to that in [4] which proves the exactness of (*) gives us the following theorem.

THEOREM 4.3. *The complex*

$$0 \longrightarrow L^{p_1\cdots(p_\ell+p_{\ell+1}+q_\ell-2)\cdots p_k}_{q_1\cdots\hat{q}_\ell\cdots q_k}F \xrightarrow{\delta} L^{p_1\cdots p_\ell}_{q_1\cdots q_{\ell-1}-1}F \otimes L^{q_\ell+p_{\ell+1}-2\cdots p_k}_{q_{\ell+1}\cdots q_k}F$$

$$\xrightarrow{C_F} \cdots \xrightarrow{C_F} L^{p_1\cdots p_\ell}_{q_1\cdots q_\ell-1}F \otimes L^{p_{\ell+1}\cdots p_k}_{q_{\ell+1}\cdots q_k}F$$

$$\xrightarrow{C_F} L^{p_1\cdots p_\ell}_{q_1\cdots q_\ell}F \otimes L^{p_{\ell+1}-1\cdots p_k}_{q_{\ell+1}\cdots q_k}F \xrightarrow{C_F} \cdots$$

$$\longrightarrow L^{p_1\cdots p_\ell}_{q_1\cdots q_\ell+p_{\ell+1}-2}F \otimes L^{1\cdots p_k}_{q_{\ell+1}\cdots q_k}F$$

$$\xrightarrow{\partial} L^{p_1\cdots\hat{p}_\ell\cdots p_k}_{q_1\cdots q_{\ell-1}(q_\ell+q_{\ell+1}+p_{\ell+1}-2)q_{\ell+2}\cdots q_k}F \longrightarrow 0$$

is exact, where δ *is the obvious diagonalization, and* ∂ *is the obvious surjection (generalizing the maps* δ *and* ∂ *in* (*)).

Remark. Just as, classically, a partition may be described as a sequence of nested hooks, the number of which is equal to the size of its Durfee square (see [3]), it may also be described as a sequence of nested skew-hooks, the number of which, again, is equal to the size of the Durfee square.

5. *Pieri formulas*

In this section we shall discuss, but not prove, the two Pieri formulas. Complete proofs of these and related formulas will be found in [1].

At the end of §2 we described the basis for $\mathbb{L}_\lambda(\phi)$ in terms of standard tableaux mod G, and indicated that, since $\mathbb{L}_\lambda(\phi)$ reduces to $L_\lambda F$ when $G = 0$, we have a description of the basis for $L_\lambda F$ in terms of standard tableaux. In particular, if $S = \{x_1,\ldots,x_n\}$ is a basis for F and T is a tableau with values in S, we say that T is *row-standard* if, for each $i = 1,\ldots,\text{length}(\lambda)$, $T(i,j) < T(i,j+1)$ for $1 \le j < \lambda_i$. To such a tableau T we assign the basis element of $\Lambda^{\lambda_1}F \otimes \cdots \otimes \Lambda^{\lambda_q}F$ given by

$$T(1,1) \wedge \cdots \wedge T(1,\lambda_1) \otimes \cdots \otimes T(q,1) \wedge \cdots \wedge T(q,\lambda_q).$$

We say that T is *column-standard* if, for each $j = 1,\ldots,\lambda_1$,

$T(i,j) \leq T(i+1,j)$ for $1 \leq i \leq \tilde{\lambda}_j$ (where $\tilde{\lambda} = (\tilde{\lambda}_1,\ldots,\tilde{\lambda}_t)$). We say that a tableau is *standard* if it is both row- and column-standard. Our assertion is that $\{d_\lambda(T) \mid T \text{ is a standard tableau}\}$ is a basis for $L_\lambda F$. Using this information, it is shown in [3] that $L_\lambda(F\oplus G)$ has a natural filtration whose associated graded module is $\Sigma\, L_\sigma F \otimes L_{\lambda/\sigma}G$, where the sum is taken over all partitions σ contained in λ. Of course, one can equally well define a natural filtration on $L_\lambda(F \oplus G)$ whose associated graded module is $\Sigma\, L_\sigma G \otimes L_{\lambda/\sigma}F$.

The proof simply makes rigorous the following heuristic argument. If we choose bases $S_1 = \{x_1,\ldots,x_n\}$ and $S_2 = \{y_1,\ldots,y_m\}$ for F and G, then their union, S, is a basis for $F \oplus G$, and we may put a total order on S by setting each element of S_1 to be less than every element of S_2. A standard tableau of shape λ with values in S must then have the property that in the ith row of λ we must have σ_i elements of S_1 arranged in strictly increasing order followed by $\lambda_i - \sigma_i$ elements of S_2 in strictly increasing order. Moreover, we must have $\sigma_i \geq \sigma_{i+1}$, for otherwise we would have an element of S_1 in the (i+1)th row lying underneath an element of S_2 in the ith row, thus contradicting column-standardness of the tableau. Consequently, $\sigma = (\sigma_1,\ldots,\sigma_q)$ is a partition contained in λ, and the tableau restricted to the diagram of σ is a standard tableau of shape σ with values in S_1, and restricted to that of λ/σ is standard of shape λ/σ with values in S_2. Knowing what we do about the bases of $L_\sigma F$ and $L_{\lambda/\sigma}G$, it is not hard to deduce the decomposition of $L_\lambda(F \oplus G)$ as $\Sigma\, L_\sigma F \otimes L_{\lambda/\sigma}G$. Clearly, if we had ordered S by putting the elements of S_2 before those of S_1, we would have obtained the other decomposition, $\Sigma\, L_\sigma G \otimes L_{\lambda/\sigma}F$.

A refinement of the above discussion, which takes into account the number of elements of S_2 which occur in the tableaux (that is, the G-content of the tableau [1]) gives us an isomorphism

$$\sum_{|\sigma|=|\lambda|-k} L_\sigma F \otimes L_{\lambda/\sigma}G \approx \sum_{|\tau|=k} L_\tau G \otimes L_{\lambda/\tau}F.$$

Suppose, now, that $G \approx R$. Then the only partition τ such that $|\tau| = k$ and for which $L_\tau G \neq 0$ is $\tau = (\underbrace{1,\ldots,1}_{k})$, and $L_\tau G \approx R$.

Thus we see that $L_{\lambda/(\underbrace{1,\ldots,1}_{k})}F$ has a filtration whose associated graded module is $\sum_{|\sigma|=|\lambda|-k} L_\sigma F \otimes L_{\lambda/\sigma}G$. However, since $G = R$, the only σ such that $L_{\lambda/\sigma}G \neq 0$ must be those for which no row of λ/σ has more than one element, and for such σ, $L_{\lambda/\sigma}G \approx R$.

Notice next that if $\lambda = (\lambda_1,\ldots,\lambda_k)$, and if we set $\lambda' = (\lambda_1+1,\ldots,\lambda_k+1,\underbrace{1,\ldots,1}_{q})$, then $S_qF \otimes L_\lambda F = L_{\lambda'/(\underbrace{1,\ldots,1}_{k})}F$. The above discussion immediately yields the following.

THEOREM 5.1. $S_qF \otimes L_\lambda F$ *admits a filtration whose associated graded module is* $\Sigma\, L_\nu F$, *where the sum is taken over all partitions* ν *containing* λ *such that* $|\nu| = |\lambda| + q$ *and for which the diagram of* ν/λ *contains at most one box in each row.* □

This theorem gives us a characteristic-free formulation of one of the Pieri formulas, but does not tell us how to calculate $\Lambda^qF \otimes L_\lambda F$. What is more, the discussion leading to Theorem 5.1 does not suggest any trivial formal modification which will lead to such a calculation. Fortunately, the machinery of Schur complexes does the job for us and, without giving formal proofs (see [1]), the following discussion shows how.

Let ϕ: $R \to F$ be any map (for example, the zero map). Then from Theorem 2.2 and the material following it, we know that a basis for $(\mathbb{L}_\lambda(\phi))_\ell$ is given by the standard tableaux mod R containing ℓ elements of the basis of R. Standardness mod R depends upon the order we impose on the union, S, of the bases of R and of F. If we order S by putting the basis element 1 of R before every element of the basis of F, and apply the definition of standard tableau mod R to this choice, we see that $(\mathbb{L}_\lambda(\phi))_\ell \approx L_{\lambda/(\ell)}F$. If, however, we order S by putting the basis element of R after every element of the basis of F, we see that $(\mathbb{L}_\lambda(\phi))_\ell$ has a filtration whose associated graded module is $\Sigma\, L_\sigma F$ where σ runs over all partitions contained in λ such that $|\sigma| = |\lambda| - \ell$ and the diagram of λ/σ has at most one box in each column (recall the definition of standardness mod R, and that R has rank 1 so that the possibility of distinct elements of R in a given column does not exist). Thus $L_{\lambda/(\ell)}F$ is isomorphic, up to filtration, to this direct sum.

Now apply these observations to the partition $\lambda' = (\lambda_1+q,\lambda_1,\lambda_2,\ldots,\lambda_k)$, observing that $L_{\lambda'/(\lambda_1)}F \approx \Lambda^q F \otimes L_\lambda F$, where $\lambda = (\lambda_1,\ldots,\lambda_k)$. We thus obtain the following.

THEOREM 5.2. $\Lambda^q F \otimes L_\lambda F$ *admits a filtration whose associated graded module is* $\Sigma\, L_\nu F$, *where the sum is taken over all partitions* ν *containing* λ *such that* $|\nu| = |\lambda| + q$ *and for which the diagram of* ν/λ *contains at most one box in each column.* □

Acknowledgements

The first author was partially supported by NSF Grant MCS-8102367, and the second author was partially supported by NSF Grant MCS-8109988.

References

1. K. Akin and D.A. Buchsbaum, "GL(n)-representation-theoretic reformulation of some classical formulas", in preparation.
2. K. Akin, D.A. Buchsbaum and J. Weyman, "Resolutions of determinantal ideals: the submaximal minors", Adv. in Math., 39 (1981), 1-30.
3. K. Akin, D.A. Buchsbaum and J. Weyman, "Schur functors and Schur complexes", Adv. in Math., 44 (1982), 207-278.
4. D.A. Buchsbaum, "Generic-free resolutions, II", Canad. J.Math., 30 (1978), 549-572.
5. D.A. Buchsbaum and D. Eisenbud, "Generic free resolutions and a family of generically perfect ideals", Adv. in Math., 18 (1975), 245-301.

Department of Mathematics,
Brandeis University,
Waltham,
Mass. 02154, U.S.A.

THE CANONICAL MODULE OF A DETERMINANTAL RING

WINFRIED BRUNS

Let R be a commutative noetherian ring. We call an R-module M a *canonical module* of R if for each prime ideal $\underline{p}$ of R the localization $M_{\underline{p}}$ is a canonical module of $R_{\underline{p}}$ in the sense of [5]. A *determinantal ring* is a residue class ring of a polynomial ring $B[X_{ij}: 1 \le i \le u, 1 \le j \le v]$ with respect to the ideal $I_{r+1}(X_{ij})$ generated by the determinants of the $(r+1,r+1)$-submatrices of the (u,v)-matrix (X_{ij}) whose entries are the algebraically independent elements X_{ij} over the commutative noetherian ring B.

Determinantal rings play an important role in various geometric and algebraic contexts. They can be considered well-understood since Hochster and Eagon proved their perfection (relative to the polynomial ring $B[X_{ij}]$) in the remarkable article [7] which furthermore contains many useful results on the ideal theory of determinantal rings. We became interested in the canonical modules of determinantal rings when their computation appeared as a rather natural step in our investigation of generic maps of a given rank and the modules associated to these maps [2]. The generic maps of rank r are the maps $\phi: S^u \to S^v$, where $S = \mathbb{Z}[X_{ij}]/I_{r+1}(X_{ij})$ and ϕ is represented by the matrix (x_{ij}) of the residue classes of the indeterminates X_{ij}. It turned out that (in the non-degenerate case where $r \ge 1$) Coker ϕ is perfect if and only if $u \ge v$. This asymmetry in the behaviour of Coker ϕ is caused by the structure of the canonical module of S which represents the asymmetry of the format of a non-square matrix in a ring-theoretic way.

THEOREM. *Let* B *be a normal Gorenstein domain, and* u,v,r *integers such that* $1 \le r \le v \le u$. *Let* X_{ij}, $1 \le i \le u$, $1 \le j \le v$, *be algebraically independent elements over* B, *and* R *the residue class ring* $B[X_{ij}]/I_{r+1}(X_{ij})$. *Let further* $\underline{p}$ *denote the ideal generated by the* r-*minors of the first* r *rows of the matrix* (x_{ij}) *of the residue classes* x_{ij} *of* X_{ij}. *Then* $\underline{p}^{u-v}$ *is a canonical module of* R.

We recently learnt from [8; p.500] that our theorem was also noted by Hochster (unpublished). As an easy corollary one obtains the following theorem of Svanes [9; Theorem (5.5.6)].

COROLLARY. R *is a Gorenstein ring if and only if* $u = v$.

In the corollary one can allow B to be an arbitrary Gorenstein ring.

The proof of the theorem consists of two steps. As a first step we compute the divisor class of the canonical module of R, that is, show that it is isomorphic to $\underline{p}^{(u-v)}$, whereas we establish in the second step the equality of the ordinary and symbolic powers of $\underline{p}$. Note that in case R is a normal domain, a canonical module of R is (isomorphic to) a divisorial ideal of R. If the natural homomorphism from the divisor class group of R into the direct product of the divisor class groups of the localizations of R is injective, then M is uniquely determined (up to isomorphism).

An important tool in the computation of the canonical modules of determinantal rings is the following lemma.

LEMMA. *Let* T *be a normal Cohen-Macaulay domain and* $\underline{a}$ *a prime ideal of height* 1 *in* T, *such that* $T/\underline{a}$ *is also normal. Suppose that the following conditions are satisfied:*

(a) *for a prime ideal* $\underline{p}$ *of height* 1, $\underline{p} \neq \underline{a}$, *the symbolic power* $\underline{p}^{(n)}$ *is a canonical module of* T;

(b) $cl(\underline{a}) = -m \cdot cl(\underline{p})$ *with* $m \geq 0$;

(c) $\mathrm{Ann}(\underline{p}^{(n+m)}/\underline{p}^{n+m}) \not\subset \underline{a} + \underline{p}$; *and*

(d) $(\underline{a}+\underline{p})/\underline{a}$ *is a prime ideal of height* 1 *in* $T/\underline{a}$.

Then $((\underline{a}+\underline{p})/\underline{a})^{(n+m)}$ *is a canonical module of* $T/\underline{a}$.

Proof. Consider the exact sequence

$$0 \to \mathrm{Hom}_T(T,\underline{p}^{(n)}) \xrightarrow{\iota} \mathrm{Hom}_T(\underline{a},\underline{p}^{(n)}) \to \mathrm{Ext}^1_T(T,\underline{p}^{(n)}) \to 0.$$

We may identify $\mathrm{Hom}_T(T,\underline{p}^{(n)})$ with $\underline{p}^{(n)}$, $\mathrm{Hom}_T(\underline{a},\underline{p}^{(n)})$ with $\underline{p}^{(n)}:\underline{a}$, and ι with the natural embedding. $\mathrm{Ext}^1_T(T,\underline{p}^{(n)})$ is a canonical module of $T/\underline{a}$. For a suitable element x in the quotient field of T we have $x\cdot(\underline{p}^{(n)}:\underline{a}) = \underline{p}^{(n+m)}$ by condition (b) and thus an exact sequence

$$0 \to x\underline{p}^{(n)} \to \underline{p}^{(n+m)} \to \operatorname{Ext}^1_T(T,\underline{p}^{(n)}) \to 0.$$

The ideal $x\underline{p}^{(n)}$ is divisorial, $\operatorname{cl}(x\underline{p}^{(n)}) = \operatorname{cl}(\underline{p}^{(n+m)} \cap \underline{a})$, $x\underline{p}^{(n)}$ contains $\underline{a}\underline{p}^{(n)}$, and $\underline{p}^{(n+m)}/x\underline{p}^{(n)}$ is isomorphic to a (divisorial) ideal of T. Then necessarily $x\underline{p}^{(n)} = \underline{p}^{(n+m)} \cap \underline{a}$. Finally it is easy to check that conditions (c) and (d) guarantee that

$$\underline{p}^{(n+m)}/(\underline{p}^{(n+m)} \cap \underline{a}) = ((\underline{a}+\underline{p})/\underline{a})^{(n+m)}. \; \square$$

Determinantal rings R are normal as soon as B is normal [7]. Their divisor class group Cl(R) was computed in [1]: $\operatorname{Cl}(R) = \operatorname{Cl}(B) \oplus \mathbb{Z}$. The second component is generated by the class of the prime ideal $\underline{p}$ of the r-minors of the first r rows of (x_{ij}) and by $-\operatorname{cl}(\underline{p}) = \operatorname{cl}(\underline{q})$, where $\underline{q}$ is the corresponding ideal for the columns. If B is a factorial domain, in particular if $B = \mathbb{Z}$ or B is a field, the natural map $\operatorname{Cl}(R) \to \operatorname{Cl}(R_{\underline{m}})$, $\underline{m}$ generated by the elements x_{ij}, is an isomorphism. We are mainly interested in these cases, and therefore we allow ourselves to speak of *the* canonical module ω_R.

PROPOSITION 1. *Let* B *be a normal Gorenstein domain. Suppose that* $u \geq v$. *Then* $\omega_R \cong \underline{p}^{(u-v)}$.

Proof. We first reduce to the case when r = 1 by a standard localization argument. Let $P = B[X_{ij}]$. Over $P[X_{11}^{-1}]$ we can transform the matrix (X_{ij}) by elementary row and column operations into

$$\begin{bmatrix} X_{11} & 0 & \cdots & 0 \\ 0 & Y_{22} & \cdots & Y_{2v} \\ \cdot & \cdot & & \cdot \\ \cdot & \cdot & & \cdot \\ \cdot & \cdot & & \cdot \\ 0 & Y_{u2} & \cdots & Y_{uv} \end{bmatrix}$$

where $Y_{ij} = X_{ij} - X_{i1}X_{1j}X_{11}^{-1}$. The elements Y_{ij} are algebraically independent over B, and the elements $X_{11},\ldots,X_{1v},X_{21},\ldots,X_{u1}$ are algebraically independent over $C := B[Y_{ij}]$. Now $R[x_{11}^{-1}]$ can be considered as a flat overring of $C/I_r(Y_{ij})$:

$$R[x_{11}^{-1}] = (C/I_r(Y_{ij}))[X_{11},\dots,X_{1v},X_{21},\dots,X_{u1}][X_{11}^{-1}].$$

The extension of the ideal $\tilde{\underline{p}}$ of $C/I_r(Y_{ij})$ generated by the (r-1)-minors of $(y_{ij}: 2 \le i \le r,\ 2 \le j \le v)$ to $R[x_{11}^{-1}]$ is just $\underline{p}R[x_{11}^{-1}]$.

Suppose that $r > 1$. By [7; Theorem 1], the element x_{11} is a prime element in R. Therefore the natural map $\mathrm{Cl}(R) \to \mathrm{Cl}(R[x_{11}^{-1}])$ is an isomorphism as is the natural map $\mathrm{Cl}(C/I_{r-1}(Y_{ij})) \to \mathrm{Cl}(R[x_{11}^{-1}])$. Now it suffices to establish the proposition for $C/I_{r-1}(Y_{ij})$, thereby reducing its proof to the case when $r = 1$.

Let $r = 1$ now and $v = 2$. In case $u = 2$, R is clearly a Gorenstein ring and $\omega_R = R = \underline{p}^{(o)}$. In case $u > 2$ let

$$R_o := B[X_{ij}: 1 \le i \le u-1,\ 1 \le j \le 2]/I_2(X_{ij})$$

and $R_1 := R_o[X_{u1},X_{u2}]$. Then

$$R = R_1/\underline{a},\quad \underline{a} = \sum_{i=1}^{u-1} R_1(x_{i1}X_{u2} - x_{i2}X_{u1}).$$

By induction on u, and since R_1 is a polynomial extension of R_o, the canonical module of R_1 is $\underline{p}_1^{(u-3)}$, $\underline{p}_1 = R_1x_{11} + R_1x_{12}$. Multiplication by $(x_{11}X_{u2} - x_{12}X_{u1})x_{11}^{-1}$ maps

$$\underline{q}_1 := \sum_{i=1}^{u-1} R_1x_{i1}$$

onto $\underline{a}$, whence $\mathrm{cl}(\underline{a}) = \mathrm{cl}(\underline{q}_1) = -\mathrm{cl}(\underline{p}_1)$. The ideal $(\underline{a}+\underline{p}_1)/\underline{a}$ is a prime ideal of height 1, and the extension of $\underline{p}_1$ in $(R_1)_{\underline{a}+\underline{p}_1}$ is principal. Therefore condition (c) of the lemma is satisfied. An application of the lemma completes the proof for $v = 2$.

Let $v > 2$ now. Because $R[x_{11}^{-1}] = B[X_{11},\dots,X_{1v},X_{21},\dots,X_{u1}][X_{11}^{-1}]$, the divisor class of ω_R is a multiple of the divisor class of $\underline{p}$. Dropping the hypothesis that $u \ge v$ momentarily and transposing (X_{ij}) if necessary we may assume that $\mathrm{cl}(\omega_R) = t\cdot\mathrm{cl}(\underline{p})$ with $t \ge 0$ and $\omega_R \cong \underline{p}^{(t)}$. All the conditions of the lemma are satisfied for

$$\underline{a} = \underline{q} = \sum_{i=1}^{u} Rx_{i1}.$$

By induction on $u+v$ the divisor class of the canonical module of $R/\underline{q}$ is $(u-v+1)\mathrm{cl}((\underline{p}+\underline{q})/\underline{q})$, and comparing this with the result of the lemma we conclude that $u - v = t \geq 0$ as desired. (Observe that $\mathrm{cl}(\underline{p}+\underline{q})/\underline{p})$ is not a torsion element in $\mathrm{Cl}(R/\underline{q})$.) □

Observe that the corollary of the theorem follows already from Proposition 1. The proof of the theorem itself is complete once we have shown that the ordinary and the symbolic powers of $\underline{p}$ coincide. This is stated in [8; Theorem 3.4] (for a field B). For an application in [2] we need a more general result, which could possibly be proved with the methods of [8]. Proposition 2 and the proof we present were found independently.

As above $\underline{q}$ denotes the prime ideal generated by the r-minors of the first r columns of (x_{ij}). For a sequence of integers $H = (i_o,\dots,i_{r-1})$, $0 \leq i_o < \dots < i_{r-1}$, $J(H)$ denotes the ideal of R which is generated by the (j+1)-minors of the first i_j rows of (x_{ij}), $j = 0,\dots,r-1$. Note that $R/J(H)$ is a normal domain by [4; Corollary 3] if B is normal.

PROPOSITION 2. *Let* B *be a normal domain and*

$\mathcal{H} := \{(i_o,\dots,i_{r-1}) : 0 \leq i_o < \dots < i_{r-1} < u\}$.

Then, for all $H \in \mathcal{H}$, *we have the following:*

(a) $J(H) + \underline{q}$ *is a prime ideal in* R;

(b) *for all* $s \geq 1$ *the ideal* $J(H) + \underline{q}^s$ *is primary.*

An application of Proposition 2 to the transpose of (X_{ij}) for $H = (0,1,\dots,r-1)$ completes the proof of the theorem. □

Proof of Proposition 2. The ideal in $B[X_{ij}]$ which defines $R/(J(H)+\underline{q})$ as a factor ring of $B[X_{ij}]$ is generated by the (j+1)-minors of the first i_j rows of (X_{ij}), $j = 0,\dots,r-1$, the (r+1)-minors of the entire matrix (X_{ij}), and, finally, the r-minors of the first r columns of (X_{ij}). Therefore part (a) is a special case of Proposition 3, (b) or (c), below.

In order to specify a minor of the matrix (X_{ij}) we introduce the following notation:

$$\Delta^{j_1\dots j_r}_{i_1\dots i_r}$$

is the minor associated with the row indices $i_1,\dots,i_r$ and the column indices $j_1,\dots,j_r$.

Part (a) is the case in which $s=1$ of the proof of part (b) by induction on s. Assume that $s>1$. We need a total order on the set of r-tuples of integers $(i_1,\ldots,i_r)$ such that $0 \le i_1 < \ldots < i_r$:

$$(i_1,\ldots,i_r) \le (j_1,\ldots,j_r)$$

if and only if $(i_1,\ldots,i_r) = (j_1,\ldots,j_r)$ or there is a $k \in \{1,\ldots,r\}$ such that $i_k < j_k$ and $i_{k+1} \le j_{k+1},\ldots,i_r \le j_r$.

To simplify the notation let $\Delta_{j_1\ldots j_r} := \Delta^{1\ldots r}_{j_1\ldots j_r}$.

For $H \in \mathcal{H}$ let R^* be the residue class ring $R/J(H)$ and $\underline{r} := \underline{q}R^*$. Since $I_{r+1}(x_{ij}) = 0$, the r-th exterior power of (x_{ij}) has rank 1, and hence

$$\Delta_{j_1\ldots j_r}\Delta^{1\ldots r-1\;r+1}_{k_1\ldots\;\;k_r} = \Delta_{k_1\ldots k_r}\Delta^{1\ldots r-1\;r+1}_{j_1\ldots\;\;j_r}$$

(minors taken from (x_{ij})). If $\Delta_{j_1\ldots j_r}$ is not zero modulo $J(H)$ then $\Delta^{1\ldots r-1\;r+1}_{j_1\ldots\;\;j_r}$ is not zero modulo $J(H) + \underline{q}$. Thus every minor $\Delta_{j_1\ldots j_r}$ which is not zero modulo $J(H)$ generates the extension of $\underline{r}$ in $R^*_{\underline{r}}$: $\underline{r}$ is a prime of height 1 in R^*, and every non-zero minor $\Delta_{j_1\ldots j_r}$ is not contained in $\underline{r}^{(2)}$. If for an element $b \in R^*$ the product $b\Delta_{j_1\ldots j_r}$ is contained in $\underline{r}^{(s)}$, we can conclude that $b \in \underline{r}^{(s-1)}$, since $R^*_{\underline{r}}$ is a discrete valuation domain.

Assume that

$$b = \sum b_{j_1\ldots j_r}\Delta_{j_1\ldots j_r} \in \underline{r}^{(s)}$$

and let $(k_1,\ldots,k_r)$ denote the maximal r-tuple relative to $\le$ such that $b_{k_1\ldots k_r}\Delta_{k_1\ldots k_r} \ne 0$. We show that $b \in \underline{r}^s$ by induction on $(k_1,\ldots,k_r)$.

If there is only one non-zero term in the representation of b, we have $b_{k_1\ldots k_r} \in \underline{r}^{(s-1)}$ by the argument just given, and $b_{k_1\ldots k_r} \in \underline{r}^{s-1}$ by induction on s. Suppose that there are at least two non-zero terms in the representation of b and let

$$\tilde{H} := (k_1-1,\dots,k_r-1).$$

Then $J(H) \subset J(\tilde{H})$ and $\Delta_{j_1\dots j_r} \in J(\tilde{H})R$ for all $(j_1,\dots,j_r) <$ $(k_1,\dots,k_r)$. This implies that

$$b_{k_1\dots k_r}\,\Delta_{k_1\dots k_r} \in J(\tilde{H})R^* + \underline{r}^{(s)}.$$

Let $\tilde{R} := R/J(\tilde{H})$ and $\tilde{\underline{r}} := \underline{q}\tilde{R}$. As above the extension of $\tilde{\underline{r}}$ in $\tilde{R}_{\tilde{\underline{r}}}$ is principal, whence

$$\mathrm{Ann}_{R^*}\underline{r}^{(s)}\tilde{R}/\tilde{\underline{r}}^s \not\subset \underline{r} + J(\tilde{H})R^*.$$

Consequently $\pi(b_{k_1\dots k_r})\pi(\Delta_{k_1\dots k_r})$ is an element of $\tilde{\underline{r}}^{(s)}$, where $\pi: R^* \to \tilde{R}$ denotes the natural epimorphism. By induction on s it follows that $\pi(b_{k_1\dots k_r}) \in \tilde{\underline{r}}^{(s-1)} = \tilde{\underline{r}}^{s-1}$ and

$$b_{k_1\dots k_r} \in \pi^{-1}(\tilde{\underline{r}}^{s-1}) = J(\tilde{H})R^* + \underline{r}^{s-1}.$$

We may assume that $b_{k_1\dots k_r} \in J(\tilde{H})R^*$. The inclusion

$$J(\tilde{H})\Delta_{k_1\dots k_r} \subset \sum_{(j_1,\dots,j_r)<(k_1,\dots,k_r)} \Delta_{j_1\dots j_r}R^*$$

finally enables us to replace $b_{k_1\dots k_r}\,\Delta_{k_1\dots k_r}$ by a linear combination of the minors $\Delta_{j_1\dots j_r}$ with $(j_1,\dots,j_r) < (k_1,\dots,k_r)$, and to complete the proof by induction on $(k_1,\dots,k_r)$.

The proof of the crucial inclusion is an exercise in expansion of determinants. Let $\Delta^{v_1\dots v_w}_{u_1\dots u_w} \in J(\tilde{H})$. One expands the (r+1)-minor $\Delta^{v_1\dots v_w t_w\dots t_r}_{u_1\dots u_w k_w\dots k_r} = 0$ with respect to the columns $v_1,\dots,v_w$, the minor $\Delta_{k_1\dots k_r}$ with respect to the rows $k_1,\dots,k_{w-1}$, and combines the two equations obtained. □

In order to prove part (a) of Proposition 2 we introduce a larger class of ideals in $B[X_{ij}]$. Throughout the rest of the article we denote the matrix (X_{ij}) simply by X. Let $H = (u_o,\dots,u_r)$ be a sequence of strictly increasing integers with $0 \le u_o$ and $u_r = u$.

Further, let p and t be integers such that $0 \le t - 1 \le p$. We consider the ideals

$$I(H,X) + I_t(X|p)$$

where $I(H,X)$ is the ideal generated by the $(j+1)$-minors of the first u_j rows of X, $j = 0,\dots,r$, and $I_t(X|p)$ is the ideal generated by the t-minors of the matrix $X|p$ consisting of the first p columns of X.

PROPOSITION 3. (a) $I(H,X) + I_t(X|p)$ *is a (strongly generically) perfect ideal of grade*

$$uv - r(u+v) + \frac{(t-2)\ (t-1)}{2} + \frac{(r-t+2p)\ (r-t+1)}{2} + \sum_{i=0}^{r-1} u_i + r.$$

(b) *If* B *is an integral domain, then* $I(H,X) + I_t(X|p)$ *is a prime ideal.*

(c) *If* B *is a normal integral domain, then*

$$B[X]/(I(H,X) + I_t(X|p))$$

is a normal integral domain.

Proof. In [4] a partial order on the set M of minors of (X_{ij}) is defined by

$$\Delta_{i_1\dots i_k}^{j_1\dots j_k} \le \Delta_{s_1\dots s_l}^{t_1\dots t_l}$$

if and only if $k \ge l$ and $i_1 \le s_1,\dots,i_k \le s_k$, $j_1 \le t_1,\dots,j_k \le t_k$. (The lower indices denote the rows, and the upper indices denote the columns, from which the minor is taken; we require that $i_1,\dots,i_k$ etc. are given in increasing order.) One easily sees that $I(H,X) + I_t(X|p)$ is generated by a poset ideal J in M. Therefore $B[X]/(I(H,X) + I_t(X|p))$ itself is an *algebra with straightening law* on the poset $P := M \setminus J$ [4; Corollary 3.5(3)]; in particular, it is a free B-module. The minor

$$\Delta_{1\ \dots\ r-1\ r+1}^{u_0+1\ \dots\ u_{r-1}+1}$$

is the single minimal element of P. Therefore P is *wonderful* in the sense of [4].

It is not hard to check that one (and, by [4; Lemma 4.3], every) maximal chain in P consists of

$$d := r(u+v) - \frac{(t-2)\ (t-1)}{2} - \frac{(r-t+2p)\ (r-t+1)}{2} - \sum_{i=0}^{r-1} u_i - r$$

elements. Furthermore

$$R := B[X]/(I(H,X) + I_t(X|p))$$

is a Cohen-Macaulay ring, once B is Cohen-Macaulay [4;Corollary 4.2].

The theory of generic perfection [3,6] shows that it suffices to prove part (a) in the case where $B = \mathbb{Z}$ or B is a field. Let $\underline{P}$ and $\underline{p}$ denote the prime ideal generated by the elements X_{ij} in B[X] and its image in R respectively. By [4;Theorem 4.1 and Proposition 3.7] a maximal $R_{\underline{p}}$-sequence in $\underline{p}R_{\underline{p}}$ consists of exactly d elements. R is a Cohen-Macaulay ring. Therefore $I := I(H,X) + I_t(X|p)$ is a perfect ideal and we obtain

$$\text{grade } I = \text{grade } I_{\underline{P}} = u.v - \text{depth } R_{\underline{p}} = u.v - d.$$

(b) As noted above, R is a free B-module. Therefore we may invert $B \setminus \{0\}$ and assume that B is a field. Now (b) is a special case of (c).

(c) We prove the assertion by induction on u+v, starting with the trivial case in which $u = v = 0$. Without restriction we may assume that $u_o = 0$. Let $\underline{a}$ denote the ideal generated by the elements

$$x_{ij},\quad 1 \le i \le u_1,\quad 1 \le j \le v,$$

in R. The preimage of $\underline{a}$ in B[X] is

$$I(\tilde{H},X) + I_t(X|p)$$

where

$$\tilde{H} = (u_1, u_1 + 1, \max(u_2, u_1 + 2), \ldots, \max(u_{r-1}, u_1 + r-1), u)$$

when $u_1 + r-1 < u$, and

$$\tilde{H} = (u_1, u_1 + 1, \max(u_2, u_1 + 2), \ldots, \max(u_{r-2}, u_1 + r-2), u)$$

when $u_1 + r-1 = u$.

By virtue of (a), $I(\tilde{H},X) + I_t(X|p)$ is a perfect ideal in B[X] and

$$\text{grade } I(\tilde{H},X) + I_t(X|p) \ge 2 + \text{grade } I(H,X) + I_t(X|p).$$

This implies that grade $\underline{a} \ge 2$. If depth $R_{\underline{q}} \le 1$ for a prime ideal $\underline{q}$ of R, then $\underline{a} \not\subset \underline{q}$. Thus it remains for us to prove that the rings

$$R[x_{ij}^{-1}],\quad 1 \le i \le u_1,\quad 1 \le j \le v,\quad x_{ij} \text{ not nilpotent},$$

are normal.

Suppose first that $j \leq p$. We again use the localization argument introduced in the proof of Proposition 1. We may assume that $i = 1$. Elementary row and column operations over $B[X][X_{11}^{-1}]$ transform the matrix X into

$$\begin{bmatrix} X_{11} & 0 & \cdots & 0 \\ 0 & Y_{22} & \cdots & Y_{2v} \\ \cdot & \cdot & & \cdot \\ \cdot & \cdot & & \cdot \\ \cdot & \cdot & & \cdot \\ 0 & Y_{u2} & \cdots & Y_{uv} \end{bmatrix}.$$

Let Y be the $(u-1,v-1)$-matrix $(Y_{i+1,j+1})$. Then

$$R[x_{11}^{-1}] \cong (B[Y]/(I(\tilde{H},Y)+I_{t-1}(Y|p-1)))[X_{11},\ldots,X_{1v},X_{21},\ldots,X_{u1}][X_{11}^{-1}]$$

with

$$\tilde{H} = (u_1-1,u_2-1,\ldots,u_r-1).$$

The ring $B[Y]/(I(\tilde{H},Y) + I_{t-1}(Y|p-1))$ is a normal domain by the induction hypothesis.

Suppose now that $p < j$. We may assume that $j = v$. Over $B[X][X_{1v}^{-1}]$ the matrix X can be transformed into

$$W = \begin{bmatrix} X_{11} & \cdots & X_{1v-1} & X_{1v} \\ Y_{21} & \cdots & Y_{1v-1} & 0 \\ \cdot & \cdots\cdot & & \cdot \\ \cdot & \cdots\cdot & & \cdot \\ \cdot & \cdots\cdot & & \cdot \\ Y_{u1} & \cdots & Y_{uv-1} & 0 \end{bmatrix}.$$

Clearly $I(H,X) + I_t(X|p) = I(H,W) + I_t(W|p)$ over $B[X][X_{1v}^{-1}]$. Let

$$Z = \begin{bmatrix} Y_{21} & \cdots & Y_{2v-1} \\ \cdot & \cdots & \cdot \\ \cdot & \cdots & \cdot \\ \cdot & \cdots & \cdot \\ Y_{u1} & \cdots & Y_{uv-1} \\ X_{11} & \cdots & X_{1v-1} \end{bmatrix}.$$

Then, over $B[X][X_{1v}^{-1}]$,

$$I(H,X) + I_t(X|p) = I(H,W) + I_t(W|p) = I(\tilde{H},Z) + I_t(Z|p)$$

where $\tilde{H} = (u_1-1,u_2-1,\ldots,u_r-1,u)$, and

$$R[X_{11}^{-1}] \cong (B[Z]/(I(\tilde{H},Z) + I_t(Z|p)))[X_{1v},\ldots,X_{uv}][X_{1v}^{-1}].$$

Again the induction hypothesis applies. □

References

1. W. Bruns, "Die Divisorenklassengruppe der Restklassenringe von Polynomenringen nach Determinantenidealen", Rev. Roumaine Math. Pures Appl., 20 (1975), 1109-1111.

2. W. Bruns, "Generic maps and modules", Compositio Math., to appear.

3. J.A. Eagon and D.G. Northcott, "Generically acyclic complexes and generically perfect ideals", Proc. Roy. Soc. London Ser. A, 299 (1967), 147-172.

4. D. Eisenbud, "Introduction to algebras with straightening laws", Ring theory and algebra III. Proceedings of the third Oklahoma Conference (ed. B.R. McDonald, Marcel Dekker, New York and Basel, 1980), pp. 243-267.

5. J. Herzog and E. Kunz, Der kanonische Modul eines Cohen-Macaulay-Rings, Lecture Notes in Mathematics 238 (Springer, Berlin-Heidelberg-New York, 1971).

6. M. Hochster, "Generically perfect modules are strongly generically perfect", Proc. London Math.Soc. (3), 23 (1971), 477-488.

7. M. Hochster and J.A. Eagon, "Cohen-Macaulay rings, invariant theory, and the generic perfection of determinantal loci", Amer. J. Math., 53 (1971), 1020-1058.

8. C. Huneke, "Powers of ideals generated by weak d-sequences", J. Algebra, 68 (1981), 471-509.

9. T. Svanes, "Coherent cohomology on Schubert subschemes of flag schemes and applications", Adv. in Math., 14 (1974), 369-453.

Fachbereich 3, Naturwissenschaften, Mathematik,
Universität Osnabrück
- Abteilung Vechta -
Driverstrasse 22,
D-2848 Vechta, West Germany.

THE MACRAE INVARIANT

HANS-BJØRN FOXBY

The aims of this note are to give an application of the MacRae ideal G(M) of a module M and to give a new description of G(M). The application is to the theory of intersection multiplicities of modules (in the sense of Serre). The new description of G(M) is as the determinant of a finite free resolution of M. The definition of the determinant of a complex of free modules relies on the description in [2] of the relative term in the localization sequence in algebraic K-theory. This gives an alternative method of computation of G(M).

O. *The basic properties of the MacRae ideals*

Throughout A *denotes a local ring,* and M *and* N *denote* A-*modules*. Here the words "local ring" include that the ring is Noetherian, commutative, and has a non-zero multiplicative identity. In addition, *all modules are supposed to be finitely generated.* Furthermore, *the projective dimension* pd M *of* M *is always supposed to be finite.*

The basic properties of the MacRae ideal G(M) of M are collected in the next lemma. (Note that pd $M < \infty$ throughout.)

(O.1) LEMMA. *If* ann $M \neq 0$ *then the following hold.*

(a) G(M) *is a principal ideal.*

(b) G(M) *is generated by* det Λ *if* M *is the cokernel of the injective map* $\Lambda : A^n \to A^n$.

(c) $G(M) = G(M')G(M'')$ *whenever there is an exact sequence of modules*

$$0 \to M' \to M \to M'' \to 0$$

with pd M′ *and* pd M″ *finite and with* ann M′ *and* ann M″ *non-zero.*

(d) $\mathrm{Supp}(A/G(M)) \subseteq \mathrm{Supp}\, M$.

(e) $G(M) \neq A$ *if and only if* $\mathrm{Ext}_A^1(M,A) \neq 0$.

Proof. This was proved by MacRae in [6]. □

1. *An application of MacRae ideals*

If $\dim(M \otimes N) = 0$ then *Serre's intersection multiplicity* is defined by

$$\chi(M,N) = \sum_i (-1)^i \ell(\mathrm{Tor}_i^A(M,N))$$

where ℓ denotes length. (Note that pd $M < \infty$ throughout.)

THE INTERSECTION CONJECTURE. *If* $\dim(M \otimes N) = 0$ *then*

(0) $\dim M + \dim N \leqq \dim A$,

(1) $\chi(M,N) = 0$ *if* $\dim M + \dim N < \dim A$, *and*

(2) $\chi(M,N) > 0$ *if* $\dim M + \dim N = \dim A$.

Serre [8] has proved that (0) holds for all regular rings A, and that (1) and (2) hold for most regular rings: consult [8; pp. 136-141] for the precise statements. Serre has conjectured that also (1) and (2) hold for all regular rings. Furthermore, Peskine and Szpiro [7] have proved that (0), (1) and (2) hold for many (non-regular) graded rings, but in general only a little is known.

If $N = k$, the residue field of A, then $\chi(M,N) = \chi(M,k)$ is the *Euler characteristic* of M; so it is known that $\chi(M,k) > 0$ if and only if $\dim M = 0$. If N has finite length, then $\chi(M,N) = \ell(N)\chi(M,k)$, since $\chi(M,-)$ is additive on short exact sequences, and so (0), (1) and (2) do certainly hold in this case.

Recall that the *grade* of M, grade M, is the least integer i such that $\mathrm{Ext}_A^i(M,A) \neq 0$, or, equivalently, grade M is the maximal length of an A-regular sequence in ann M, the annihilator of M. Recall also that grade $M \leqq$ height ann M, and so $\dim M +$ grade $M \leqq \dim A$. (It has been conjectured that equality holds always in both places.) If grade $M = 0$, that is if ann $M \subseteqq z(A)$, the zero-divisors on A, then it is known that ann $M = 0$, so that (0), (1) and (2) hold for trivial reasons.

(1.1) THEOREM. (0), (1) *and* (2) *hold if either* grade $M = 1$ *or* $\dim N = 1$.

That (0) and (1) hold when grade $M = 1$ is probably well known and should be contained in [3] (of which, however, I have not yet seen a copy).

Proof. First assume that grade M = 1. Then

$\dim N - 1 \leqq \dim (N/G(M)N) \leqq \dim (M \otimes N)$

by (0.1)(a),(d), so that $\dim N \leqq 1$. Similarly $\dim A - 1 \leqq \dim M$, and so $\dim M = \dim A - 1$. Thus (0), and thereby (1), have been proved. (2) follows from the discussion below.

Next assume that $\dim N = 1$ (and grade M is arbitrary). Then necessarily $\dim M \leqq \dim A - 1$, that is, (0) holds and $\chi(M,k) = 0$.

Since N has a filtration with factors of the form $A/\underline{p}$ with $\underline{p} \in \operatorname{Supp} N$, and since $\chi(M, -)$ is additive on short exact sequences, it suffices to assume that $N = A/\underline{p}$, and the desired assertions follow from the lemma below and (0.1)(e). □

(1.2) LEMMA. *Let* $\underline{p} \in \operatorname{Spec} A$ *be such that* $\ell(M \otimes \bar{A}) < \infty$ *and* $\dim \bar{A} = 1$ *where* $\bar{A} = A/\underline{p}$. *Then* $\chi(M,\bar{A}) = \ell(\bar{A}/G(M)\bar{A})$.

Proof. This is by induction on $p = \operatorname{pd} M$.

$p = 0$ is impossible.

$p = 1$. In this case M is the cokernel of some injective linear map $\Lambda : A^n \to A^n$. Now $\chi(M,\bar{A}) = \ell(M \otimes \bar{A})$ since $\operatorname{Tor}_1^A(M,\bar{A}) = 0$. This follows from the well-known general fact that if L is a module such that $\ell(M \otimes L) < \infty$ then

(1.3) $\sup\{i \mid \operatorname{Tor}_i^A(M,L) \neq 0\} = \operatorname{pd} M - \operatorname{depth} L.$

(This is easily established by induction on depth L.) The equality now follows, since it is also well known that $\ell(\bar{A}/\det \bar{\Lambda}) = \ell(\operatorname{Coker} \bar{\Lambda})$ for any injective linear map $\Lambda : \bar{A}^n \to \bar{A}^n$: see for example (2.6).

$p > 1$. Choose a free module F_0, a surjective linear map $F_0 \to M$, and an element $a \in \operatorname{ann} M - z(A) \cup \underline{p}$. Let K denote the kernel of the induced map $F_0/aF_0 \to M$. It follows from the exact sequence

$$0 \to K \to F_0/aF_0 \to M \to 0$$

that $\operatorname{pd} K = p - 1$, since $\operatorname{pd}(F_0/aF_0) = 1$. Now the identity follows from the inductive hypothesis, since $\chi(-,A)$ and $\ell(\bar{A}/G(-)\bar{A})$ both are additive on short exact sequences, the latter by (0.1)(c). □

(1.4) COROLLARY. *If* A *is regular and* $\dim A \leqq 4$, *then* (1) *and* (2) *hold.*

This was first proved by Hochster [5; Corollary 2.11] (by different methods).

Proof. Directly from (1.1) it follows that (1) holds. Directly, or from (1.1), it follows that the only remaining case in (2) is that when $\dim A = 4$, $\dim M = 2$ and $\dim N = 2$, and in this case (2) can be proved by standard methods. (Namely, it suffices to assume that $M = A/\underline{p}$ and $N = A/\underline{q}$ where $\underline{p}, \underline{q} \in \operatorname{Spec} A$. Write $\Omega = \operatorname{Ext}_A^2(M,A)$ and $\Delta = \operatorname{Ext}_A^2(N,A)$. Then depth $\Omega = 2$, and hence pd $\Omega = 4 - 2 = 2$. (If $B = A/(a_1,a_2)$ where $a_1,a_2 \in \underline{p}$ is an A-regular sequence, then $\Omega \simeq \operatorname{Hom}_B(M,B)$, a second syzygy B-module.) Also depth $\Delta = 2$, so the fact (1.3) shows that $\chi(\Omega,\Delta) = \ell(\Omega \otimes \Delta) > 0$. On the other hand, since Ω (respectively Δ) has a filtration with factors $A/\underline{r}$ with $\underline{r} \in \operatorname{Spec} A$ and $\underline{r} \supseteq \underline{p}$ (respectively $\underline{r} \supseteq \underline{q}$) and since (1) already holds, $n\chi(M,N) = \chi(\Omega,\Delta)$ for some non-negative integer n (which actually is 1).) □

(1.5) *Remark.* If A is regular and $\dim A = 5$, then (1.1) shows that the only remaining case in (1) is that when $\dim M = 2$ and $\dim N = 2$. It is possible to prove that (1) also holds in this case (using some techniques from [2]). However this is already known; a (different) proof should be in Dutta's thesis [3].

2. *A new description of the MacRae ideal*

First recall that there is an exact sequence of abelian groups, the so-called *localization sequence*

$$(2.1) \quad K_1(A) \to K_1(S^{-1}A) \xrightarrow{\partial} K_0(A;S) \to K_0(A) \to K_0(S^{-1}A).$$

Here and in all that follows S *denotes a multiplicatively closed subset of* A *with* $S \cap z(A) = \emptyset$ (that is, S contains no zerodivisor on A). For the localization sequence consult [1; §10]. The basic definitions can be found in [1; §§1, 4 and 10], where the group $K_0(A;S)$ is denoted by $K_0(\mathscr{P}_{<\infty}(A)_S)$. However, the following description of $K_0(A;S)$ is the one that will be needed later. Details can be found in Fossum, Foxby, Iversen [2] (see, in particular, Proposition 4.8, where the group is denoted by $K_0(\operatorname{Hot}(\underline{\underline{P}}(A),S))$).

Description of $K_0(A;S)$. The abelian group $K_0(A;S)$ is presented by generators $[P.]$, only depending on the isomorphism class of the bounded complex $P.$ of projective modules with $S^{-1}P.$ exact, subject to the relation $[P.] = 0$ if $P.$ is exact, and to the relation $[P.] = [P'_.] + [P''_.]$ whenever there is an exact sequence of bounded complexes

$$0 \longrightarrow P'_. \longrightarrow P. \longrightarrow P''_. \longrightarrow 0$$

of projective modules with $S^{-1}P'_.$, $S^{-1}P.$ and $S^{-1}P''_.$ exact.

Definitions. If P_0 and P_1 are projective A-modules and $\phi : P_1 \to P_0$ is A-linear and such that $S^{-1}\phi : S^{-1}P_1 \to S^{-1}P_0$ is an isomorphism, then $[\phi]$ denotes the class in $K_0(A;S)$ of the complex

$$0 \longrightarrow P_1 \xrightarrow{\phi} P_0 \longrightarrow 0$$

concentrated in degrees 0 and 1.

If Λ is an $n \times n$-matrix with entries in A such that the matrix $S^{-1}\Lambda$ (with entries in $S^{-1}A$) is invertible, then $[\Lambda]$ denotes the class in $K_0(A;S)$ induced by the linear map $\Lambda : A^n \to A^n$.

In particular, if $s \in S$ then $[s]$ denotes the class in $K_0(A;S)$ induced by $s : A \to A$ (multiplication by s).

Now the homomorphism ∂ of the localization sequence (3) can be described: see [1] and [2].

(2.2) LEMMA. *If* $s \in S$ *and* Λ *is an* $n \times n$-*matrix with entries in* A *and such that* $S^{-1}\Lambda$ *is invertible then*

$$\partial([s^{-1}\Lambda]) = [\Lambda] - n[s] \quad (\in K_0(A;S)). \quad \square$$

Definition. Let $U(A;S) = U(S^{-1}A)/U(A)$, where for any ring R the notation U(R) denotes the multiplicative group of units. An element of $U(A;S)$ will be thought of as a cyclic A-submodule Av of $S^{-1}A$ generated by a unit v of $S^{-1}A$.

(2.3) LEMMA. *There is a commutative diagram with exact rows*

$$\begin{array}{ccccccc}
K_1(A) & \longrightarrow & K_1(S^{-1}A) & \longrightarrow & K_0(A;S) & \longrightarrow & 0 \\
\downarrow{\scriptstyle \det_A} & & \downarrow{\scriptstyle \det_{S^{-1}A}} & & \downarrow{\scriptstyle \det_S} & & \\
U(A) & \longrightarrow & U(S^{-1}A) & \longrightarrow & U(A;S) & \longrightarrow & 0.
\end{array}$$

Proof. The homomorphisms $\det_A$ and $\det_{S^{-1}A}$ are just the usual determinants (see [1;§1]) and so the left rectangle is certainly commutative.

Thus to define $\det_S$ it suffices to prove that ∂ is surjective, and this follows easily from the sequence (2.1), since the homomorphism $K_0(A) \to K_0(S^{-1}A)$ is injective. Here, for the first time, the fact that A is local has been used: for any ring R there is an injective homomorphism $\rho_R : \mathbb{Z} \to K_0(R)$ given by $\rho_R(1) = [R]$; since each projective A-module is free ρ_A is an isomorphism. The homomorphism $K_0(A) \to K_0(S^{-1}A)$ is the composite $\rho_{S^{-1}A}\rho_A^{-1}$, and thus injective. □

(2.4) PROPOSITION. $G(M) = \det_S[F.]$ *if* $S^{-1}M = 0$ *whenever* F. *is a finite free resolution of* M.

Proof. This is by induction on p = pd M, but first notice that $\det_S[F.] = \det_S[P.]$ if P. is another finite free resolution of M: see [2; Proposition 3.5].

If p = 1 the identity follows directly from (0.1)(b) and (2.2).

If p > 1 choose a, F_0 and K as in the proof of (1.2) and use (0.1)(c),(d). □

(2.5) *Remark.* The above result gives an alternative method of computation of G(M). Namely, since $S^{-1}F.$ is exact it is possible to choose linear maps $s_i : F_i \to F_{i+1}$ for all i such that all the maps $a_i = s_{i-1}d_i + d_{i+1}s_i$ are injective (namely with $\det a_i \notin z(A)$). Define a matrix Σ as follows.

$$\Sigma = \begin{pmatrix} d_1 & & & \\ s_1 & d_3 & & \\ & s_3 & \ddots & \\ & & & \ddots \end{pmatrix} : \begin{matrix} F_1 \\ \oplus \\ F_3 \\ \oplus \\ \vdots \end{matrix} \longrightarrow \begin{matrix} F_0 \\ \oplus \\ F_2 \\ \oplus \\ \vdots \end{matrix} .$$

Then

$$G(M)\Pi_i(\det a_{2i})^i = (\det \Sigma)\Pi_i(\det a_{2i+1})^i .$$

This follows from [2; Theorem 2.1 and Theorem 4.4]. (We adopt the convention whereby the determinant of a 0×0-matrix is 1.)

(2.6) *Remark.* Let dim A = 1 and S = A - z(A), the set of non-zerodivisors on A. Then the ring $S^{-1}A$ is semilocal and so $\det_{S^{-1}A} : K_1(S^{-1}A) \to U(S^{-1}A)$ is an isomorphism: see [1; Corollary (2.8)]. Now let $\Lambda : A^n \to A^n$ be an injective linear map. Then $\det \Lambda \in S$ (McCoy's Theorem), and it follows from the above that $[\det \Lambda] = [\Lambda]$ in $K_1(S^{-1}A)$. The homomorphism χ: $K_0(A;S) \to \mathbb{Z}$ is defined by $\chi([P.]) = \Sigma(-1)^i \ell(H_i(P.))$. It follows that

$$\ell(A/(\det \Lambda)) = \chi(\partial([\det \Lambda])) = \chi(\partial([\Lambda])) = \ell(\text{Coker } \Lambda).$$

This is however known: see [4; Lemme (21.10.17.3), p.298] (where the proof is somewhat different).

Acknowledgement

The author has been supported, in part, by the Danish Natural Science Research Council.

References

1. H. Bass, Introduction to some methods of algebraic K-theory, C.B.M.S. Regional Conference Series in Mathematics 20 (American Mathematical Society, Providence, 1974).

2. R. Fossum, H.-B. Foxby and B. Iversen, "A characteristic class in algebraic K-theory", to appear.

3. S. Dutta, Thesis, University of Michigan, Ann Arbor, 1981.

4. A. Grothendieck, Élements de géométrie algébrique, IV, Publications Mathématiques 32 (Institut des Hautes Études Scientifiques, Paris, 1967).

5. M. Hochster, "Cohen-Macaulay modules", Conference on commutative algebra, Lecture Notes in Mathematics 311 (eds. J.W.Brewer and E.A. Rutter, Springer, Berlin, Heidelberg, New York, 1973), pp. 120-152.

6. R.E. MacRae, "On an application of the Fitting invariants", J. Algebra, 2 (1965), 153-169.

7. C. Peskine and L. Szpiro, "Syzygies et multiplicités", C. R. Acad. Sci. Paris Sér. A, 278 (1974), 1421-1424.

8. J.-P. Serre, Algèbre locale: multiplicités, Lecture Notes in Mathematics 11 (Springer, Berlin, Heidelberg, New York, 1965).

Department of Mathematics,
University of Oklahoma,
Norman, Oklahoma 73019,
U.S.A.

and (from 1982)

Matematisk Institut,
Københavns Universitet,
Universitetsparken 5,
DK 2100 København Ø, Denmark.

FINITE FREE RESOLUTIONS AND SOME BASIC CONCEPTS OF COMMUTATIVE ALGEBRA

D.G.NORTHCOTT

This paper deals with certain developments in the theory of finite free resolutions. The developments in question are of interest not only for their own sake, but also because they have led to a reappraisal of some basic concepts of commutative algebra. It appears that in the absence of Noetherian conditions, certain very familiar and fundamental concepts let us down rather badly, but happily there have been found ways of dealing with the resulting problems. This in itself, it seems to the author, may have more than a passing interest.

1. *Modules with Euler characteristic zero*

Suppose that R is a commutative ring with a non-zero identity element, and let $\mathbb{F}$ denote the class of modules that have a resolution, of finite length, by means of free R-modules of finite rank. Thus an R-module E belongs to $\mathbb{F}$ precisely when there exists an exact sequence

$$0 \to F_n \to F_{n-1} \to \ldots \to F_1 \to F_0 \to E \to 0, \qquad (1.1)$$

where each F_i is a free module with a finite base. The aim of the theory of finite free resolutions is to study such modules (and their resolutions) preferably without imposing any unnecessary conditions on the ring R.

Let E belong to $\mathbb{F}$ and suppose that (1.1) is a finite free resolution of E. The *Euler characteristic*, $\mathrm{Char}_R(E)$, of E is defined by

$$\mathrm{Char}_R(E) = \sum_{\nu=0}^{n} (-1)^{\nu} \mathrm{rank}_R(F_{\nu}),$$

where $\mathrm{rank}_R(F_{\nu})$ denotes the number of elements in a base of F_{ν}. This is an *invariant* of E, that is to say it does not depend on the chosen resolution, and it has many properties. The particular property that is relevant here states that $\mathrm{Char}_R(E) \geq 0$. There are, of course, plenty of modules whose Euler characteristic is zero. If therefore we put

$$\mathbb{F}_0 = \{E \in \mathbb{F} : \mathrm{Char}_R(E) = 0\},$$

then $\mathbb{F}_0$ is a subclass of $\mathbb{F}$ which is likely to be especially interesting.

It is known that when R is *Noetherian* there are two contrasting characterizations of $\mathbb{F}_0$. These may be stated as in (A) and (B) below.

(A) *For* E *in* $\mathbb{F}$ *we have* $E \in \mathbb{F}_0$ *if and only if the annihilator* $\mathrm{Ann}_R(E)$, *of* E, *is non-zero.*

(B) *For* E *in* $\mathbb{F}$ *we have* $E \in \mathbb{F}_0$ *if and only if* $\mathrm{Ann}_R(E)$ *contains a non-zerodivisor.*

However the effect of dropping the Noetherian condition is disconcerting for, as was shown by W.V.Vasconcelos, (A) remains true whereas (B) becomes false. Nevertheless, as Vasconcelos pointed out, (B) is not altogether false.

It has taken some time for the full significance of the last remark to become clear. In order to explain what is involved, let us suppose that I is an ideal of R and X is an indeterminate. If $r \in I$ and is a non-zerodivisor in R, then $r \in IR[X]$ and it remains a non-zerodivisor in R[X]. On the other hand it is perfectly possible for I to be composed of zerodivisors and yet for IR[X] to contain a non-zerodivisor. Let us agree to say that I contains a *latent* non-zerodivisor whenever IR[X] contains a non-zerodivisor. We can then rescue (B) by changing it to read as follows.

(B)′ *For* E *in* $\mathbb{F}$ *we have* $E \in \mathbb{F}_0$ *if and only if* $\mathrm{Ann}_R(E)$ *contains a latent non-zerodivisor.*

All this suggests that we should modify our attitude to non-zerodivisors and to concepts which involve them. One obvious concept which comes into this category is that of *grade*. Classically the grade of an ideal I is the upper bound of the lengths of all sequences $r_1, r_2, \ldots, r_s$ in I such that, for each i, r_i is a non-zerodivisor on $R/(r_1, r_2, \ldots, r_{i-1})$. This number will be denoted by $gr_R(I)$. In order to take account of possible latent non-zero-divisors, we introduce an infinite sequence $X_1, X_2, X_3, \ldots$ of different indeterminates and put

$$Gr_R(I) = \lim_{n\to\infty} gr_{R[X_1,\ldots,X_n]}(IR[X_1,\ldots,X_n]).$$

(The limit exists because the right hand side increases with n.) Let us call $Gr_R(I)$ the *true* grade of I. If $n \geq 2$, then in a suitable ring R there exists an ideal I, generated by n elements, such that $gr_R(I) = 0$ and $Gr_R(I) = n$. Thus the two grades can differ considerably although it is not difficult to see that they always coincide when R is Noetherian.

Let us return to (B). In its original form it asserts that a module E, in $\mathbb{F}$, belongs to $\mathbb{F}_0$ when and only when $gr_R(Ann_R(E)) > 0$, and we have already noted that this may be false when R is not Noetherian. However we now have the following theorem which holds without any such condition on R.

THEOREM. *Let* E *belong to* $\mathbb{F}$. *Then* $E \in \mathbb{F}_0$ *if and only if* $Gr_R(Ann_R(E)) > 0$.

The moral seems to be that we should use true grade rather than classical grade, and presumably this should apply generally and not just in the context of the theory of finite free resolutions. (It was, I believe, M.Hochster who first questioned the appropriateness of the traditional definition of grade.) As I shall try to show presently, the consequences are unexpectedly far-reaching, but for the moment let us stay with the subject of resolutions.

Denote by $\mathbb{P}$ the class of modules which have resolutions, of finite length, by means of finitely generated *projective* modules. Thus for E to belong to $\mathbb{P}$ there must exist an exact sequence

$$0 \to \Pi_n \to \Pi_{n-1} \to \ldots \to \Pi_1 \to \Pi_0 \to E \to 0,$$

where Π_j is a finitely generated projective module. The class $\mathbb{P}$ is somewhat larger than the class $\mathbb{F}$ and our previous discussion now enables us to identify the subclass $\mathbb{P}_0$ that generalizes $\mathbb{F}_0$. We do this by putting

$$\mathbb{P}_0 = \{E \in \mathbb{P} : Gr_R(Ann_R(E)) > 0\}.$$

So far we have been operating in something of a vacuum. To get the feel of what is happening let us take a quick look at the case where $R = \mathbb{Z}$. For this situation there is no distinction between free and projective modules, and so $\mathbb{F} = \mathbb{P}$ and $\mathbb{F}_0 = \mathbb{P}_0$. It is easy to see that $\mathbb{P}$ is the class of finitely generated $\mathbb{Z}$-modules, that is to say it is the class of *finitely generated abelian groups*. With this interpretation $\mathbb{P}_0$ is the subclass of *finite abelian groups*. For E in $\mathbb{P}_0$ let O(E) denote the *order* of the finite abelian group. Then O(E) is *multiplicative*, that is if $0 \to E' \to E \to E'' \to 0$ is an exact sequence in $\mathbb{P}_0$, then $O(E) = O(E')O(E'')$. This suggests that for a general R there may be a similar multiplicative invariant associated with $\mathbb{P}_0$. Indeed, when R is Noetherian, R.E.MacRae has identified this invariant. But when we try to drop the Noetherian condition we immediately encounter the kind of difficulty I have been discussing. In this instance we are not yet equipped to deal with the problems that arise, so let me prepare the way by following up a previously mentioned clue.

2. *The theory of attached prime ideals*

It was stated earlier that the principle of replacing classical grade by true grade has some far-reaching consequences. This will now be illustrated by means of the theory of associated prime ideals. As is well known, if E is an R-module and P is a prime ideal, then P is said to be *associated* with E if $P = \mathrm{Ann}_R(Re)$ for some $e \in E$. The set of such prime ideals is denoted by $\mathrm{Ass}_R(E)$.

There is an extensive theory surrounding this concept. For example, if R is Noetherian and E is finitely generated, then the following three statements all hold.

(I) $\mathrm{Ass}_R(E)$ *is empty if and only if* $E = 0$.

(II) *An element* $r \in R$ *is a zerodivisor on* E *if and only if* $r \in P$ *for some* $P \in \mathrm{Ass}_R(E)$.

(III) *If* S *is a multiplicatively closed subset of* R, *then* $\mathrm{Ass}_{R_S}(E_S) = \{PR_S : P \in \mathrm{Ass}_R(E)$ *and* $P \cap S = \emptyset\}$.

However when the finiteness conditions on R and E are removed all three statements can be false. Since the failure of (I) is particularly inconvenient, an example to show how easily it can happen may be of interest. Suppose then that $X_1, X_2, X_3, \ldots$ are indeterminates, let $R = \mathbb{Q}[X_1, X_2, X_3, \ldots]$ and let I be the R-ideal $(X_1^2, X_2^2, X_3^2, \ldots)$. Then R/I is a non-zero R-module and it is easy to check that $\mathrm{Ass}_R(R/I)$ is empty.

It is here that our earlier remarks about grade can help. First we note that if $\mathrm{gr}_R(I;E)$ denotes the classical grade of an ideal I on a module E, then, using the same device as before, we can define the true grade $\mathrm{Gr}_R(I;E)$ of I on E. This said, assume for the moment that R is Noetherian and E is finitely generated. Then a prime R-ideal P is associated with E if and only if the R_P-ideal PR_P is associated with E_P, and this happens if and only if every element of PR_P is a zerodivisor on E_P. Consequently $P \in \mathrm{Ass}_R(E)$ when and only when

$$\mathrm{gr}_{R_P}(PR_P;E_P) = 0.$$

The way ahead is now clear. For arbitrary R and E we now put

$$\mathrm{Att}_R(E) = \{P \in \mathrm{Spec}(R) : \mathrm{Gr}_{R_P}(PR_P;E_P) = 0\},$$

and we say that the prime ideals in $\mathrm{Att}_R(E)$ are *attached* to E. With this definition we have $\mathrm{Ass}_R(E) \subseteq \mathrm{Att}_R(E)$ and the inclusion may be strict. Every finitely generated attached prime ideal turns out to be an associated prime ideal, and so $\mathrm{Ass}_R(E) = \mathrm{Att}_R(E)$ whenever R is Noetherian. But we have gained considerably because, for example, assertions (I), (II) and (III) are now true quite generally provided we use attached rather than associated prime ideals.

The theory of attached prime ideals has been well developed largely through the efforts of P.Dutton. Almost all the properties of associated prime ideals generalize and the more general theory seems to provide all that one could reasonably expect.

3. *Quasi-invertible ideals*

There is one more concept that needs to be modified before we can continue with the theory of resolutions. This is the notion of an invertible ideal. Let Σ be the full ring of fractions of R so that a typical member of Σ has as numerator an element of R and as demoninator a non-zerodivisor of R. If M and N are R-submodules of Σ, then we can form their product MN just as we form the product of two ideals. The submodules then form a commutative semi-group with R as its neutral element. The ideals of R are members of this semi-group (because R is a subring of Σ) and an ideal is called *invertible* if it has a semi-group inverse.

But what has this to do with grade? To answer this question suppose that I is an ideal of R. Then there is a result which states that for I to be invertible it is necessary and sufficient that I be projective (as a module) and contain a non-zerodivisor. Thus I is invertible if and only if I is projective and $gr_R(I) > 0$. This prompts the following.

Definition. The ideal I is said to be *quasi-invertible* if I is projective and $Gr_R(I) > 0$.

Quasi-invertible ideals enjoy many of the properties of invertible ideals. For example, they are always finitely generated. Again if I,J are ideals, $I \subseteq J$ and J is quasi-invertible, then $I = JA$ for a unique ideal A. Finally we mention that the product of two ideals is quasi-invertible when and only when both the factors are quasi-invertible.

4. *The MacRae invariant*

We are now ready to resume the discussion of free and projective resolutions. Let $E \in \mathbb{P}_0$. Then, as is easily seen, E is finitely presented and therefore there exists an exact sequence

$$G \xrightarrow{f} F \to E \to 0,$$

where F and G are free modules of let us say ranks p and q respectively. Suppose we choose a base for each of F and G. Then f is represented by a $p \times q$ matrix A, and the $p \times p$ minors of A generate an ideal $\mathscr{F}_R(E)$ that depends only on E. This is the well-known initial *Fitting invariant* of E. $\mathscr{F}_R(E)$ is closely allied to $Ann_R(E)$ and in particular they coincide when E is *cyclic*.

The vital fact here is that, because E is in $\mathbb{P}_0$, it can be shown that there is a smallest quasi-invertible ideal containing

$\mathcal{F}_R(E)$. This quasi-invertible ideal will be denoted by $\mathcal{G}_R(E)$. It is essentially MacRae's generalization of the order of a finite group, but now we have removed the restriction that R has to be Noetherian. Among the properties of the new invariant we select two for mention because they are striking in themselves and particularly relevant here. The properties in question are enshrined in the next two theorems.

THEOREM. *If* $0 \to E' \to E \to E'' \to 0$ *is an exact sequence in* $\mathbb{P}_0$, *then* $\mathcal{G}_R(E) = \mathcal{G}_R(E')\mathcal{G}_R(E'')$.

THEOREM. *If* E *belongs to* $\mathbb{F}_0$, *then* $\mathcal{G}_R(E)$ *is a principal ideal generated by a non-zerodivisor.*

For applications of MacRae's invariant it is often important to know when $\mathcal{G}_R(E)$ is a *proper* ideal, that is to say different from the whole ring. This happens, of course, when $R/\mathcal{G}_R(E) \neq 0$ or equivalently when $\mathrm{Att}_R(R/\mathcal{G}_R(E))$ is not empty. Now it is known that

$\mathrm{Att}_R(R/\mathcal{G}(E)) = \{P \in \mathrm{Att}_R(E) : \mathrm{Gr}_{R_P}(PR_P) = 1\}$ and indeed I suspect that this can be simplified to

$$\mathrm{Att}_R(R/\mathcal{G}(E)) = \{P \in \mathrm{Att}_R(E) : \mathrm{Gr}_R(P) = 1\},$$

although I have not been able to find a proof. In fact this seems to be connected with an open problem concerned with the phenomenon of grade stability, and, as the problem may be unfamiliar, it perhaps merits a digression.

Suppose that P is a prime ideal. It is easy to see that

$$\mathrm{Gr}_R(P) \leq \mathrm{Gr}_{R_P}(PR_P).$$

Should it happen that

$$\mathrm{Gr}_R(P) = \mathrm{Gr}_{R_P}(PR_P),$$

then P is said to be *grade stable*.

Problem. Is it true that whenever $E \in \mathbb{P}$ all the prime ideals in $\mathrm{Att}_R(E)$ are grade stable?

It seems very likely that the answer to the question posed here is 'Yes'. For example, if $E \in \mathbb{P}$ then every finitely generated member of $\mathrm{Att}_R(E)$ is grade stable. Consequently, as was observed by MacRae, the answer is affirmative whenever R is Noetherian. Of course, because we are striving for full generality, this doesn't help here. However there is one additional piece of evidence. It can be shown that if $E \in \mathbb{P}$ and $E \neq 0$, then at least one prime ideal in $\mathrm{Att}_R(E)$ is grade stable. Fortunately this is sufficient for the applications described below, but an affirmative answer, if correct, would add a finishing touch to what is already an elegant theory.

5. *Applications*

We conclude by indicating what can be achieved by way of applications. First we have the following.

THEOREM. *An ideal* I, *of* R, *can be generated by a non-zero-divisor if and only if*

(i) $I \in \mathbb{F}$, *and*

(ii) $\mathrm{Gr}_R(P) = 1$ *for all* P *in* $\mathrm{Att}_R(R/I)$.

For Noetherian *domains* this too goes back to MacRae. Note that in the form just stated, where there are no extra conditions on R at all, there is a satisfying economy in the hypotheses. It is also worthwhile noting in passing that the theorem serves to pinpoint the origin of the connection between unique factorization and finite homological dimension.

As the proof illustrates the interconnections between the various topics described above it will be given in outline. However the interesting part of the demonstration consists in showing that conditions (i) and (ii) imply that I is generated by a non-zerodivisor, and so only this aspect will be considered.

Condition (i) ensures that $R/I \in \mathbb{F}$ and from (ii) it is easily deduced that $\mathrm{Gr}_R(\mathrm{Ann}_R(R/I))$ is greater than zero. Consequently $R/I \in \mathbb{F}_0$. Thus $\mathscr{G}_R(R/I)$ is defined and it is moreover an ideal generated by a non-zerodivisor. We also have

$$I = \mathrm{Ann}_R(R/I) = \mathscr{F}_R(R/I) \subseteq \mathscr{G}_R(R/I),$$

whence, because $\mathscr{G}_R(R/I)$ is a principal ideal,

$$I = \mathscr{G}_R(R/I)\mathrm{Ann}_R(\mathscr{G}_R(R/I)/I). \qquad (5.1)$$

Put $E = \mathscr{G}_R(R/I)/I$. It will suffice to show that $E = 0$, for then we shall have $I = \mathscr{G}_R(R/I)$ and it has already been noted that the latter is generated by a non-zerodivisor. In any event

$\mathcal{G}_R(R/I)$, and hence E as well, is cyclic and therefore

$$\mathcal{F}_R(E) = \mathrm{Ann}_R(E) = \mathrm{Ann}_R(\mathcal{G}_R(R/I)/I).$$

Thus (5.1) can be rewritten as

$$I = \mathcal{F}_R(E)\mathcal{G}_R(R/I). \tag{5.2}$$

By (i), $I \in \mathbb{F}$. We also have $\mathcal{G}_R(R/I) \in \mathbb{F}$ because $\mathcal{G}_R(R/I)$ is a free module of rank one. Consequently $E = \mathcal{G}_R(R/I)/I$ belongs to $\mathbb{F}$; indeed $E \in \mathbb{F}_0$ because $\mathrm{Ann}_R(E)$ contains I and so is not zero. Thus $\mathcal{G}_R(E)$ is defined and now (5.2) yields

$$I \subseteq \mathcal{G}_R(E)\mathcal{G}_R(R/I) \subseteq \mathcal{G}_R(R/I)$$

because $\mathcal{F}_R(E) \subseteq \mathcal{G}_R(E)$. But, by definition, $\mathcal{G}_R(R/I)$ is the *smallest* quasi-invertible ideal containing $\mathcal{F}_R(R/I) = I$, and $\mathcal{G}_R(E)\mathcal{G}_R(R/I)$, because it is the product of two quasi-invertible ideals, is itself quasi-invertible. It follows that $\mathcal{G}_R(E)\mathcal{G}_R(R/I) = \mathcal{G}_R(R/I)$ and therefore $\mathcal{G}_R(E) = R$ by another of the properties of quasi-invertible ideals. We have now shown that $\mathcal{G}_R(E)$ is an improper ideal and this, as we saw earlier, means that *there can be no prime ideal* P *in* $\mathrm{Att}_R(E)$ *for which* $\mathrm{Gr}_{R_P}(PR_P) = 1$.

On the other hand $E = \mathcal{G}_R(R/I)/I$ is a submodule of R/I from which it follows that

$$\mathrm{Att}_R(E) \subseteq \mathrm{Att}_R(R/I).$$

Accordingly, by condition (ii), $\mathrm{Gr}_R(P) = 1$ for every P in $\mathrm{Att}_R(E)$. Now we know that if $E \neq 0$, then there will be at least one P in $\mathrm{Att}_R(E)$ which is grade stable, and for such a P we would have

$$\mathrm{Gr}_{R_P}(PR_P) = \mathrm{Gr}_R(P) = 1.$$

But we have just established that this situation cannot arise. Consequently E = 0, as we wished to prove. □

Let us return to the statement of the last theorem. An ideal which can be generated by a non-zerodivisor is the same as a non-zero free ideal. It is therefore to be expected that there will be a related result concerning projective ideals. In fact it is better to state the companion theorem in terms of Picard modules.

An R-module M will be called a *Picard module* if there exists a second module N such that $M \otimes_R N$ is isomorphic to R. If [M] denotes the isomorphism class of M, then the isomorphism classes of Picard modules form the *Picard group*, Pic(R), of R when composition is defined by

$$[M_1] + [M_2] = [M_1 \otimes_R M_2].$$

The Picard modules are just the rank one projective modules, but the definition just given is more succinct.

We now have sufficient terminology to state a final result.

THEOREM. *Let* K *be a submodule of a Picard module* M. *Then* K *is also a Picard module if and only if*

(i) $K \in \mathbb{P}$, *and*

(ii) $\mathrm{Gr}_R(P) = 1$ *for all* $P \in \mathrm{Att}_R(M/K)$.

To obtain from this a result concerning projective ideals all that is necessary is to replace the Picard module M by the ring R.

References

1. P.Dutton, "Prime ideals attached to a module", Quart.J.Math. Oxford (2), 29 (1978), 403-413.

2. R.E.MacRae, "On the homological dimension of certain ideals", Proc.Amer.Math.Soc., 14 (1963), 746-750.

3. R.E.MacRae, "On an application of the Fitting invariants", J.Algebra, 2 (1965), 153-169.

4. D.G.Northcott, Finite free resolutions, Cambridge Tracts in Mathematics 71 (Cambridge University Press, Cambridge, 1976).

5. D.G.Northcott, "Projective ideals and MacRae's invariant", J.London Math.Soc. (2), 24 (1981), 211-226.

Department of Pure Mathematics,
University of Sheffield,
Hicks Building,
Sheffield S3 7RH, U.K.

PART III

MULTIPLICITY THEORY, HILBERT AND POINCARÉ SERIES, ASSOCIATED GRADED RINGS, AND RELATED TOPICS

BLOWING-UP OF BUCHSBAUM RINGS

SHIRO GOTO

1. *Introduction*

The purpose of this paper is to give a characterization of Buchsbaum rings in terms of blowing-up.

Let A be a Noetherian local ring with maximal ideal $\underline{m}$ and $\dim A = d$. Then A is called *Buchsbaum* if the difference

$$I(A) = \ell_A(A/\underline{q}) - e_{\underline{q}}(A)$$

is an invariant which does not depend on the choice of the parameter ideal $\underline{q}$ of A. (Here $\ell_A(A/\underline{q})$ and $e_{\underline{q}}(A)$ denote, respectively, the length of $A/\underline{q}$ and the multiplicity of $\underline{q}$.) This is equivalent to the condition that every system $a_1, a_2, \ldots, a_d$ of parameters for A is a weak sequence, that is the equality

$$(a_1, \ldots, a_i) : a_{i+1} = (a_1, \ldots, a_i) : \underline{m}$$

holds for every $0 \leq i < d$ (c.f. [21; Satz 10]). Thus a Cohen-Macaulay ring A is a Buchsbaum ring with $I(A) = 0$ and vice versa. In this sense the concept of Buchsbaum ring is an extension of that of Cohen-Macaulay ring and the theory of Buchsbaum rings has started from an answer of Vogel [24] to a problem of Buchsbaum [3; p. 228].

Let $\underline{q} = (a_1, a_2, \ldots, a_d)$ be a parameter ideal of A and put $R(\underline{q}) = \bigoplus_{n \geq 0} \underline{q}^n$. Then the canonical morphism $f : \operatorname{Proj} R(\underline{q}) \to \operatorname{Spec} A$ is said to be the *blowing-up of* Spec A *with centre* Spec $A/\underline{q}$. Recall that

$$\operatorname{Proj} R(\underline{q}) = \bigcup_{i=1}^{d} \operatorname{Spec} A[x/a_i \mid x \in \underline{q}]$$

and that the fibre of f over Spec $A/\underline{q}$ is given by Proj $G(\underline{q})$ where $G(\underline{q}) = \bigoplus_{n \geq 0} \underline{q}^n/\underline{q}^{n+1}$.

Let $H^i_{\underline{m}}(.)$ denote the i-th local cohomology functor.

With this notation the main result of this paper is stated as follows.

THEOREM (1.1). *Suppose that* dim A > 0. *Then the following conditions are equivalent:*

(1) $A/H^0_{\underline{m}}(A)$ *is a Buchsbaum ring;*

(2) Proj $R(\underline{q})$ *is a locally Cohen-Macaulay scheme for every parameter ideal* $\underline{q}$ *of* A.

The theory of Buchsbaum singularities is now developing very rapidly (see [5,6,7,8,9,10,17,18,22]) and it is known that they enjoy pretty good properties. For example suppose that A is Buchsbaum. Then for every prime ideal $\underline{p}$ of A such that $\underline{p} \neq \underline{m}$ the local ring $A_{\underline{p}}$ is a Cohen-Macaulay ring with $\dim A_{\underline{p}} = d - \dim A/\underline{p}$. Moreover the local cohomology modules $H^i_{\underline{m}}(A)$ $(i \neq d)$ are vector spaces, that is $\underline{m}.H^i_{\underline{m}}(A) = 0$, and we have

$$I(A) = \sum_{i=0}^{d-1} \binom{d-1}{i} . h^i(A)$$

[16; Satz 2] where, for each i, $h^i(A)$ denotes the dimension of $H^i_{\underline{m}}(A)$ as a vector space over $A/\underline{m}$. It was shown in [17] (see also [23]) that Buchsbaum rings may be characterized in terms of Koszul homology relative to systems of parameters. There was given in [22] and [29] a very powerful criterion, the so-called surjectivity criterion, for Buchsbaum rings in terms of local cohomology, and subsequently, using this, a lot of examples of non-Cohen-Macaulay Buchsbaum and normal rings were discovered [5] . It was also pointed out in [6] and [10] that certain Buchsbaum rings are characterized by the behaviour of the Rees algebras $R(\underline{q})$ of parameter ideals $\underline{q}$ of them. Nevertheless in spite of the importance of the theory of Buchsbaum rings there has been established no definitive characterization which really clarifies them as singularities. From this point of view our theorem (1.1) may have some interest.

As a consequence of (1.1) one has the following.

COROLLARY (1.2). *Suppose that* dim A > 0. *Then the following conditions are equivalent:*

(1) $A/H^0_{\underline{m}}(A)$ *is a Gorenstein ring;*

(2) Proj $R(\underline{q})$ *is locally Gorenstein for every parameter ideal* $\underline{q}$ *of* A.

We shall prove (1.2) in Section 4. A similar characterization of complete intersections may be found also in that section. (Further applications of (1.1) will be discussed in a subsequent paper.) Theorem (1.1) itself will be proved in Section 3. Section 2 is devoted to some remarks on rings with finite local cohomology which we shall often need in the proof of (1.1).

Finally the author wishes to thank Y.Shimoda for helpful discussion during this research. Lemma (3.4) was suggested to the author by him.

Throughout this paper A always denotes a Noetherian local ring with maximal ideal $\underline{m}$ and dim A = d. Also $H^i_{\underline{m}}(.)$ will stand for the i-th local cohomology functor.

2. *Preliminaries*

We say that A has finite local cohomology if the local cohomology modules $H^i_{\underline{m}}(A)$ are finitely generated (that is the lengths $\ell_A(H^i_{\underline{m}}(A))$ are finite) for all $i \neq d$.

First of all we note the following.

PROPOSITION (2.1) [19, 22]. *The following conditions are equivalent:*

(1) A *has finite local cohomology;*

(2) *there exists an* $\underline{m}$*-primary ideal* I *of* A *such that for every system* $a_1, a_2, \ldots, a_d$ *of parameters contained in* I *and for every integer* $0 \leq i < d$ *we have*

$$(a_1, \ldots, a_i) : a_{i+1} = (a_1, \ldots, a_i) : I.$$

When this is the case, $I.H^i_{\underline{m}}(A) = (0)$ *for every* $i \neq d$ *and* $A_{\underline{p}}$ *is a Cohen-Macaulay local ring with* $\dim A_{\underline{p}} = d - \dim A/\underline{p}$ *for every prime ideal* $\underline{p}$ *of* A *such that* $\underline{p} \neq \underline{m}$.

Proof. See [19]. See also [22; Lemma 3]. □

Let J be an ideal of A and let

$$J = \bigcap_{\underline{p} \in \mathrm{Ass}\, A/J} J(\underline{p})$$

denote a primary decomposition of J in A. We put

$$\operatorname{Assh} A/J = \{\underline{p} \in \operatorname{Ass} A/J \mid \dim A/\underline{p} = \dim A/J\}$$

and

$$U(J) = \bigcap_{\underline{p}\in \operatorname{Assh} A/J} J(\underline{p}).$$

As every element of Assh A/J is a minimal prime divisor of J this definition of U(J) does not depend on the choice of primary decomposition of J. Notice that Ass A/U(J) = Assh A/J and dim A/U(J) = dim A/J.

For the rest of this section we assume that our ring A has finite local cohomology and we let I be an ideal of A obtained by (2.1). Let $a_1, a_2, \dots, a_d$ be a fixed system of parameters for A contained in I. We put $\underline{q}_i = (a_1, \dots, a_i)$ $(0 \le i \le d)$.

LEMMA (2.2). $H^0_{\underline{m}}(A) = U(0) = (0) : I = (0) : a_1$.

Proof. As I is $\underline{m}$-primary, $H^0_{\underline{m}}(A) \supset (0) : I$ and so we have that $H^0_{\underline{m}}(A) = (0) : I$ since $I.H^0_{\underline{m}}(A) = 0$. Because Assh A = Ass A $\setminus \{\underline{m}\}$ (c.f. (2.1)), U(0) has finite length, whence $H^0_{\underline{m}}(A) \supset U(0)$. Since depth A/U(0) > 0 we get the opposite inclusion $H^0_{\underline{m}}(A) \subset U(0)$. Thus $H^0_{\underline{m}}(A) = U(0)$. The equality $(0) : a_1 = (0) : I$ follows from the choice of I. □

COROLLARY (2.3). $U(\underline{q}_i) = \underline{q}_i : I = \underline{q}_i : a_{i+1}$ *for every* $0 \le i < d$. □

THEOREM (2.4) (c.f. [10; (4.2)]). *Let* $0 \le i \le j \le d$ *be integers. Then*

$$U(\underline{q}_i) \cap \underline{q}_j^n = \underline{q}_i \underline{q}_j^{n-1}$$

for every $n > 0$.

Proof. We use induction on i. If i = j there is nothing to prove. Assume that i < j and that our equality holds for i + 1. First of all notice that $U(\underline{q}_i) \cap \underline{q}_j^n \subset U(\underline{q}_{i+1}) \cap \underline{q}_j^n$. In fact if j = i + 1 the $U(\underline{q}_{i+1})$ contains $\underline{q}_j^n$ and the assertion is clearly true. Suppose that i + 1 < j. Then we get, by (2.3), that

$$U(\underline{q}_i) = \underline{q}_i : I \subset U(\underline{q}_{i+1}) = \underline{q}_{i+1} : I.$$

Hence in any case we have the required inclusion.

Now let $f \in U(\underline{q}_i) \cap \underline{q}_j^n$ and write $f = g + a_{i+1}h$ with $g \in \underline{q}_i\underline{q}_j^{n-1}$ and $h \in \underline{q}_j^{n-1}$. Then as $a_{i+1}h \in U(\underline{q}_i)$ and a_{i+1} is $A/U(\underline{q}_i)$-

regular we see that $h \in U(\underline{q}_i)$. If $n = 1$ this implies that $a_{i+1}h \in \underline{q}_i$, whence $f \in \underline{q}_i$. Suppose that $n \geq 2$ and that

$$U(\underline{q}_i) \cap \underline{q}_j^{n-1} = \underline{q}_i\underline{q}_j^{n-2}.$$

Then $h \in \underline{q}_i\underline{q}_j^{n-2}$ as $h \in U(\underline{q}_i) \cap \underline{q}_j^{n-1}$. Hence we have that $a_{i+1}h \in \underline{q}_i\underline{q}_j^{n-1}$, which yields $f \in \underline{q}_i\underline{q}_j^{n-1}$ as required. Therefore $U(\underline{q}_i) \cap \underline{q}_j^{n} \subset \underline{q}_i\underline{q}_j^{n-1}$. The opposite inclusion is trivial. □

COROLLARY (2.5). $U(0) \cap \underline{q}_i = (0)$ *for every* $0 \leq i \leq d$. □

We close this section with the following three remarks. They are known. However we shall need them so frequently that we give proofs for completeness.

PROPOSITION (2.6). *Let* a *be an element of* $\underline{m}$ *and assume that* $\dim A/aA = d - 1$. *Then*

(1) *the length* $\ell_A((0) : a)$ *is finite;*

(2) *there is an exact sequence*

$$\begin{aligned} 0 \longrightarrow (0) : a \longrightarrow H^0_{\underline{m}}(A) &\xrightarrow{a} H^0_{\underline{m}}(A) \longrightarrow H^0_{\underline{m}}(A/aA) \\ \longrightarrow H^1_{\underline{m}}(A) &\xrightarrow{a} H^1_{\underline{m}}(A) \longrightarrow H^1_{\underline{m}}(A/aA) \longrightarrow \cdots \\ \longrightarrow H^i_{\underline{m}}(A) &\xrightarrow{a} H^i_{\underline{m}}(A) \longrightarrow H^i_{\underline{m}}(A/aA) \longrightarrow \cdots \end{aligned}$$

of local cohomology modules;

(3) *the local ring* A/aA *again has finite local cohomology.*

Proof. As $\operatorname{Assh} A = \operatorname{Ass} A \setminus \{\underline{m}\}$ the element a is contained in no associated prime ideal $\underline{p}$ of A such that $\underline{p} \neq \underline{m}$; hence $(0) : a$ has finite length. Now consider the exact sequence

$$0 \longrightarrow (0) : a \longrightarrow A \xrightarrow{a} A \longrightarrow A/aA \longrightarrow 0$$

and split it into the two short exact sequences

$$0 \longrightarrow (0) : a \longrightarrow A \xrightarrow{f} aA \longrightarrow 0, \qquad \text{(a)}$$

$$0 \longrightarrow aA \xrightarrow{i} A \longrightarrow A/aA \longrightarrow 0. \qquad \text{(b)}$$

Apply the functors $H^i_{\underline{m}}(.)$ to (a) and obtain isomorphisms

$$H^i_{\underline{m}}(A) \overset{f}{\cong} H^i_{\underline{m}}(aA) \quad (i \geq 1)$$

and a short exact sequence

$$0 \longrightarrow (0) : a \longrightarrow H^0_{\underline{m}}(A) \xrightarrow{f} H^0_{\underline{m}}(aA) \longrightarrow 0 \qquad \text{(c)}$$

because the length of $(0) : a$ is finite. Similarly we get a long exact sequence

$$\begin{aligned} 0 \longrightarrow H^0_{\underline{m}}(aA) &\xrightarrow{i} H^0_{\underline{m}}(A) \longrightarrow H^0_{\underline{m}}(A/aA) \\ \longrightarrow H^1_{\underline{m}}(aA) &\xrightarrow{i} H^1_{\underline{m}}(A) \longrightarrow H^1_{\underline{m}}(A/aA) \longrightarrow \cdots \qquad\qquad (d) \\ \longrightarrow H^i_{\underline{m}}(aA) &\xrightarrow{i} H^i_{\underline{m}}(A) \longrightarrow H^i_{\underline{m}}(A/aA) \longrightarrow \cdots \end{aligned}$$

of local cohomology modules which comes from the sequence (b). Therefore, replacing $H^i_{\underline{m}}(aA)$ in the sequence (d) by $H^i_{\underline{m}}(A)$ for $i \geq 1$ and combining the sequence (c) with the resulting one, we obtain the required exact sequence

$$\begin{aligned} 0 \longrightarrow (0) : a \longrightarrow H^0_{\underline{m}}(A) &\xrightarrow{a} H^0_{\underline{m}}(A) \longrightarrow H^0_{\underline{m}}(A/aA) \\ \longrightarrow H^1_{\underline{m}}(A) &\xrightarrow{a} H^1_{\underline{m}}(A) \longrightarrow H^1_{\underline{m}}(A/aA) \longrightarrow \cdots \\ \longrightarrow H^i_{\underline{m}}(A) &\xrightarrow{a} H^i_{\underline{m}}(A) \longrightarrow H^i_{\underline{m}}(A/aA) \longrightarrow \cdots . \end{aligned}$$

(Note that the triangle

$$\begin{array}{ccc} H^i_{\underline{m}}(aA) & \xrightarrow{\quad i \quad} & H^i_{\underline{m}}(A) \\ {\scriptstyle f}\nwarrow & & \nearrow{\scriptstyle a} \\ & H^i_{\underline{m}}(A) & \end{array}$$

is commutative for every i.) The last assertion (3) follows from this sequence. □

COROLLARY (2.7). *Suppose that* $d = 2$ *and that* depth $A > 0$. *Let* $J = (0) : H^1_{\underline{m}}(A)$. *Then*

$$(a) : b = (a) : J$$

for every system a,b *of parameters contained in* J. *(Hence one may take* J *to be an ideal* I *obtained from* (2.1)(2).)

Proof. By the sequence of (2.6)(2) we have that $H^0_{\underline{m}}(A/aA) \cong H^1_{\underline{m}}(A)$. On the other hand because $\ell_A((a) : b/(a))$ is finite by (2.6)(1) we get that $(a) : b/(a) \subset H^0_{\underline{m}}(A/aA)$; hence $J.[(a) : b] \subset (a)$, that is $(a) : b$ is contained in $(a) : J$. The opposite inclusion is trivial. □

COROLLARY (2.8). *Let* a *be a regular element of* A *and assume that* $d \geq 3$. *Then* A *is a Cohen-Macaulay ring if*

$$H^i_{\underline{m}}(A/aA) = (0)$$

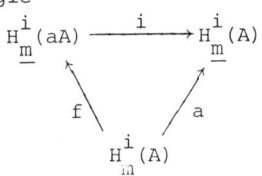

for every $1 \leq i \leq d - 2$.

Proof. Considering the exact sequence in (2.6)(2) we find that the homomorphism $H^1_{\underline{m}}(A) \xrightarrow{a} H^1_{\underline{m}}(A)$ is onto and that a is a non-zerodivisor on $H^i_{\underline{m}}(A)$ for all $2 \leq i \leq d - 1$. Because $H^i_{\underline{m}}(A)$ has finite length for $i \neq d$ by our standard assumption, these facts yield that $H^i_{\underline{m}}(A) = (0)$ for all $i \neq d$. Thus A is a Cohen-Macaulay ring. □

3. *Proof of Theorem* (1.1)

In this section assume that $d = \dim A > 0$ and let $\underline{q} = (a_1, a_2, \ldots, a_d)$ be a parameter ideal of A. We shall maintain the following notation:

$R = \bigoplus_{n \geq 0} \underline{q}^n$, the Rees algebra of $\underline{q}$;

$G = \bigoplus_{n \geq 0} \underline{q}^n/\underline{q}^{n+1}$, the associated graded ring of $\underline{q}$;

$M = \underline{m}R + R_+$, the unique graded maximal ideal of R;

$N = \underline{m}G + G_+$, the unique graded maximal ideal of G.

Also, we shall denote by $H^i_M(R)$ (respectively $H^i_N(G)$) the local cohomology modules of R (respectively G) relative to M (respectively N).

We note the following.

LEMMA (3.1). *Let* P *be a prime ideal of a Noetherian graded ring* $S = \bigoplus_{n \in \mathbb{Z}} S_n$ *and let* P* *denote the largest graded ideal of* S *contained in* P. *Then* P* *is again a prime ideal of* S *and* S_P *is a Cohen-Macaulay (respectively Gorenstein) local ring if and only if* S_{P^*} *is.*

Proof. See [11; (1.1.3)]. □

PROPOSITION (3.2). *Suppose that* Proj G *is Cohen-Macaulay. Then the length* $\ell_G(H^i_N(G))$ *of the local cohomology module is finite for all* $i \neq d$. *Moreover* A *has finite local cohomology.*

Proof. First of all notice that the local ring G_P is Cohen-Macaulay for every prime ideal P of G such that $P \neq N$. (This follows, by (3.1), from our assumption that Proj G is Cohen-Macaulay.) Then we get that the length $\ell_G(H^i_N(G))$ is finite for $i \neq d$ because $\underline{m}G$ is a unique minimal prime ideal of G and $\dim G/\underline{m}G = d$ (c.f. [19; (2.5) and (3.8)]; recall that

$$H^i_N(G) = H^i_{NG_N}(G_N)).$$

Let $f_i = a_i \bmod \underline{q}^2$ $(1 \le i \le d)$. Then by virtue of (3.3) in [19] we see that

$$\sup_{n_1,\ldots,n_d > 0}\left[\ell_G(G/(f_1^{n_1},\ldots,f_d^{n_d}) - e_{(f_1^{n_1},\ldots,f_d^{n_d})}(G)\right] < \infty \qquad (\neq)$$

where $e_{(f_1^{n_1},\ldots,f_d^{n_d})}(G)$ denotes the multiplicity of the ideal $(f_1^{n_1},\ldots,f_d^{n_d})$ in G. On the other hand we have that

$$\ell_G(G/(f_1^{n_1},\ldots,f_d^{n_d}) \ge \ell_A(A/(a_1^{n_1},\ldots,a_d^{n_d}))$$

and that

$$\begin{aligned} e_{(f_1^{n_1},\ldots,f_d^{n_d})}(G) &= \prod_{i=1}^{d} n_i \cdot e_{(f_1,\ldots,f_d)}(G) \\ &= \prod_{i=1}^{d} n_i \cdot e_{\underline{q}}(A) = e_{(a_1^{n_1},\ldots,a_d^{n_d})}(A) \end{aligned}$$

for all integers $n_1, n_2, \ldots, n_d > 0$. Hence by the inequality $(\neq)$ we see that

$$\sup_{n_1,\ldots,n_d > 0}\left[\ell_A(A/(a_1^{n_1},\ldots,a_d^{n_d}) - e_{(a_1^{n_1},\ldots,a_d^{n_d})}(A)\right] < \infty$$

which yields, again by (3.3) of [19], that A has finite local cohomology. □

COROLLARY (3.3). *Suppose that* Proj R *is Cohen-Macaulay. Then* A *has finite local cohomology.* □

LEMMA (3.4). *Suppose that* A *has finite local cohomology and let* I *be an ideal of* A *satisfying condition* (2) *of* (2.1). *Assume that* $a_1, a_2, \ldots, a_d$ *are contained in* I. *Then*

$$a_1, a_2/a_1, \ldots, a_d/a_1$$

is a regular sequence in $A[x/a_1 \mid x \in \underline{q}]$.

Proof. We put $B = A[x/a_1 \mid x \in \underline{q}]$. Let f be an element of B and assume that

$$a_{k+1}/a_1 \cdot f = a_1 \cdot f_1 + a_2/a_1 \cdot f_2 + \ldots + a_k/a_1 \cdot f_k$$

for some $f_i \in B$. First of all let us write $f = x/a_1^n$ and $f_i = x_i/a_1^n$

where $x, x_i \in \underline{q}^n$ with n a sufficiently large integer. Then we have an equation

$$a_{k+1}x = a_1^2x_1 + a_2x_2 +\ldots+ a_kx_k \qquad (\neq)$$

in B, whence

$$a_{k+1}x - (a_1^2x_1 + a_2x_2 +\ldots+ a_kx_k) \in (0) : a_1^r$$

for some $r > 0$. Because $((0) : a_1^r) \cap (a_1,\ldots,a_{k+1}) = (0)$ by (2.2) and (2.5) we actually get the equation ($\neq$) in A. Therefore $x_1 \in (a_2,\ldots,a_{k+1}) : a_1^2$ and hence we may write

$$x_1 = a_2y_2 +\ldots+ a_{k+1}y_{k+1} \qquad (\neq\neq)$$

with $y_i \in \underline{q}^{n-1}$, because $((a_2,\ldots,a_{k+1}) : a_1^2) \cap \underline{q}^n = (a_2,\ldots,a_{k+1})\underline{q}^{n-1}$ by (2.4). Now consider both the equations ($\neq$) and ($\neq\neq$). Then we find that

$$a_{k+1}(x - a_1^2y_{k+1}) \in (a_2,\ldots,a_k),$$

which allows us to write

$$x = a_1^2y_{k+1} + a_2z_2 +\ldots+ a_kz_k$$

with $z_i \in \underline{q}^{n-1}$ since $((a_2,\ldots,a_k) : a_{k+1}) \cap \underline{q}^n = (a_2,\ldots,a_k)\underline{q}^{n-1}$. Thus

$$f = a_1.y_{k+1}/a_1^{n-1} + a_2/a_1.z_2/a_1^{n-1} +\ldots+ a_k/a_1.z_k/a_1^{n-1}$$

in B, whence $f \in (a_1, a_2/a_1,\ldots,a_k/a_1)B$. Therefore $a_1, a_2/a_1,\ldots,a_d/a_1$ is a B-regular sequence. □

PROPOSITION (3.5). *Let* $k = A/\underline{m}$ *and suppose that* A *is a Buchsbaum ring. Let* $\bar{k}/k$ *be an extension of fields. Then there exists a Buchsbaum local* A-*algebra* $\bar{A}$ *with maximal ideal* $\bar{\underline{m}}$ *such that* (a) $\bar{A}$ *is* A-*flat*, (b) $\bar{\underline{m}} = \underline{m}\bar{A}$ *and* (c) $\bar{k} \cong \bar{A}/\bar{\underline{m}}$ *as* k-*algebras*.

Proof. Passing to the completion of A we may assume that A is a homomorphic image of a regular local ring R, say A = R/I. Choose a Noetherian local R-algebra S such that (a) S is R-flat, (b) $\underline{m}_S = \underline{m}_R S$ (here $\underline{m}_R$ (respectively $\underline{m}_S$) denotes the maximal ideal of R (respectively S)) and (c) $\bar{k} \cong S/\underline{m}_S$ as k-algebras. (Such an R-algebra S must exist: see, for example, [12; Chapter 0, (10.3.1)].) Let $\bar{A} = S \otimes_R A$. Then because A is Buchsbaum we see by Satz 1 of [20] that the canonical homomorphisms

$$\mathrm{Ext}_R^i(R/\underline{m}_R, A) \to H^i_{\underline{m}_R}(A)$$

are surjective for all $i \neq d$; hence so also are the homomorphisms

$$\mathrm{Ext}^i_S(S/\underline{m}_S,\overline{A}) = S \otimes_R \mathrm{Ext}^i_R(R/\underline{m}_R,A) \longrightarrow H^i_{\underline{m}_S}(\overline{A}) = S \otimes_R H^i_{\underline{m}_R}(A)$$

for all $i \neq d$. Therefore we get by Theorem 1 of [22] that $\overline{A}$ is a Buchsbaum ring and it is clear that $\overline{A}$ satisfies all the requirements (a), (b) and (c). □

Proof of Theorem (1.1) ((1) ⇒ (2)). Passing to $A/H^0_{\underline{m}}(A)$ we may assume that A is a Buchsbaum ring. Let $\overline{k}$ denote the algebraic closure of $k = A/\underline{m}$ and choose a Buchsbaum overring $\overline{A}$ of A so that $\overline{A}$ satisfies the conditions in (3.5). Let $\overline{R} = \overline{A} \otimes_A R$. Then because the induced morphism $\mathrm{Proj}\ \overline{R} \longrightarrow \mathrm{Proj}\ R$ is flat, to prove that Proj R is Cohen-Macaulay it is enough to show that $\mathrm{Proj}\ \overline{R}$ is Cohen-Macaulay. Therefore we may assume that the residue class field k of A is algebraically closed. As

$$\mathrm{Proj}\ R = \bigcup_{i=1}^{d} \mathrm{Spec}\ A[x/a_i \mid x \in \underline{q}]$$

it suffices to prove that $A[x/a_i \mid x \in \underline{q}]$ is a Cohen-Macaulay ring for every $1 \leq i \leq d$. Of course one may assume without loss of generality that $i = 1$.

Now let M be a maximal ideal of $B = A[x/a_1 \mid x \in \underline{q}]$. We put $\underline{p} = M \cap A$. If $a_1 \notin \underline{p}$ then $A_{\underline{p}} = B_M$, whence B_M is a Cohen-Macaulay ring. Suppose that $a_1 \in \underline{p}$. Then $M \supset \underline{m}B$ as $\underline{q}B = a_1B$ and therefore M determines a maximal ideal $M/\underline{m}B$ of the k-algebra $B/\underline{m}B$. Hence we may write

$$M = \underline{m}B + \left(\frac{a_2 - c_2a_1}{a_1}, \frac{a_3 - c_3a_1}{a_1}, \ldots, \frac{a_d - c_da_1}{a_1}\right)B$$

with $c_i \in A$ $(2 \leq i \leq d)$ because the field k is algebraically closed. Let $b_i = a_i - c_ia_1$ for $2 \leq i \leq d$. Then by (3.4) we see that $a_1, b_2/a_1, \ldots, b_d/a_1$ is a B-regular sequence because

$$\underline{q} = (a_1, b_2, \ldots, b_d)$$

by the choice of c_i. As $M = \sqrt{Q}$ where $Q = (a_1, b_2/a_1, \ldots, b_d/a_1)B$, this guarantees that B_M is a Cohen-Macaulay ring. In any case the local ring B_M is Cohen-Macaulay and we have proved the implication (1) ⇒ (2) in Theorem (1.1). □

In order to prove the converse we need further preliminaries. We put

$$B = A[x/a_1 \mid x \in \underline{q}],$$
$$Q = (a_1, a_2/a_1, \ldots, a_d/a_1)B,$$
$$M = \underline{m}B + Q.$$

Now M is a maximal ideal of B with $A/\underline{m} = B/M$. Note that $M = \sqrt{Q}$. We denote by $e_Q(B)$ the multiplicity of Q in B.

LEMMA (3.6). *Suppose that* A *has finite local cohomology and let* I *be an ideal obtained by* (2.1). *Assume that* $a_1, a_2, \ldots, a_d$ *are contained in* I. *Then*

$$Q \cap A = a_1A + U(a_2, \ldots, a_d)$$

and hence

$$A/a_1A + U(a_2, \ldots, a_d) \cong B/Q$$

as A-*algebras.*

Proof. Let $x \in U(a_2, \ldots, a_d)$ and write $a_1x = a_2x_2 + \ldots + a_dx_d$ with $x_i \in A$. (Recall that $U(a_2, \ldots, a_d) = (a_2, \ldots, a_d) : a_1$. See (2.3).) Then as $x = a_2/a_1.x_2 + \ldots + a_d/a_1.x_d$ in B we have that x is in $Q \cap A$ and therefore $a_1A + U(a_2, \ldots, a_d) \subset Q \cap A$.

Conversely let $f \in Q \cap A$. Let us write

$$f = a_1f_1 + a_2/a_1.f_2 + \ldots + a_d/a_1.f_d$$

with $f_i \in B$ and subsequently let us choose $x_i \in \underline{q}^n$ $(1 \le i \le d)$ so that $f_i = x_i/a_1^n$ where n is a sufficiently large integer. Then

$$a_1^{n+1}f = a_1^2x_1 + a_2x_2 + \ldots + a_dx_d \qquad (\neq)$$

in B, whence

$$a_1^{n+1}f - (a_1^2x_1 + a_2x_2 + \ldots + a_dx_d) \in (0) : a_1^r$$

in A for some integer $r \ge 0$. As $((0) : a_1^r) \cap \underline{q} = (0)$ by (2.5) we find that the equation $(\neq)$ actually holds in A. Now let us write

$$x_1 = a_1^ny_1 + a_2y_2 + \ldots + a_dy_d \qquad (\neq\neq)$$

with $y_i \in \underline{q}^{n-1}$ $(2 \le i \le d)$ and $y_1 \in A$. (This is possible since $x_1 \in \underline{q}^n$ by our choice and $\underline{q}^n = (a_1^n) + (a_2, \ldots, a_d)\underline{q}^{n-1}$.) Then we get a new expression

$$a_1^{n+1}f = a_1^{n+2}y_1 + a_2z_2 + \ldots + a_dz_d$$

for $a_1^{n+1}f$ where $z_i = x_i + a_1^2a_iy_i$ $(2 \le i \le d)$, which yields that $f - a_1y_1$ is in $U(a_2, \ldots, a_d)$ (c.f. (2.3)). Thus $f \in a_1A + U(a_2, \ldots, a_d)$. This completes the proof of the inclusion $Q \cap A \subset a_1A + U(a_2, \ldots, a_d)$. □

PROPOSITION (3.7). *Suppose that* A *has finite local cohomology. Then*

(1) $\dim B_M = d$;

(2) $e_Q(B) = e_{\underline{q}}(A) = \ell_A(A/a_1A + U(a_2,\dots,a_d))$.

Proof. Let I be an ideal of A obtained by (2.1). Choose an integer $n > 0$ such that $\underline{q}^n \subset I$. We put $b_i = a_i^n$ $(1 \le i \le d)$ and $\underline{q}' = (b_1,b_2,\dots,b_d)A$. Further let

$$B' = A[x/b_1 \mid x \in \underline{q}'],$$

$$Q' = (b_1,b_2/b_1,\dots,b_d/b_1)B',$$

$$M' = \underline{m}B' + Q'.$$

Then clearly $B' \subset B$ and B is module-finite over B'. Note that M is the unique prime ideal of B lying over M'. Since $B'_{M'}$ is a Cohen-Macaulay local ring with $\dim B'_{M'} = d$ (c.f. (3.4)), this implies that $\dim B_M = d$. Moreover as $B'/M' = B/M$ $(= A/\underline{m})$ and as B and B' are birational we see that

$$n^d.e_Q(B) = e_{Q'B}(B) = e_{Q'}(B').$$

Thus

$$n^d.e_Q(B) = \ell_{B'}(B'/Q')$$

since $\ell_{B'}(B'/Q') = e_{Q'}(B')$. Therefore by (3.6) we have that

$$n^d.e_Q(B) = \ell_A(A/(b_1) + U(b_2,\dots,b_d)).$$

Because

$$\ell_A(A/(b_1) + U(b_2,\dots,b_d))$$

$$= \ell_A(A/\underline{q}') - \ell_A(U(b_2,\dots,b_d)/(b_2,\dots,b_d))$$

(note that $U(b_2,\dots,b_d) \cap \underline{q}' = (b_2,\dots,b_d)$; see (2.4)) and

$$\ell_A(A/\underline{q}') = e_{\underline{q}'}(A) + \ell_A(U(b_2,\dots,b_d)/(b_2,\dots,b_d))$$

by Corollary 4.8 of [1] (recall that $b_1,b_2,\dots,b_d$ is a reducing system of parameters (see (2.1) or (2.6)) and note that $U(b_2,\dots,b_d) = (b_2,\dots,b_d) : b_1$ by (2.3)), we finally get that

$$n^d.e_Q(B) = e_{\underline{q}'}(A) = n^d.e_{\underline{q}}(A).$$

Thus $e_Q(B) = e_{\underline{q}}(A)$ as required.

Now let us prove that $e_{\underline{q}}(A) = \ell_A(A/a_1A + U(a_2,\dots,a_d))$. We put $\underline{q}_1 = (a_2,\dots,a_d)$. Then because $a_1,a_2,\dots,a_d$ is a reducing system of parameters (c.f. (2.6)) we see that

$$\ell_A(A/\underline{q}) = e_{\underline{q}}(A) + \ell_A(\underline{q}_1 : a_1/\underline{q}_1). \qquad (\neq)$$

CLAIM. $\ell_A(A/\underline{q}) = \ell_A(A/(a_1) + U(\underline{q}_1)) + \ell_A(\underline{q}_1 : a_1/\underline{q}_1)$.

Proof. Note that

$$(a_1A + U(\underline{q}_1))/\underline{q} \cong U(\underline{q}_1)/\underline{q} \cap U(\underline{q}_1) = U(\underline{q}_1)/(a_1A \cap U(\underline{q}_1) + \underline{q}_1)$$

and that $a_1A \cap U(\underline{q}_1) = a_1U(\underline{q}_1)$ as a_1 is $A/U(\underline{q}_1)$-regular. Then we get an isomorphism

$$a_1A + U(\underline{q}_1)/\underline{q} \cong U(\underline{q}_1)/a_1U(\underline{q}_1) + \underline{q}_1.$$

Hence

$$\begin{aligned}\ell_A(A/\underline{q}) &= \ell_A(A/a_1A + U(\underline{q}_1)) + \ell_A(a_1A + U(\underline{q}_1)/\underline{q}) \\ &= \ell_A(A/a_1A + U(\underline{q}_1)) + \ell_A(U(\underline{q}_1)/a_1U(\underline{q}_1) + \underline{q}_1) \\ &= \ell_A(A/a_1A + U(\underline{q}_1)) + \ell_A(U(\underline{q}_1)/\underline{q}_1) + \ell_A(a_1U(\underline{q}_1) + \underline{q}_1/\underline{q}_1).\end{aligned}$$

On the other hand we see that

$$a_1U(\underline{q}_1) + \underline{q}_1/\underline{q}_1 \cong a_1U(\underline{q}_1)/a_1U(\underline{q}_1) \cap \underline{q}_1 \cong U(\underline{q}_1)/\underline{q}_1 : a_1.$$

Therefore we get the required equality

$$\ell_A(A/\underline{q}) = \ell_A(A/a_1A + U(\underline{q}_1)) + \ell_A(\underline{q}_1 : a_1/\underline{q}_1). \ \square$$

It follows from this claim and the equality $(\neq)$ that $\ell_A(A/a_1A + U(\underline{q}_1)) = e_{\underline{q}}(A)$, which completes the proof of (3.7). $\square$

COROLLARY (3.8). *Suppose that* A *has finite local cohomology. Then the local ring* B_M *is Cohen-Macaulay if and only if*

$$U(a_2,\ldots,a_d) \subset (a_2/a_1,\ldots,a_d/a_1)B \cap A.$$

(When this is the case $U(a_2,\ldots,a_d) = (a_2/a_1,\ldots,a_d/a_1)B \cap A$.*)*

Proof. Let $Q' = (a_2/a_1,\ldots,a_d/a_1)B$ and $J = Q' \cap A$. Note that $A/J \cong B/Q'$. Now suppose that B_M is a Cohen-Macaulay ring. Then because $a_2/a_1,\ldots,a_d/a_1$ is a B_M-regular sequence (recall that $a_1, a_2/a_1,\ldots,a_d/a_1$ is a system of parameters of B_M (c.f. (3.7))) we see that A/J is a Cohen-Macaulay ring of dimension 1. Let $\underline{p} \in \mathrm{Supp}_A J/(a_2,\ldots,a_d)$. If $\underline{p} \neq \underline{m}$ then $a_1 \notin \underline{p}$ as $\underline{p} \supset (a_2,\ldots,a_d)$, whence $A_{\underline{p}} = B_{\underline{p}}$. Therefore $A_{\underline{p}} \otimes_A J/(a_2,\ldots,a_d) = (0)$ because $Q'B_{\underline{p}} = (a_2,\ldots,a_d)B_{\underline{p}}$ in our case. This is a contradiction and we conclude that $\underline{p} = \underline{m}$. Hence the length $\ell_A(J/(a_2,\ldots,a_d))$ is finite and we get that

$$H^0_{\underline{m}}(A/(a_2,\ldots,a_d)) = J/(a_2,\ldots,a_d) \qquad (\neq)$$

since depth $A/J = 1$ as we have remarked above. Because

$$H^0_{\underline{m}}(A/(a_2,\ldots,a_d)) = U(a_2,\ldots,a_d)/(a_2,\ldots,a_d)$$

by (2.2) we get that $J = U(a_2,\ldots,a_d)$ as required.

Conversely suppose that $U(a_2,\ldots,a_d) \subset Q' \cap A$. Then B/Q is a homomorphic image of $A/a_1A + U(a_2,\ldots,a_d)$ and hence we have the inequality

$$\ell_B(B/Q) \leq \ell_A(A/a_1A + U(a_2,\ldots,a_d));$$

this implies that $\ell_B(B/Q) = e_Q(B)$ as $\ell_A(A/a_1A + U(a_2,\ldots,a_d)) = e_Q(B)$ by (3.7). Thus B_M is a Cohen-Macaulay ring. □

LEMMA (3.9). *Suppose that* $d = 2$. *Then* $A/H^0_{\underline{m}}(A)$ *is a Buchsbaum ring if* Proj $R(\underline{q})$ *is Cohen-Macaulay for every parameter ideal* $\underline{q}$ *of* A.

Proof. We may assume that depth $A > 0$. Note that $H^1_{\underline{m}}(A)$ has finite length (c.f. (3.3)). We put $I = (0) : H^1_{\underline{m}}(A)$. (By virtue of (2.7) we may consider this ideal I to be an ideal obtained by (2.1).) To prove that A is Buchsbaum it is enough to show that $I \supset \underline{m}$ (c.f. [16; Satz 3]). Assume the contrary and choose a system a,b of parameters for A such that $a \notin I$ but $b \in I$. Let $\underline{q} = (a,b)$ and $B = A[b/a]$. Then applying (3.8) to this situation we get $aU(bA) \subset bB$. Let x be an element of $U(bA)$ and write $ax = bf$ with $f \in B$. Choose an integer $n > 0$ and $y \in \underline{q}^n$ such that $f = y/a^n$ and $a^n \in I$. Because $\underline{q}^n = (a^n) + b\underline{q}^{n-1}$ we may write

$$y = a^n z + bv$$

for some $z \in A$ and $v \in \underline{q}^{n-1}$. Since

$$a^{n+1}x = a^n bz + b^2 v \tag{$\neq$}$$

we see that $v \in U(a^n)$ by (2.3), which implies that $bv \in (a^n)$ (again by (2.3)). Let $bv = a^n w$ with $w \in A$. Then by the equation ($\neq$) we get $a^{n+1}x = a^n(bz + bw)$, whence $ax = b(z + w)$ as a^n is A-regular. Therefore $ax \in bA$ and so we have that $aU(bA) \subset bA$. On the other hand by virtue of (2.2) (respectively (2.6)(2)) we know that

$$U(bA)/bA = H^0_{\underline{m}}(A/bA) \quad \text{(respectively } H^0_{\underline{m}}(A/bA) \cong H^1_{\underline{m}}(A)\text{)},$$

whence $U(bA)/bA \cong H^1_{\underline{m}}(A)$. Thus $a.H^1_{\underline{m}}(A) = (0)$ because $aU(bA) \subset bA$ as we have proved above, that is $a \in I$ - this contradicts the choice of a. Therefore we conclude that $\underline{m}.H^1_{\underline{m}}(A) = (0)$ and hence that A is a Buchsbaum ring. □

Proof of Theorem (1.1) ((2) ⇒ (1)). We may assume that depth $A > 0$. By (3.9) we may further assume that $d \geq 3$ and that our assertion holds for $d - 1$. Let $a_1, a_2, \ldots, a_d$ be a system of parameters for A and put $\underline{q} = (a_1, a_2, \ldots, a_d)$. Let $\bar{A} = A/U(a_d A)$ and denote by $\bar{.}$ reduction mod $U(a_d A)$. Note that $\bar{A}$ has finite local cohomology because A has, by (3.3) (see also (2.6)).

CLAIM 1. $\bar{A}$ *is a Buchsbaum ring.*

Proof. Let $\bar{\underline{q}} = (\bar{a}_1, \bar{a}_2, \ldots, \bar{a}_{d-1})\bar{A}$ and $\bar{B} = \bar{A}[x/\bar{a}_1 \mid x \in \bar{\underline{q}}]$. Then by (3.8) and the induction hypothesis on d, in order to prove that $\bar{A}$ is Buchsbaum it suffices to show that $U(\bar{a}_2, \ldots, \bar{a}_{d-1}) \subset (\bar{a}_2/\bar{a}_1, \ldots, \bar{a}_{d-1}/\bar{a}_1)\bar{B}$. Let x be an element of A and assume that $\bar{x} \in U(\bar{a}_2, \ldots, \bar{a}_{d-1})$. Then as $x \in U(a_2, \ldots, a_d)$ we see by (3.8) that $x \in (a_2/a_1, \ldots, a_d/a_1)B$, whence we may write $a_1^{n+1}x = a_2x_2 + \ldots + a_dx_d$ with $x_i \in \underline{q}^n$. Because

$$\bar{x} = \bar{a}_2/\bar{a}_1 . \bar{x}_2/\bar{a}_1^{n} + \ldots + \bar{a}_{d-1}/\bar{a}_1 . \bar{x}_{d-1}/\bar{a}_1^{n}$$

in $\bar{B}$ we find that $\bar{x} \in (\bar{a}_2/\bar{a}_1, \ldots, \bar{a}_{d-1}/\bar{a}_1)\bar{B}$, as required. □

CLAIM 2. $\underline{m}.H^i_{\underline{m}}(A) = (0)$ $(i \neq d)$ *and*

$$\ell_A(H^i_{\underline{m}}(\bar{A})) = \ell_A(H^i_{\underline{m}}(A)) + \ell_A(H^{i+1}_{\underline{m}}(A)) \quad (1 \leq i \leq d-2).$$

Moreover $U(a_dA)/a_dA \cong H^1_{\underline{m}}(A)$.

Proof. Let I be an ideal satisfying the condition in (2.1) and let $b \in I$ such that $\dim A/bA = d - 1$. Then by (2.6)(2) and the choice of I we get exact sequences

$$0 \longrightarrow H^i_{\underline{m}}(A) \longrightarrow H^i_{\underline{m}}(A/bA) \longrightarrow H^{i+1}_{\underline{m}}(A) \longrightarrow 0 \quad (0 \leq i \leq d-2) \qquad (\neq)$$

of local cohomology modules. Now recall that $\underline{m}.H^i_{\underline{m}}(A/bA) = (0)$ for every $1 \leq i \leq d - 2$ as $H^i_{\underline{m}}(A/bA) \cong H^i_{\underline{m}}(A/U(bA))$ and as $A/U(bA)$ is a Buchsbaum ring by Claim 1. Then we find that $\underline{m}.H^i_{\underline{m}}(A) = (0)$ $(i \neq d)$ by the sequences $(\neq)$. Hence again by (2.6)(2) we obtain exact sequences

$$0 \longrightarrow H^i_{\underline{m}}(A) \longrightarrow H^i_{\underline{m}}(A/a_dA) \longrightarrow H^{i+1}_{\underline{m}}(A) \longrightarrow 0 \quad (0 \leq i \leq d-2);$$

these imply that

$$\ell_A(H^i_{\underline{m}}(A/U(a_dA))) = \ell_A(H^i_{\underline{m}}(A)) + \ell_A(H^{i+1}_{\underline{m}}(A))$$

for every $1 \leq i \leq d - 2$. It is clear that $U(a_dA)/a_dA = H^0_{\underline{m}}(A/a_dA) \cong H^1_{\underline{m}}(A)$. This completes the proof of all the assertions in Claim 2.□

CLAIM 3. $\underline{q} \cap U(a_1A) = a_1A$.

Proof. Let $\underline{q}' = (a_2, \ldots, a_d)$. It suffices to show that $\underline{q}' \cap U(a_1A) \subset a_1A$. Let $x \in \underline{q}' \cap U(a_1A)$. Then

$$\bar{x} \in (\bar{a}_2, \ldots, \bar{a}_{d-1}) \cap U(\bar{a}_1\bar{A})$$

clearly, whence $\bar{x} \in (\bar{a}_1)$ by (2.4). (Recall that $\bar{A}$ is a Buchsbaum ring by Claim 1.) Thus $x \in a_1A + U(a_dA)$. Let us write $x = y + z$ with $y \in a_1A$ and $z \in U(a_dA)$. Then as $z = x - y \in U(a_1A) \cap U(a_dA)$ we may, by Claim 2, write $a_1z = a_du$ and $a_dz = a_1v$ with $u, v \in A$. Thus

$$a_1a_dz = a_1^2v = a_d^2u,$$

whence $u \in (a_1^2) : a_d^2$. Recalling that $(a_1^2) : a_d^2 \subset U(a_1^2A)$ and that $U(a_1^2A) \subset (a_1^2) : \underline{m}$ by Claim 2 we find that $a_du = a_1^2w$ for some $w \in A$. Therefore $a_1z = a_1^2w$ and so $z = a_1w$ as a_1 is A-regular. Thus $x = y + z$ is in a_1A and hence we get that $\underline{q}' \cap U(a_1A) \subset a_1A$ as required. □

Now let us finish the proof of the implication (2) ⇒ (1). First of all consider the exact sequence

$$0 \longrightarrow U(a_dA)/a_dA \longrightarrow A/\underline{q} \longrightarrow \bar{A}/\bar{\underline{q}} \longrightarrow 0$$

which follows from the fact that $U(a_dA) \cap \underline{q} = a_dA$ (see Claim 3 and note the symmetry between a_1 and a_d): we get that

$$\ell_A(A/\underline{q}) = \ell_{\bar{A}}(\bar{A}/\bar{\underline{q}}) + \ell_A(H^1_{\underline{m}}(A))$$

as $U(a_dA)/a_dA \cong H^1_{\underline{m}}(A)$ by Claim 2. On the other hand we already know from Claim 1 that $\bar{A}$ is a Buchsbaum ring. Hence

$$\ell_{\bar{A}}(\bar{A}/\bar{\underline{q}}) = e_{\bar{\underline{q}}}(\bar{A}) + I(\bar{A})$$

where $I(\bar{A})$ denotes the Buchsbaum invariant of $\bar{A}$. Recalling that $e_{\underline{q}}(A) = e_{\bar{\underline{q}}}(\bar{A})$, we conclude that the difference

$$\ell_A(A/\underline{q}) - e_{\underline{q}}(A) = I(\bar{A}) + \ell_A(H^1_{\underline{m}}(A))$$

does not depend on the choice of $\underline{q}$ because

$$I(\bar{A}) = \sum_{i=1}^{d-2} \binom{d-2}{i} \cdot \ell_A(H^i_{\underline{m}}(\bar{A}))$$

by Satz 2 of [16] and because

$$\ell_A(H^i_{\underline{m}}(\bar{A})) = \ell_A(H^i_{\underline{m}}(A)) + \ell_A(H^{i+1}_{\underline{m}}(A))$$

$(1 \le i \le d-2)$ by Claim 2. Thus A is a Buchsbaum ring by definition. This completes the proof of Theorem (1.1). □

4. *Proof of Corollary* (1.2)

As in Section 3 assume that $d = \dim A > 0$ and let

$\underline{q} = (a_1, a_2, \ldots, a_d)$

be a parameter ideal of A. We shall preserve the notation of Section 3 and identify the Rees algebra $R = R(\underline{q})$ with the A-subalgebra $A[a_1X, a_2X, \ldots, a_dX]$ of $A[X]$ where X denotes an indeterminate over A.

First of all let us recall the following (known) fact [15]. We shall give a proof since we shall use this proof once more.

PROPOSITION (4.1). *Suppose that* A *is a Cohen-Macaulay ring. Then* A *is a Gorenstein ring if and only if the ring*

$B = A[x/a_1 \mid x \in \underline{q}]$

is Gorenstein.

Proof. Let $T = A[X_1, X_2, \ldots, X_d]$ denote a polynomial ring and let $f : T \to R$ be the A-algebra map defined by $f(X_i) = a_iX$ for $1 \le i \le d$. Then $K = \mathrm{Ker}\, f$ is a perfect ideal of grade $d - 1$ and generated by 2×2 minors of the matrix

$$\begin{pmatrix} a_1 & a_2 & \cdots & a_d \\ X_1 & X_2 & \cdots & X_d \end{pmatrix}$$

(c.f. [2; p. 93]). We put $R' = R[1/a_1X]$ and $T' = T[1/X_1]$. Then

$KT' = (a_1X_i - a_iX_1 \mid 2 \le i \le d)$

clearly and therefore $\{a_1X_i - a_iX_1\}_{2 \le i \le d}$ forms a T'-regular sequence. (Recall that KT' is still a perfect ideal of grade $d - 1$.) Thus $R' = T'/KT'$ is a complete intersection in T' and so, recalling that $R' = B[a_1X, 1/a_1X]$ and that a_1X is algebraically independent over B, we conclude that B is Gorenstein if and only if A is.

We note the following.

PROPOSITION (4.2). *The following conditions are equivalent:*

(1) Proj R *is Gorenstein;*

(2) $A[x/a_i \mid x \in \underline{q}]$ *are Gorenstein rings for all* $1 \le i \le d$;

(3) R_P *is a Gorenstein local ring for every prime ideal* P *of* R *such that* $P \ne M$.

When this is the case $A_{\underline{p}}$ *is a Gorenstein local ring for every prime ideal* $\underline{p}$ *of* A *such that* $\underline{p} \neq \underline{m}$.

Proof. (1) ⇔ (2) ⇐ (3) These are well known.

(1) ⇒ (3) See (3.1).

The last assertion follows from the fact that Spec A \ Spec A/$\underline{q}$ ≅ Proj R \ Proj G. □

For the moment assume that A is complete and let E denote the injective envelope of A/$\underline{m}$. We put

$$K_A = \mathrm{Hom}_A(H^d_{\underline{m}}(A), E)$$

and call it the canonical module of A. Various properties of canonical modules are discussed in [13] and [14]. Here let us summarize some of them which we shall need in the sequel.

PROPOSITION (4.3). (1) $(0) : K_A = U(0)$.

(2) $\mathrm{depth}_A K_A \geq 2$ *if* $\dim A \geq 2$.

(3) *Suppose that* A *is Cohen-Macaulay. Then* A *is a Gorenstein ring if and only if* $K_A \cong A$.

Proof. (1) See [13; Proposition 6.6(7)].

(2) See, for example, [4; proof of Lemma 2.4].

(3) See [14; Satz 5.9].

Proof of Corollary (1.2). We may assume that depth A > 0.

(1) ⇒ (2) This follows from (4.1) and (4.2).

(2) ⇒ (1) We use induction on d. By virtue of (4.1) and (4.2) it is enough to show that A is a Cohen-Macaulay ring. Notice that A is at least Buchsbaum (c.f. (1.1)).

If d = 1 we have nothing to prove. Suppose that d = 2 and let a,b be a system of parameters for A. Then by (3.4) and (3.6) we see that A/aA + U(bA) is a Gorenstein ring as also is the ring B = A[b/a]. Hence A/U(bA) is Gorenstein because a is regular on A/U(bA). Let E (respectively $\hat{A}$) denote the injective envelope of A/$\underline{m}$ (respectively the completion of A) and apply the functor $\mathrm{Hom}_A(\cdot, E)$ to the exact sequence

$$0 \longrightarrow H^1_{\underline{m}}(A) \longrightarrow H^1_{\underline{m}}(A/bA) \longrightarrow H^2_{\underline{m}}(A) \xrightarrow{\ b\ } H^2_{\underline{m}}(A) \longrightarrow 0$$

of local cohomology modules, which comes from the short exact sequence

$$0 \longrightarrow A \xrightarrow{b} A \longrightarrow A/bA \longrightarrow 0$$

and the fact that $b.H^1_{\underline{m}}(A) = (0)$. Then we get an exact sequence

$$0 \longrightarrow K_{\hat{A}}/bK_{\hat{A}} \longrightarrow K_{\hat{A}/b\hat{A}} \longrightarrow \mathrm{Hom}_A(H^1_{\underline{m}}(A),E) \longrightarrow 0 \qquad (\neq)$$

of $\hat{A}$-modules. Recall that $H^1_{\underline{m}}(A/bA) \cong H^1_{\underline{m}}(A/U(bA))$; and we see by (4.3)(3) that

$$K_{\hat{A}/b\hat{A}} \cong \hat{A}/U(bA)\hat{A}$$

because $A/U(bA)$ is a Gorenstein ring as we have remarked above.

Now assume that A is not Cohen-Macaulay. Then since $\underline{m}.H^1_{\underline{m}}(A) = (0)$ and $K_{\hat{A}/b\hat{A}}$ is cyclic, we obtain from the sequence $(\neq)$ that

$$\mathrm{Hom}_A(H^1_{\underline{m}}(A),E) \cong A/\underline{m},$$

which implies, again by the sequence $(\neq)$, that

$$K_{\hat{A}}/bK_{\hat{A}} \cong \underline{m}\hat{A}/U(bA)\hat{A} \qquad (\neq\neq)$$

as $\hat{A}$-modules. Hence the embedding dimension $v(A/U(bA))$ of $A/U(bA)$ does not depend on the choice of b. This is impossible. (For example, if b is chosen so that $b \notin \underline{m}^2$ then $v(A/U(bA)) < v(A)$ clearly (here $v(A)$ denotes the embedding dimension of A). On the other hand as $U(b^2A) = bU(bA)$, we must have $v(A/U(b^2A)) = v(A)$.) Hence A is a Cohen-Macaulay ring.

Suppose that $d \geq 3$ and that our assertion holds for $d - 1$. Let $\underline{q} = (a_1, a_2, \ldots, a_d)$ be a parameter ideal of A and put $a = a_1$. We put $\bar{A} = A/U(aA)$, $\bar{\underline{q}} = \underline{q}\bar{A}$ and $\bar{R} = R(\bar{\underline{q}})$ (the Rees algebra of $\bar{\underline{q}}$). Then by virtue of (4.4) in [9] we have an exact sequence

$$0 \longrightarrow {}_hU(aA) \longrightarrow R/(aX) \longrightarrow \bar{R} \longrightarrow 0 \qquad (\neq\neq\neq)$$

of R-modules, where ${}_hU(aA)$ denotes $U(aA)$ considered as an R-module via the canonical projection $h : R \to A$.

Now let us show that $\mathrm{Proj}\,\bar{R}$ is Gorenstein. Let $f : R \to \bar{R}$ denote the canonical homomorphism of rings. Let $\bar{Q}$ be a prime ideal of $\bar{R}$ such that $\bar{Q} \neq \bar{M}$ (here $\bar{M} = \underline{m}\bar{R} + (\bar{R})_+$ denotes the unique graded maximal ideal of $\bar{R}$) and put $Q = f^{-1}(\bar{Q})$. Then $Q \neq M$, whence $R_Q/(aX)R_Q$ is a Gorenstein ring as aX is R_Q-regular. Therefore, because $[{}_hU(aA)]_Q = (0)$ if $Q \not\supset R_+$, we get from the sequence $(\neq\neq\neq)$ that $\bar{R}_{\bar{Q}}$ is a Gorenstein ring in this case. If $Q \supset R_+$ then $\bar{Q} \not\supset \underline{m}\bar{A}$

and therefore $\bar{R}_{\bar{Q}}$ is obtained by localization from $\bar{R}_{\bar{p}} = \bar{A}_{\bar{p}}[X]$ where $\bar{p} = \bar{Q} \cap \bar{A}$. Let $\underline{p} = Q \cap A$. Then as $\bar{A}_{\bar{p}} = A_{\underline{p}}/aA_{\underline{p}}$ is a Gorenstein ring we get that the ring $\bar{R}_{\bar{Q}}$ is also Gorenstein in this case. Thus Proj $\bar{R}$ is Gorenstein (c.f. (4.2)) whence $\bar{A}$ is a Gorenstein local ring by the induction hypothesis on d. In particular $\bar{A}$ is a Cohen-Macaulay ring and so we find that

$$H^i_{\underline{m}}(A/bA) = (0)$$

for $1 \leq i \leq d - 2$ since $H^i_{\underline{m}}(A/bA) \cong H^i_{\underline{m}}(\bar{A})$ for every $i > 0$. Thus by (2.8) we get that A itself is a Cohen-Macaulay ring and this completes the proof of Corollary (1.2). □

COROLLARY (4.4). *Suppose that* A *is a homomorphic image of a regular ring. Then the following conditions are equivalent*:

(1) $A/H^0_{\underline{m}}(A)$ *is a complete intersection*;

(2) Proj $R(\underline{q})$ *is locally a complete intersection for every parameter ideal* $\underline{q}$ *of* A.

Proof. By passing to $A/H^0_{\underline{m}}(A)$ we may assume that depth $A > 0$. By virtue of (1.2) we may further assume that A is a Cohen-Macaulay ring. Therefore the assertion follows from the proof of (4.1). □

Remark (4.5). The implication (1) ⇒ (2) in (4.4) appeared in [15] (c.f. Theorem 2).

We shall close this paper with the next remark.

PROPOSITION (4.6). *The following conditions are equivalent*:

(1) $A/H^0_{\underline{m}}(A)$ *is a regular local ring*;

(2) Proj $R(\underline{q})$ *is smooth for some parameter ideal* $\underline{q}$ *of* A.

When this is the case $\underline{m} = \underline{q} + H^0_{\underline{m}}(A)$ *if* $d \geq 2$.

To prove this we need one more lemma.

LEMMA (4.7). *Suppose that* A *has finite local cohomology and that* depth $A > 0$. *Then* A *is a regular local ring if* $A/U(a_2,\ldots,a_d)$ *is a DVR for some subsystem* $a_2, a_3, \ldots, a_d$ *of parameters for* A.

Proof. We use induction on d. We may assume that A is complete. It is enough to show that A is Cohen-Macaulay. If $d = 1$ there is nothing to prove. Consider the case in which $d = 2$ and put $a = a_2$. Let E denote the injective envelope of $A/\underline{m}$ and take the E-dual of the exact sequence

$$H^1_{\underline{m}}(A/aA) \longrightarrow H^2_{\underline{m}}(A) \xrightarrow{a} H^2_{\underline{m}}(A) \longrightarrow 0$$

of local cohomology modules which is induced by the exact sequence

$$0 \longrightarrow A \xrightarrow{a} A \longrightarrow A/aA \longrightarrow 0.$$

Then we get an embedding

$$0 \longrightarrow K_A/aK_A \longrightarrow K_{A/aA}. \qquad (\neq)$$

On the other hand, recalling that $H^1_{\underline{m}}(A/aA) \cong H^1_{\underline{m}}(A/U(aA))$, we see that $K_{A/aA} \cong A/U(aA)$ because $A/U(aA)$ is a Gorenstein ring by our assumption. Therefore K_A/aK_A is isomorphic to some ideal of $A/U(aA)$, whence K_A must be cyclic as $A/U(aA)$ is a DVR. Thus by (4.3)(1),(2) we find that A is Cohen-Macaulay in this case.

Now assume that $d \geq 3$ and that our assertion holds for $d - 1$. Let $a = a_2$ and put $\bar{A} = A/U(aA)$. Let $\bar{\cdot}$ denote reduction mod $U(aA)$. Then since

$$\bar{A}/U(\bar{a}_3,\ldots,\bar{a}_d) \cong A/U(a_2,\ldots,a_d)$$

(note that $U(a_2,\ldots,a_d) \supset U(aA)$; see (2.2) and (2.6)) and since $\bar{A}$ also has finite local cohomology we get by the induction hypothesis on d that $\bar{A}$ is regular. Therefore $H^i_{\underline{m}}(A/aA) = (0)$ for $1 \leq i \leq d - 2$ as $H^i_{\underline{m}}(A/aA) \cong H^i_{\underline{m}}(\bar{A})$ for all $i \geq 1$. Hence A is a Cohen-Macaulay ring by (2.8) and this completes the proof of (4.7). □

Proof of (4.6). We may assume that depth $A > 0$.

(1) ⇒ (2) Take $\underline{q} = \underline{m}$. (It is well known that Proj $R(\underline{m})$ is smooth.)

(2) ⇒ (1) Let $\underline{q} = (a_1,a_2,\ldots,a_d)$ be a parameter ideal of A such that Proj $R(\underline{q})$ is smooth. Notice that A has finite local cohomology (c.f. (3.3)). We put $B = A[x/a_1 \mid x \in \underline{q}]$ and $Q' = (a_2/a_1,\ldots,a_d/a_1)B$. Let $M = \underline{m}B + Q'$. Then it is easy to check that $a_2/a_1,\ldots,a_d/a_1$ mod M^2 are linearly independent over B/M. Hence $B_M/Q'B_M$ is a DVR and therefore, recalling that $A/U(a_2,\ldots,a_d) \cong B/Q'$ (c.f. (3.8)), we see that $A/U(a_2,\ldots,a_d)$ is a DVR. Thus the assertion follows from (4.7).

Now consider the last assertion. Since $A/(a_j \mid j \neq i)$ is a DVR for every $1 \leq i \leq d$ we get that $a_i \notin \underline{m}^2$ for all $1 \leq i \leq d$. Choose $c \in \underline{m}$ such that $\underline{m} = (c) + (a_2,\ldots,a_d)$ and let us write

$a_1 = cx + \sum_{i=2}^{d} a_i x_i$ with $x, x_i \in A$. Then $\underline{q} = (cx, a_2, \ldots, a_d)$. If $x \in \underline{m}$ then $a_1' = cx \in \underline{m}^2$ - this contradicts the fact that no member of a minimal system of generators of $\underline{q}$ is contained in $\underline{m}^2$. Thus $x \notin \underline{m}$ whence $\underline{q} = \underline{m}$ as claimed. □

Acknowledgement

The author was supported by a Grant-in-Aid for Co-operative Research.

References

1. M.Auslander and D.A.Buchsbaum, "Codimension and multiplicity", Ann. of Math., 68 (1958), 625-657.

2. J.Barshay, "Graded algebras of powers of ideals generated by A-sequences", J. Algebra, 25 (1973), 90-99.

3. D.A.Buchsbaum, "Complexes in local ring theory", Some aspects of ring theory, (C. I. M. E., Rome, 1965).

4. R.Fossum, H.-B.Foxby, P.Griffith and I.Reiten, "Minimal injective resolutions with applications to dualizing modules and Gorenstein modules", Publications Mathématiques 45 (Institut des Hautes Études Scientifiques, Paris, 1975), pp.193-213.

5. S.Goto, "On Buchsbaum rings", J. Algebra, 67 (1980), 272-279.

6. S.Goto, "On the Cohen-Macaulayfication of certain Buchsbaum rings", Nagoya Math. J., 80 (1980), 107-116.

7. S.Goto, "On Buchsbaum rings obtained by gluing", Nagoya Math. J., 83 (1981), 123-135.

8. S.Goto, "Buchsbaum rings with multiplicity 2", J. Algebra, 74 (1982), 494-508.

9. S.Goto, "Buchsbaum rings of maximal embedding dimension", J. Algebra, to appear.

10. S.Goto and Y.Shimoda, "On Rees algebras over Buchsbaum rings", J. Math. Kyoto Univ., 20 (1980), 691-708.

11. S.Goto and K.Watanabe, "On graded rings, I", J. Math. Soc. Japan, 30 (1978), 179-213.

12. A.Grothendieck, Éléments de géométrie algébrique, Publications Mathématiques 11 (Institut des Hautes Études Scientifiques, Paris, 1965).

13. A.Grothendieck, Local cohomology, Lecture Notes in Mathematics 41 (Springer, Berlin, Heidelberg, New York, 1967).

14. J.Herzog and E.Kunz, Der kanonische Modul eines Cohen-Macaulay-Rings, Lecture Notes in Mathematics 238 (Springer, Berlin, Heidelberg, New York, 1971).

15. T.Matsuoka, "Some remarks on a certain transformation of Macaulay rings", J. Math. Kyoto Univ., 11 (1971), 301-309.

16. B.Renschuch, J.Stückrad and W.Vogel, "Weitere Bemerkungen zu einem der Schnittheorie und über ein Maß von A.Seidenberg für die Imperfektheit", J. Algebra, 37 (1975), 447-471.

17. P.Schenzel, "Applications of dualizing complexes to Buchsbaum rings", Adv. in Math., 44 (1982), 61-77.

18. P.Schenzel, "On Buchsbaum rings and their canonical modules", preprint, Martin Luther Universität, 1981.

19. P.Schenzel, N.V.Trung and N.T.Cuong, "Verallgemeinerte Cohen-Macaulay-Moduln", Math. Nachr., 85 (1978), 57-73.

20. J.Stückrad, "Über die kohomologische Charakterisierung von Buchsbaum-Moduln", Math. Nachr., 95 (1980), 265-272.

21. J.Stückrad and W.Vogel, "Eine Verallgemeinerung der Cohen-Macaulay Ringe und Anwendungen auf ein Problem der Multiplizitätstheorie", J. Math. Kyoto Univ., 13 (1973), 513-528.

22. J.Stückrad and W.Vogel, "Toward a theory of Buchsbaum singularities", Amer. J. Math., 100 (1978), 727-746.

23. N.Suzuki, "On the Koszul complex generated by a system of parameters for a Buchsbaum module", Science Reports of Shizuoka College of Pharmacy, Department of General Education, 8 (1979), 27-35.

24. W.Vogel, "Über eine Vermutung von D.A.Buchsbaum", J. Algebra, 25 (1973), 106-112.

Department of Mathematics,
Nihon University,
Sakurajosui 3-25-40,
Setagaya-ku,
Tokyo, Japan.

NECESSARY CONDITIONS FOR AN ANALYTIC ALGEBRA TO BE STRICT

J.HERZOG

In [3] we introduced the notion of a strict local ring. Here we slightly modify the definition and generalize the result of [3]. For technical reasons we restrict ourselves to consideration of only analytic k-algebras $(A, \underline{m}_A)$, that is residue class rings of formal power series rings over a field k.

Definition. We call an analytic k-algebra B *strict* if for all k-algebras A and all regular A-sequences $\underline{t}$ with $B \simeq A/\underline{t}A$, the sequence $\underline{t}$ is super-regular.

Recall that a regular sequence $\underline{t}$ is called *super-regular* if the sequence of initial forms $\underline{t}^*$ is a regular $gr_{\underline{m}_A}(A)$-sequence.

If, for example, A is CM (Cohen-Macaulay) and $\underline{t}$ is a system of parameters of A such that $A/\underline{t}A$ is strict, then the tangent cone $gr_{\underline{m}_A}(A)$ is again CM. Hence it is interesting to find strict artinian algebras. The artinian ring $(B, \underline{m}_B)$ with $\underline{m}_B^2 = 0$ is one prominent algebra which is strict: see [2] or [3].

Each CM-ring with minimal multiplicity specializes to such an algebra and hence has CM-tangent cone. This is a result of J.Sally [4].

To come back to the general situation, suppose that we are given analytic k-algebras A and B, and a regular A-sequence $\underline{t}$ such that $A/\underline{t}A \simeq B$. We choose a presentation $B \simeq P/I$, where $P = k[[X_1,\dots,X_m]]$. Then A can be written as Q/J, where $Q = k[[X_1,\dots,X_m,T_1,\dots,T_n]]$ and where $t_i = T_i \bmod J$ for $i = 1,\dots,m$. We are going to look for intrinsic properties of B which guarantee that $\underline{t}$ is super-regular. To this end, pick a system of generators $f_1,\dots,f_k$ of I. Let H_1 denote the first Koszul homology of this sequence. There is a natural exact sequence of B-modules

$$0 \longrightarrow (I/I^2)^* \xrightarrow{\varepsilon^*} E^* \xrightarrow{j^*} H_1^*, \qquad (1^*)$$

which is obtained by dualizing the natural sequence

$$H_1 \xrightarrow{j} E \xrightarrow{\varepsilon} I/I^2 \longrightarrow 0. \qquad (1)$$

In this sequence E is a free module with basis $e_1,\ldots,e_k$. Given an element $[z] \in H_1$, $z = \Sigma r_i e_i$, $r_i \in P$, we define $j([z]) = \Sigma \bar{r}_i e_i$, where $\bar{r}_i = r_i \bmod I$. The map ε simply assigns to each e_i the element $f_i + I^2$.

The module $(I/I^2)^*$ contains as a submodule the module D of those homomorphisms which are induced by derivations. Thus D is generated by the elements $\partial_i : I/I^2 \to B$, $i = 1,\ldots,m$, with $\partial_i(f+I^2) = \frac{\partial f}{\partial X_i} + I$. The quotient module $T^1(B/k,B) = (I/I^2)^*/D$ is known to be an invariant of B, that is it does not depend on the presentation of B. In fact, T^1 classifies the first order analytic deformations of B. In this context there is another intrinsic B-module assigned to B, namely $T^2(B/k,B)$, the cokernel of $E^* \xrightarrow{j^*} H_1^*$. In deformation theory T^2 is the module of obstructions for lifting a given deformation. For details about the modules T^1 and T^2 the reader can consult [1] and [5].

The analytic k-algebras are $\underline{m}$-adically filtered. Thus I/I^2 has a natural filtration as a subquotient of P. Let $d_i = \deg f_i$ be the degree of the leading form of f_i. We define a natural filtration on E, given by

$$F_j E = \bigoplus_{i=1}^{k} \underline{m}_B^{j-d_i} e_i.$$

Here $\underline{m}_B^{\ell}$ is understood to be B if $\ell \le 0$. Now H_1 also admits a natural filtration: we set $x \in F_j H_1$ if x can be represented by a cycle $z = \Sigma r_i e_i$, with $r_i \in \underline{m}_P^{j-d_i}$ for $i = 1,\ldots,k$. It is clear that, with these filtrations, (1) is a sequence of filtered modules.

Whenever there is given a filtered B-module M, there exists a natural filtration on the dual module M^*, given by

$$F_i M^* = \{\phi \in \mathrm{Hom}(M,B) \mid \phi F_j M \subseteq F_{j+i} M \text{ for all } j\}.$$

Thus (1*) is in a natural way an exact sequence of filtered modules. It is clear that the modules $T^1(B/k,B)$ and $T^2(B/k,B)$ have induced filtrations which depend only on B, and not on the presentation of B nor on the choice of the generators of I.

Recall that a homomorphism of filtered modules $f : M \to N$ is called *strict* if

$$f(F_j M) = f(M) \cap F_j N$$

for all j. Provided the filtration is separated, there is a convenient characterization of strict homomorphisms. Let L be the cokernel of f with its induced filtration; then f is strict if and only if the induced sequence

$$\mathrm{gr}(M) \xrightarrow{\mathrm{gr}(f)} \mathrm{gr}(N) \longrightarrow \mathrm{gr}(L) \longrightarrow 0$$

of graded modules is exact.

We are now ready to formulate our main result. In the terminology we used so far we have the following.

THEOREM. *Let* $B = k[[X_1,\dots,X_n]]/I$ *be an analytic* k-*algebra and let* $j^* : E^* \to H_1^*$ *be the canonical homomorphism associated to a set of generators of* I. *Then* B *is strict if* $F_j T^1(B/k,B) = T^1(B/k,B)$ *for* $j < 0$, *and if* j^* *is a strict homomorphism.*

Remark 1. The condition "j^* is a strict homomorphism" does not depend on the particular choice of the generators of I. To see this, let $\underline{f}$ and $\underline{g}$ be two sets of generators of I. Let j_1^*, j_2^* and j^* be the homomorphisms corresponding to $\underline{f}, \underline{g}$ and the composed sequence $\underline{f} \cup \underline{g}$. Since $H_1(\underline{f},\underline{g};P) \simeq H_1(\underline{f};P) \oplus F$ where F is a free B-module, one sees that $j^* \simeq j_1^* \oplus \mathrm{id}_{F^*}$. Thus j_1^* is strict if and only if j^* is strict. The same argument applies for j_2^*, and the assertion follows.

Remark 2. Suppose that I is generated by homogeneous forms $f_1,\dots,f_k$. It is immediate that j^* with respect to this sequence is a strict homomorphism. Furthermore it is clear that in this case

$$\mathrm{gr}\, T^1(B/k,B) \simeq T^1(\mathrm{gr}(B)/k, \mathrm{gr}(B)).$$

Thus if I is generated by homogeneous forms, our theorem asserts that B is strict if $T^1(\mathrm{gr}(B)/k,\mathrm{gr}(B))_\nu = 0$ for $\nu < -1$. Here M_ν denotes the ν-th homogeneous component of a graded module M. This is exactly the result in [3].

Proof of the theorem. As before let $P = k[[X_1,\dots,X_n]]$, $B = P/I$, $Q = k[[X_1,\dots,X_n,T_1,\dots,T_m]]$, $A = Q/J$ and $t_i = T_i+J$ for $i = 1,\dots,m$.

Suppose that $\underline{t} = t_1,\dots,t_m$ is a regular A-sequence with $B \simeq A/\underline{t}A$. Choose generators $f_1,\dots,f_\ell$ of I whose leading forms generate $\mathrm{Ker}(\mathrm{gr}(P) \to \mathrm{gr}(B))$. Let

$$P^\ell \xrightarrow{\phi} P \longrightarrow B \longrightarrow 0$$

be the corresponding presentation of B, equipped with the natural filtrations. Then, by the choice of the f_i, the associated complex

$$\mathrm{gr}(P)^\ell \xrightarrow{\mathrm{gr}(\phi)} \mathrm{gr}(P) \longrightarrow \mathrm{gr}(B) \longrightarrow 0$$

is exact.

In a later step of the proof we are going to show that we can find $F_1,\dots,F_\ell \in J$ such that for $i = 1,\dots,\ell$ we have

(a) $f_i = F_i \bmod I$,

(b) $\deg F_i = \deg f_i$.

Let $Q^\ell \xrightarrow{\Phi} Q \longrightarrow A \longrightarrow 0$ be the presentation of A corresponding to the generators $F_1,\dots,F_\ell$ of J satisfying (a) and (b). We then obtain a commutative diagram

$$\begin{array}{ccccccc} Q^\ell & \xrightarrow{\Phi} & Q & \longrightarrow & A & \longrightarrow & 0 \\ \downarrow & & \downarrow & & \downarrow & & \\ P^\ell & \xrightarrow{\phi} & P & \longrightarrow & B & \longrightarrow & 0 \end{array} \qquad (2)$$

of filtered modules, where the vertical arrows correspond to reduction modulo I. Now (2) induces the commutative diagram

$$\begin{array}{ccccccc} \mathrm{gr}(Q)^\ell & \longrightarrow & \mathrm{gr}(Q) & \longrightarrow & \mathrm{gr}(A) & \longrightarrow & 0 \\ \downarrow & & \downarrow & & \downarrow & & \\ \mathrm{gr}(P)^\ell & \longrightarrow & \mathrm{gr}(P) & \longrightarrow & \mathrm{gr}(B) & \longrightarrow & 0 \end{array} \qquad (3)$$

of filtered modules. We claim that

(i) the sequence

$$\mathrm{gr}(Q)^\ell \xrightarrow{\mathrm{gr}(\Phi)} \mathrm{gr}(Q) \longrightarrow \mathrm{gr}(A) \longrightarrow 0$$

is exact, and

(ii) the induced homomorphism $\psi : \mathrm{Ker}\,\mathrm{gr}(\Phi) \longrightarrow \mathrm{Ker}\,\mathrm{gr}(\phi)$ is surjective.

If $\underline{t}^*$ denotes the sequence of leading forms of $\underline{t}$, then by (i) and (ii) we have

$$0 = \text{Coker}\ \psi \simeq \text{Tor}_1^{gr(Q)}(gr(P), gr(A)) \simeq H_1(\underline{t}^*; gr(A)).$$

This means that $\underline{t}$ is a super-regular A-sequence.

Thus the proof of the theorem will be completed once we have established (i) and (ii), and once we have found generators of J satisfying (a) and (b).

To prove (i) we assume for simplicity that the sequence $\underline{T}$ has length 1. The general case is achieved by induction on the length of the sequence. As before G^* denotes the leading form of a polynomial G, and $\bar{G}$ denotes its residue modulo T_1.

Now pick $x \in \text{Ker}(gr(Q) \to gr(A))$. By diagram chasing one finds that $x = A^*.T_1^*$, up to an element of Im $gr(\Phi)$. Hence we can find $B \in Q$ with deg B > deg A and such that $AT_1 - B \in J$. It follows that $\bar{B} \in I$. Since we are assuming that J is generated by $F_1, \ldots, F_\ell$ satisfying (a) and (b), we can find $C \in J$ such that deg C = deg $\bar{B}$ and $\bar{C} = \bar{B}$. It follows that $C - B = D.T_1$, and hence $(A-D)T_1 = (AT_1 - B) + C \in J$, and $(A-D)^* = A^*$. Thus we may assume from the beginning that $AT_1 \in J$. Since T_1 is regular modulo J, it follows that $A \in J$. Hence we have $x = A^*.T_1^*$ with $A^* \in \text{Ker}(gr(R) \to gr(A))$. Since deg A^* < deg x, the assertion follows by induction on the degree.

In order to prove (ii), we pick an element $x \in \text{Ker}\ gr(\phi)$. Since ϕ is strict, there exists $r \in \text{Ker}\ \phi$ such that $r^* = x$. We prove in a moment that we can find $R \in \text{Ker}\ \Phi$ with $\bar{R} = r$ and deg R = deg r. It is then clear that R^* is mapped onto r^*.

We find such an element $R \in \text{Ker}\ \Phi$ step by step, as follows. Denote by K the ideal generated by $T_1, \ldots, T_n$ and let $\Phi_i = \Phi \otimes Q^\ell / K^{i+1} Q^\ell$. Suppose we have found $R^{(i-1)} \in \text{Ker}\ \Phi_{i-1}$ with $r = R^{(i-1)}$ mod K and deg $R^{(i-1)}$ = deg r. Consider the commutative diagram of canonically filtered modules

$$\begin{array}{ccccccc}
 & & 0 & & 0 & & 0 \\
 & & \downarrow & & \downarrow & & \downarrow \\
K^iQ^\ell/K^{i+1}Q^\ell & \xrightarrow{\ \pi\ } & K^i/K^{i+1} & \xrightarrow{\ \varepsilon\ } & K^iA/K^{i+1}A & \longrightarrow & 0 \\
{\scriptstyle\tau}\downarrow & & \downarrow & & \downarrow & & \\
Q^\ell/K^{i+1}Q^\ell & \xrightarrow{\ \Phi_i\ } & Q/K^{i+1} & \longrightarrow & A/K^{i+1}A & \longrightarrow & 0. \\
{\scriptstyle\sigma}\downarrow & & \downarrow & & \downarrow & & \\
Q^\ell/K^iQ^\ell & \xrightarrow{\ \Phi_{i-1}\ } & Q/K^i & \longrightarrow & A/K^iA & \longrightarrow & 0 \\
\downarrow & & \downarrow & & \downarrow & & \\
0 & & 0 & & 0 & &
\end{array} \qquad (4)$$

All columns and the second and third row are trivially exact. The top row is exact since $\underline{t}$ is a regular A-sequence. Let $d = \deg R^{(i-1)}$, and choose $\tilde{R}^{(i)} \in F_dQ^\ell/K^{i+1}Q^\ell$ such that $\sigma(\tilde{R}^{(i)}) = R^{(i-1)}$. Then $\Phi_i(\tilde{R}^{(i)}) \in F_dK^i/K^{i+1}$ and $\varepsilon\Phi_i(\tilde{R}^{(i)}) = 0$. Thus we can find $P \in F_dK^iQ^\ell/K^{i+1}Q^\ell$ with $\pi(P) = \Phi_i(\tilde{R}^{(i-1)})$. Let $R^{(i)} = \tilde{R}^{(i)}-P$; then $R^{(i)} \in \operatorname{Ker}\Phi_i$ and $\deg R^{(i)} = \deg R^{(i-1)}$. The relation R is obtained as the "limit" of the $R^{(i)}$.

To complete the proof of the theorem it now suffices to find the $F_i \in J$ satisfying (a) and (b). Pick any $F_i \in J$ with $\bar{F}_i = f_i$ for $i = 1,\ldots,\ell$. Write

$$F_i = \sum_\nu f_i^{(\nu)}T^\nu$$

with $f_i^{(\nu)} \in P$ and $f_i^{(0)} = f_i$. We claim that $\deg \tilde{f}_i^{(\nu)} \geq \deg f_i - |\nu|$ for all i and $|\nu|$. Here ~ denotes the I-residue of an element of P. It is then clear that the F_i may be modified in such a way that they satisfy $\deg F_i = \deg f_i$ for $i = 1,\ldots,\ell$, as well.

To prove the claim assume to the contrary that for some i, say i = 1, and for some $\nu \neq 0$ we have $d = \deg \tilde{f}_1^{(\nu)} < \deg f_1 - |\nu|$, and $\deg f_i^{(\mu)} \geq \deg f_i - |\mu|$ for all $|\mu| < |\nu|$. In the exact sequence (1*) let $\alpha \in E^*$ be the element with $\alpha(e_i) = \tilde{f}_i^{(\nu)}$ for $i = 1,\ldots,\ell$.

Let $j+1 = |\nu|$ and $F_i^{(j)} = F_i \bmod K^{j+1}$; then by the choice of ν

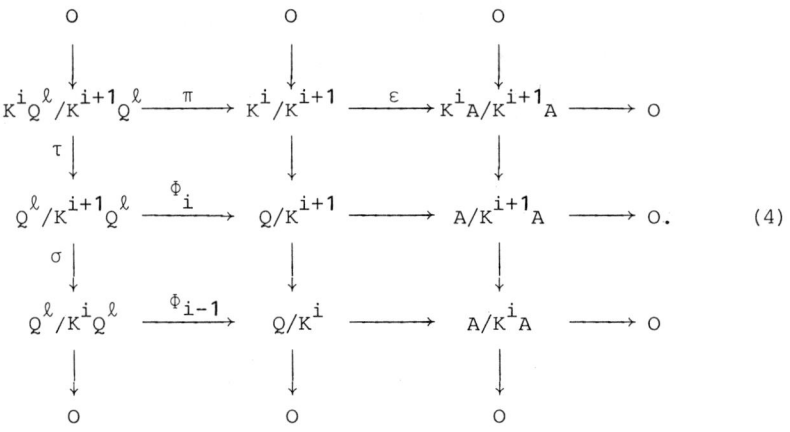

we have $\deg F_i^{(j)} = \deg f_i$ for $i = 1,\dots,\ell$. Using diagram (4) we see, as before, that any relation $r = (r_i)$ can be lifted to a relation $(R_i^{(j)}) \in Q^\ell/K^{j+1}Q^\ell$ of $(F_1^{(j)},\dots,F_\ell^{(j)})$ with $\deg R_i^{(j)} = \deg r_i$. Write $R_i^{(j)} = \sum_{|\mu|<j+1} r_i^{(\mu)} T^{(\mu)}$. We claim that

$$j^*(\alpha)[r] = - \sum_{\substack{\mu+\sigma=\nu \\ \mu\neq 0,\sigma\neq 0}} \sum_i r_i^{(\mu)} f_i^{(\sigma)} \mod I.$$

In fact we can find $r_i^{(\mu)} \in P$ with $|\mu| = j+1$ such that

$$\sum_i (R_i^{(j)} + \sum_\mu r_i^{(\mu)} T_i^\mu) F_i^{(j+1)} = 0 \mod K^{j+1}.$$

Consider the coefficient of T^ν in this sum. We get

$$\sum_i r_i^{(\nu)} f_i + \sum_{\substack{\mu+\sigma=\nu \\ \mu\neq 0,\sigma\neq 0}} \sum_i r_i^{(\mu)} f_i^{(\sigma)} + \sum_i r_i f_i^{(\nu)} = 0.$$

On the other hand $j^*(\alpha)[r] = \sum_i r_i f_i^{(\nu)} \mod I$, and the above equality follows. We have $\alpha \notin F_{-(j+1)}E^*$, but $j^*(\alpha) \in F_0 H_1^*$, since, for $\mu \neq 0$, $\sigma \neq 0$, $\deg r_i^{(\mu)} f_i^{(\sigma)} \geq \deg r_i + \deg f_i \geq \deg r$. Since, by assumption, j^* is strict there exists $\beta \in F_0 E^*$ such that $j^*(\beta) = j^*(\alpha)$. Since (1*) is exact it follows that $\alpha-\beta \in (I/I^2)^* - F_{-(j+1)}(I/I^2)^*$. This contradicts our assumption that $F_j T^1(B/k,B) = T^1(B/k,B)$ for $j < 0$.

References

1. M.Artin, Lectures on deformations of singularities, Tata Institute of Fundamental Research, Bombay, 1976.

2. J.Herzog, "When is a regular sequence super-regular?", Nagoya Math. J., 83 (1981), 183-195.

3. J.Herzog, "Strict local rings", Proc. Amer. Math. Soc., 84 (1982), 165-172.

4. J.Sally, "On the associated graded ring of a local Cohen-Macaulay ring", J. Math. Kyoto Univ., 17 (1977), 19-21.

5. M.Schlessinger, "Functors of Artin rings", Trans. Amer. Math. Soc., 130 (1968), 208-222.

Universität Essen-Gesamthochschule,
Fachbereich 6-Mathematik,
Universitätsstrasse 3,
Postfach 103764,
D-4300 Essen 1, West Germany.

MULTIPLICITIES, HILBERT FUNCTIONS AND DEGREE FUNCTIONS

D.REES

When I was asked to give a talk at this colloquium, I decided to attempt a historical survey of the theory of multiplicities of ideals. When I began to think about it in more detail, I realised that to cover the topic in an hour, I would have to impose some limitations. What I am going to do is to consider three papers and a book of lecture notes which exemplify the early history of multiplicities and then jump forward to the last 10 years and consider two papers which to me seem to indicate an interesting area for further work. My choice of material is inevitably personal. The first two papers played a large part in influencing me to change to commutative algebra as a field of research. The others have taught me new ways of looking at commutative algebra.

Let me start with the first two papers on my list.

[C] C.Chevalley, "On the theory of local rings", Ann.of Math., 44 (1943), 690-708.

[Sa] P.Samuel, "La notion de multiplicité en algèbre et en géométrie algébrique", J.Math.Pures Appl., 30 (1951), 159-274.

Chevalley's paper contains a great deal of material we take for granted now, but its importance for this talk lies in the fact that it contains, as far as I know, the first definition of multiplicity as we know it. He did impose limitations on his local rings, and on the ideals to which the definition applied. He takes a complete local domain Q of Krull dimension d which contains a field k over which $Q/\underline{m}$ is a finite extension. Now let $x_1,\dots,x_d$ be elements such that $x_1Q+\dots+x_dQ$ is $\underline{m}$-primary. Then Q contains the regular local ring $R = k[[x_1,\dots,x_d]]$ and Q is a finite R-module. Chevalley showed that

$$[Q : R] = e[Q/\underline{m} : k]$$

where e is independent of the choice of k. In fact, e was, in modern notation, $e(x_1,\ldots,x_d;Q)$. Chevalley applied this in later papers to give a definition of intersection multiplicity for sub-varieties of an algebraic variety at a simple point of the ambient variety. He also proved under suitable restrictions one result whose proof in general was to challenge those of us working in the early 1950's. This is the associativity formula which I will come to later.

The drawback to Chevalley's definition was the restrictions that were imposed. The removal of these was one, but only one, of Samuel's contributions. (Perhaps I should say here that Samuel's results had been announced from 1947 onwards.) Samuel's great idea was the introduction of Hilbert functions. Basically, if Q is any local ring of Krull dimension d and $\underline{q}$ is an $\underline{m}$-primary ideal of Q, then, for large n, $\ell(Q/\underline{q}^n)$ is a polynomial S(n) in n of degree d and

$$S(n) = \frac{1}{d!}e(\underline{q})n^d + \ldots ,$$

where $e(\underline{q})$ is a positive integer. If $\underline{q} = x_1Q + \ldots + x_dQ$, then

$$e(\underline{q}) = e(x_1,\ldots,x_d;Q)$$

under Chevalley's restrictions. We can go further. If $x_1,\ldots,x_d \in \underline{q}$, then

$$e(x_1Q + \ldots + x_dQ) \geq e(\underline{q})$$

and, provided $Q/\underline{m}$ is infinite, we can choose $x_1,\ldots,x_d$ so that equality holds. In fact it is sufficient to choose $(x_1,\ldots,x_d)$ to be a reduction of $\underline{q}$.

Samuel's approach applied to local rings in general and one major aim of that period was to place in a general context the associativity formula of Chevalley. This was achieved by Christer Lech in the third paper I want to mention.

[L] C.Lech, "On the associativity formula for multiplicities", Ark.Mat., 3 (1956), 301-314.

This is a very elegant paper indeed. First could I state the associativity formula in the form Lech proved it.

Let Q be a local ring of Krull dimension d, let $(x_1,\ldots,x_d)$ generate an $\underline{m}$-primary ideal $\underline{q}$ and let $\underline{a} = (x_{m+1},\ldots,x_d)$. Then, if

$\underline{p}$ ranges over those minimal prime ideals of $\underline{a}$ for which $\dim Q/\underline{p} + \operatorname{ht}\underline{p} = d$,

$$e(\underline{q}) = \sum_{\underline{p}} e(\underline{q}+\underline{p}/\underline{p})\, e(\underline{a}Q_{\underline{p}}).$$

Lech's approach depends on a new formula for the multiplicity of an ideal $(x_1,\ldots,x_d) = \underline{q}$:

$$e(\underline{q}) = \lim_{\min n_i \to \infty} \frac{\ell(Q/x_1^{n_1}Q + \ldots + x_d^{n_d}Q)}{n_1\ldots n_d}.$$

At the end of Lech's paper is another result which I will refer to later. Suppose that $\underline{q} = (x_1,x_2,\ldots,x_d)$ and $\underline{q}' = (x_1',x_2,\ldots,x_d)$ are $\underline{m}$-primary. Then, if $\underline{q}'' = (x_1x_1',x_2,\ldots,x_d)$, $\underline{q}''$ is $\underline{m}$-primary and

$$e(\underline{q}'') = e(\underline{q}) + e(\underline{q}').$$

The year 1956 in which Lech's paper appeared was a major turning-point in multiplicity theory. In fact, in a similar lecture to this one, given in December 1956, Pierre Samuel was able to indicate accurately the direction it was to take in the following years. Reference to Samuel's lecture enables me to correct an injustice in that I have not mentioned two names, Nagata and Northcott: both made, as Samuel says, many important contributions before 1956, and both have made many such contributions since then.

Samuel referred in his lecture particularly to two new directions of attack. The first, and I use his word, is Modulation. If $\underline{q}$ is any $\underline{m}$-primary ideal and M is a finitely generated Q-module we can define a non-negative integer $e(\underline{q};M)$ by a Hilbert function

$$\ell(M/\underline{q}^nM) = \frac{1}{d!}e(\underline{q};M)n^d + \ldots$$

for large n, and it is not difficult to deduce from this definition that if

$$0 \to M' \to M \to M'' \to 0$$

is exact, then

$$e(\underline{q};M) = e(\underline{q};M') + e(\underline{q};M'');$$

this leads to simplified proofs of, for example, the associativity formula. However it is with another development of this period that my fourth reference is mainly concerned.

[Se] J.-P.Serre, Algèbre locale: multiplicités, Cours au Collège de France, 1957-1958, Lecture Notes in Mathematics 11 (Springer, Berlin, 1965).

In the hands of Auslander, Buchsbaum, Serre and others, the introduction of the methods of homological algebra into commutative ring theory quickly led to the solution of long-standing problems and opened up new fields of research. In particular it provided a new definition of multiplicity via the Koszul complex. I will assume you are familiar with the structure of the Koszul complex $K(x_1,\ldots,x_d;Q)$ associated with an $\underline{m}$-primary ideal $\underline{q} = (x_1,\ldots,x_d)$. If M is a finitely generated Q-module, then we write $K(x_1,\ldots,x_d;M)$ for the complex $K(x_1,\ldots,x_d;Q) \otimes M$ and $H_i(x_1,\ldots,x_d;M)$, $i = 0,\ldots,d$, for its homology groups. These are annihilated by $\underline{q}$ and so have finite length. We now define

$$\chi(\underline{q},M) = \sum_{i=0}^{d} (-1)^i \ell(H_i(x_1,\ldots,x_d;M)).$$

Now I want to sketch a proof of the equality

$$\chi(\underline{q},M) = e(\underline{q};M)$$

and, in fact, a stronger result, which I will use again later.

Could I first recall that by a *good* $\underline{q}$-*filtration of* M is meant a sequence $\{M_r\}$ of submodules of M ($-\infty < r < \infty$) such that

(1) $M_r = M$ for r sufficiently large and negative,

(2) $M_r \supseteq M_{r+1} \supseteq \underline{q}M_r$, and

(3) $M_{r+1} = \underline{q}M_r$, if r is large and positive.

Now let $R = Q[tx_1,\ldots,tx_d,u]$, where t is an indeterminate and $u = t^{-1}$, and let $\mathcal{M}$ be the graded R-module $\sum_{-\infty}^{\infty} t^r M_r$, which is finitely generated. We now consider the graded Koszul Complex

$$K(tx_1,\ldots,tx_d,\mathcal{M})$$

adjusted to make the boundary operators have degree 0. Its homology groups are annihilated by $(tx_1,\ldots,tx_d)$, which is irrelevant, that is contains all elements of positive degree. It follows that the component $K_n(tx_1,\ldots,tx_d,\mathcal{M})$ of degree n is acyclic if n is large and positive. Now K_{ni} is the direct sum of $\binom{d}{i}$ copies of M_{n-i} and K_n is a subcomplex of $K(x_1,\ldots,x_d;M)$. The module $(K/K_n)_i$ is isomorphic to M/M_{n-i} and if we write $\ell(n) = \ell(M/M_n)$ and $\Delta f(n) = f(n) - f(n-1)$

for any function f on $\mathbb{Z}$, we obtain immediately

$$\chi(x_1,\ldots,x_d,M) = \chi(K_n(tx_1,\ldots,tx_d,\mathcal{M})) + \Delta^d \ell(n)$$

for all n, whence, as $\chi(K_n(tx_1,\ldots,tx_d,\mathcal{M})) = 0$ for n large, $\ell(n)$ is a polynomial of degree d in n and the coefficient of n^d is $\chi(x_1,\ldots,x_d,M)/d!$. This completes the link between the Hilbert Function and Homological Algebra approaches to multiplicities.

As I said at the beginning, at this stage I am going to pass over a great deal. Serre's notes contain a great deal more than I have had time to mention. For example, I think that the section on partial Euler-Poincaré characteristics and the later paper of Stephen Lichtenbaum call out for further application. The extensive simplifications of the earlier theory arising from the work of the Sheffield school, documented in Douglas Northcott's "Lessons on rings, modules and multiplicities", and the beautiful work of Joseph Lipman also deserve mention, but so do many other developments. I have chosen just two papers, published in 1973 and 1977 by Bernard Teissier, to comment on in more detail, because they impressed me particularly. These papers are as follows.

[T_1] B.Teissier, "Cycles évanescents, sections planes et conditions de Whitney", Singularités à Cargèse 1972, Astérisque, 7-8 (1973).

[T_2] B.Teissier, "Sur une inégalité à la Minkowski pour les multiplicités", (Appendix to a paper by D.Eisenbud and H.I.Levine), Ann. of Math., 106 (1977), 38-44.

Let me deal with the first of these papers. It is really concerned with singularities of complex varieties and of maps of one onto another, but imbedded in it is a short section on Hilbert functions and multiplicities, due to Teissier and J.J.Risler, which is inspired. It was well known that if $\underline{q}_1,\ldots,\underline{q}_s$ are $\underline{m}$-primary ideals of Q and M is a finitely generated Q-module, then for large $n_1,\ldots,n_s$, $\ell(M/\underline{q}_1^{n_1}\ldots\underline{q}_s^{n_s}M)$ is a polynomial of total degree d in $n_1,\ldots,n_s$, and this leads to a formula for the multiplicity of $\underline{q}_1^{m_1}\ldots\underline{q}_s^{m_s}$:

$$e(\underline{q}_1^{m_1}\ldots\underline{q}_s^{m_s};M) = \sum_{|k|=d} \frac{d!}{k_1!\ldots k_s!}\, e(\underline{q}_1^{k_1}|\ldots|\underline{q}_s^{k_s};M)\, m_1^{k_1}\ldots m_s^{k_s},$$

the sum being over all non-negative $k_1,\ldots,k_s$ such that $|k_1| = k_1+\ldots+k_s = d$. The first major contribution of Risler and Teissier is to identify the coefficients $e(\underline{q}_1^{k_1}|\ldots|\underline{q}_s^{k_s};M)$ as follows. Let x_{ij}, $i = 1,\ldots,s$, $j = 1,\ldots,k_i$, be chosen from $\underline{q}_i$, $i = 1,\ldots,s$, "sufficiently generally". Then

$$e(\underline{q}_1^{k_1}|\ldots|\underline{q}_s^{k_s};M) = e(x_{11},\ldots,x_{sk_s};M).$$

Let me give a brief sketch of a proof of this result which also gives a meaning to the words "sufficiently generally". First let us term a set of elements x_{ij}, $i = 1,\ldots,s$, $j = 1,\ldots,k_i$, a *joint reduction* of $\underline{q}_1,\ldots,\underline{q}_s$ of type $(k_1,\ldots,k_s)$ if

$$\bigcup_{i=1}^{s}\bigcup_{j=1}^{k_i} x_{ij}\underline{q}_1^{n_1}\ldots\underline{q}_i^{n_i-1}\ldots\underline{q}_s^{n_s} = \underline{q}_1^{n_1}\ldots\underline{q}_s^{n_s}$$

whenever $n_1,\ldots,n_s$ are all sufficiently large.

Now we consider the multigraded ring R consisting of all finite sums $\Sigma c(n_1,\ldots,n_s)t_1^{n_1}\ldots t_s^{n_s}$ with $t_1,\ldots,t_s$ indeterminates, and $c(n_1,\ldots,n_s) \in \underline{q}_1^{n_1}\ldots\underline{q}_s^{n_s}$. (Note that we allow the exponents $n_1,\ldots,n_s$ to be negative, but, if $n_i \le 0$, then $\underline{q}_i^{n_i} = Q$.)

Now suppose that $M(n_1,\ldots,n_s)$ is a good $\underline{q}_1,\ldots,\underline{q}_s$ filtration of M, that is

(i) $M(n_1,\ldots,n_s) = M$ if $n_1,\ldots,n_s$ are all large and negative,

(ii) $M(n_1,\ldots,n_i,\ldots,n_s) \supset M(n_1,\ldots,n_i+1,\ldots,n_s) \supseteq \underline{q}_i M(n_1,\ldots,n_s)$, and

(iii) $M(n_1,\ldots,n_i+1,\ldots,n_s) = \underline{q}_i M(n_1,\ldots,n_s)$

for all $n_1,\ldots,n_s$ such that $n_i \ge$ a certain integer n_i^0, $i = 1,\ldots,s$. Then we can define a multigraded module $\mathcal{M}$ consisting of all finite sums $\Sigma m(n_1,\ldots,n_s)t_1^{n_1}\ldots t_s^{n_s}$ with $m(n_1,\ldots,n_s) \in M(n_1,\ldots,n_s)$. $\mathcal{M}$ is finitely generated. We can now copy my earlier proof to obtain the following result. Let $\ell(n_1,\ldots,n_s) = \ell(M/M(n_1,\ldots,n_s))$ and let

$$\Delta_i f(n_1,\ldots,n_s) = f(n_1,\ldots,n_s) - f(n_1,\ldots,n_i-1,\ldots,n_s).$$

Then, for $n_1,\ldots,n_s$ sufficiently large,

$$\Delta_1^{k_1}\ldots\Delta_s^{k_s}\ell(n_1,\ldots,n_s) = e(x_{11},\ldots,x_{sk_s};M),$$

which yields the Risler-Teissier result.

Now could I turn to the second of Teissier's papers. In the first he conjectured that

$$e(\underline{q}_1\underline{q}_2\cdots\underline{q}_s)^{1/d} \le e(\underline{q}_1)^{1/d} + \dots + e(\underline{q}_s)^{1/d}$$

for any s $\underline{m}$-primary ideals of a d-dimensional local ring. The second paper contains the proof, admittedly under restrictions later to be removed by Rodney Sharp by a most elegant argument. The inequality above has found, already, many applications, particularly in relation to theorems of the Bonnesen type on convex bodies.

To finish off this talk I want to discuss the third topic of my title, degree functions. This topic is much older than Teissier's papers. In fact, degree functions were first introduced by Samuel in about 1952, but they began to have real consequence as a result of Teissier's first paper.

Degree functions are easily defined. Let Q be a local ring, let $\underline{q}$ be an $\underline{m}$-primary ideal and let x range over the multiplicatively closed set of elements of $\underline{m}$ such that Q/xQ has dimension d-1 and x $\notin$ any minimal prime ideal of Q of dimension d-1. Then we set $d(\underline{q},x) = e(\underline{q}+xQ/xQ)$. It is easy to see that

$$d(\underline{q},xy) = d(\underline{q},x) + d(\underline{q},y).$$

Now we must relate degree functions to $\underline{m}$-valuations, which I shall now define. An $\underline{m}$-valuation v of Q is a mapping from Q to $\mathbb{Z} \cup \{\infty\}$ which is, for some minimal prime ideal $\underline{p}$ of Q such that $\dim Q/\underline{p} = d$, the composition of the natural homomorphism $Q \to Q/\underline{p}$ and a valuation $w : K(\underline{p}) \to \mathbb{Z} \cup \{\infty\}$ of $K(\underline{p})$ such that

$$w(y) \ge 0 \text{ on } Q/\underline{p},$$

$$w(y) > 0 \text{ on } \underline{m}/\underline{p},$$

and the residue field K_w of w is a finitely generated extension of $k = Q/\underline{m}$ of transcendence degree d-1.

Now we can express $d(\underline{q},x)$ in terms of $\underline{m}$-valuations:

$$d(\underline{q},x) = \sum_v d(\underline{q},v)v(x),$$

$d(\underline{q},v)$ being a non-negative integer which is clearly zero for all save a finite set of $\underline{m}$-valuations. Now $d(\underline{q},v)$ can be described more closely. First, we can extend v to the completion $\bar{Q}$ of Q. This extension will take the value ∞ on a minimal prime ideal $\underline{p}_v$ of zero in $\bar{Q}$ such that $\dim \bar{Q}/p_v = d$. Let δ_v denote the length of the

$\underline{p}_v$-primary component of (O) in $\bar{Q}$. Next suppose that $(x_1,\ldots,x_d)$ is a reduction of $\underline{q}$ (I am assuming that $Q/\underline{m}$ is infinite). Then if $d(\underline{q},v) \neq 0$ we will have, first, that $v(x_1) = \ldots = v(x_d)$ and, secondly, that the images $\xi_1,\ldots,\xi_{d-1}$ of $x_1/x_d,\ldots,x_{d-1}/x_d$ in K_v are algebraically independent. Then

$$d(\underline{q},v) = \delta_v[K_v : k(\xi_1,\ldots,\xi_{d-1})].$$

In fact the coefficients $d(\underline{q},v)$ are uniquely determined by the equation

$$d(\underline{q},x) = \sum_v d(\underline{q},v)v(x).$$

Now we come to the relation with Teissier's paper. First suppose that $\underline{q}_1,\ldots,\underline{q}_d$ are d $\underline{m}$-primary ideals, where $d = \dim Q$. Then $e(\underline{q}_1|\ldots|\underline{q}_d)$ is a "multilinear" function in the sense that $e(\underline{q}_1|\ldots|\underline{q}_i\underline{q}_i'|\ldots|\underline{q}_d) = e(\underline{q}_1|\ldots|\underline{q}_i|\ldots|\underline{q}_d) + e(\underline{q}_1|\ldots|\underline{q}_i'|\ldots|\underline{q}_d)$.

This follows from the result quoted from Lech's paper earlier. Next the uniqueness of the coefficients $d(\underline{q},v)$ in the degree formula

$$d(\underline{q},x) = \sum_v d(\underline{q},v)v(x)$$

leads very easily to a proof that $d(\underline{q}_1^{n_1}\ldots\underline{q}_s^{n_s},v)$ is a homogeneous polynomial

$$\sum_{|k|=d-1} \frac{(d-1)!}{k_1!\ldots k_s!} d(\underline{q}_1^{k_1}|\ldots|\underline{q}_s^{k_s},v)n_1^{k_1}\ldots n_s^{k_s},$$

and we may similarly define a "multilinear" function $d(\underline{q}_1|\ldots|\underline{q}_{d-1},v)$.

Now Teissier and Risler proved a result which we can write in the form

$$e(\underline{q}_1|\ldots|\underline{q}_d) = d(\underline{q}_1|\ldots|\underline{q}_{i-1}|\underline{q}_{i+1}|\ldots|\underline{q}_d,x)$$

provided x is a sufficiently general element of $\underline{q}_i$, and this yields d expressions for $e(\underline{q}_1|\ldots|\underline{q}_d)$:

$$e(\underline{q}_1|\ldots|\underline{q}_d) = \sum_v d(\underline{q}_1|\ldots|\underline{q}_{i-1}|\underline{q}_{i+1}|\ldots|\underline{q}_d)v(\underline{q}_i),\ i = 1,\ldots,d,$$

where $v(\underline{q}_i) = \min_{x\in\underline{q}_i} v(x)$, and the numbers $d(\underline{q}_1|\ldots|\underline{q}_{i-1}|\underline{q}_{i+1}|\ldots|\underline{q}_d)$ are non-negative integers. The fact that this holds for each i yields a number of relations for the numbers

$$d(\underline{q}_1|\ldots|\underline{q}_{i-1}|\underline{q}_{i+1}|\ldots|\underline{q}_d)$$

and with this I will finish.

References

[C] C.Chevalley, "On the theory of local rings", Ann.of Math., 44 (1943), 690-708.

[L] C.Lech, "On the associativity formula for multiplicities", Ark.Mat., 3 (1956), 301-314.

[Sa] P.Samuel, "La notion de multiplicité en algèbre et en géométrie algébrique", J.Math.Pures Appl., 30 (1951), 159-274.

[Se] J.-P.Serre, Algèbre locale: multiplicités, Cours au Collège de France, 1957-1958, Lecture Notes in Mathematics 11 (Springer, Berlin, 1965).

[T_1] B.Teissier, "Cycles évanescents, sections planes, et conditions de Whitney", Singularités à Cargèse 1972, Astérisque, 7-8 (1973).

[T_2] B.Teissier, "Sur une inégalité à la Minkowski pour les multiplicités", (Appendix to a paper by D.Eisenbud and H.I. Levine), Ann.of Math., 106 (1977), 38-44.

Department of Mathematics,
University of Exeter,
North Park Road,
Exeter EX4 4QE, U.K.

FINITENESS CONDITIONS IN COMMUTATIVE ALGEBRA AND SOLUTION OF A PROBLEM OF VASCONCELOS

JAN-ERIK ROOS

Introduction

It is fair to say that a major part of current commutative algebra is a theory about noetherian rings. However, even if one tries hard to restrict oneself to noetherian rings, one is inevitably led to study non-noetherian commutative rings. Here is a simple example (it will be dealt with in more detail in §6).

Let R be a commutative noetherian domain and let K be its field of fractions. It is well known (cf. [13; Theorem 93 (Krull, Akizuki and Cohen)] and [10]) that the following assertions are equivalent:

(i) every ring S between R and K,

$R \subset S \subset K$,

is noetherian;

(ii) the Krull dimension of R is ≤ 1.

Thus if $\dim R \geq 2$, there is always a non-noetherian domain between R and K.

PROBLEM 1. *Which finiteness conditions do the rings between a commutative noetherian domain and its field of fractions satisfy?*

PROBLEM 2. *Let* R *be a noetherian domain and let* $\bar{R}$ *be the integral closure of* R *in its field of fractions. Which finiteness conditions does* $\bar{R}$ *satisfy?* (It is known [17; Appendix, Example 5] that $\bar{R}$ can be non-noetherian if $\dim R \geq 3$.)

A natural generalization of noetherian rings are coherent rings.

Definition 1. A (not necessarily commutative) ring S is called (left) *coherent* if each finitely generated left ideal of S is finitely presented (that is can be presented by a finite number of generators and a finite number of relations).

Thus if $\underline{a}$ is an ideal as in Definition 1, then we have an exact sequence of left S-modules

$$F_2 \longrightarrow F_1 \longrightarrow S \longrightarrow S/\underline{a} \longrightarrow 0$$

where the F_i are finitely generated free S-modules.

Here are some examples of coherent rings.

(a) The polynomial ring in any number of variables (not necessarily finite!) over a noetherian ring is coherent.

(b) $S = \prod_{\mathbb{N}} \mathbb{Q}[X,Y]$ is coherent, but $S' = \prod_{\mathbb{N}} \mathbb{C}[X,Y,Z]$ is *not* coherent. For more details about example (b) we refer the reader to [24; p.90], where also the interesting fact that S[X] is *not* coherent is mentioned (Soublin).

(c) Here is a non-commutative example. The group ring $\mathbb{Z}[G]$ of a finitely generated solvable group G is coherent if and only if G is polycyclic *or* an ascending HNN-group over a polycyclic base group. This is proved in [6], where the terminology is explained. For a partial generalization, see [1].

Returning to Problem 1, it is however known that coherence is not enough to describe those finiteness conditions that the rings of Problem 1 must satisfy. Indeed, Papick has shown [19] that if dim R ≥ 2, there is always a *non*-coherent ring S between R and K. However, we shall prove (cf. §6 below) that the ring constructed by Papick has λ-dimension 2 in the sense of the following definition.

Definition 2. A ring R has (left) *λ-dimension* ≤ n if for each exact sequence of left R-modules

$$F_n \xrightarrow{\phi_n} F_{n-1} \longrightarrow \cdots \longrightarrow F_1 \longrightarrow F_0 \longrightarrow M \longrightarrow 0,$$

where the F_i are finitely generated and free, we have that Ker ϕ_n is a finitely generated R-module.

It follows that

λ-dim R ≤ 0 if and only if R is left noetherian,

and that

λ-dim R ≤ 1 if and only if R is left coherent,

this last assertion being an easy consequence of the general theory of coherent rings (cf. [24] or the exercises of Chapitre 1, §2 of [7]).

The first time λ-dimension occurs explicitly in the literature seems to be in Vasconcelos' book [24], but this notion seems to have been referred to as r-coherence earlier (cf. [24; p. 90]) and the notion is at least implicit in the exercises of Bourbaki [7] mentioned above (see also the exercises of §3 of [8]). It is noted in [24; p. 92] that

$$\lambda\text{-dim } S[X] \le \lambda\text{-dim } S + 1 \tag{1}$$

if S is coherent, or more generally if flat $\dim_S \prod_I S + 1 = \lambda\text{-dim } S$, (I big), but we shall see in §2 below that this last equality does not hold in general for non-coherent S.

It follows from the inequality (1) that for the S of (b) above we have $\lambda\text{-dim } S[X] = 2$.

Until now, *among commutative rings*, only rings with λ-dimension 0, 1 and 2 were known. This led Vasconcelos to state the following problem.

PROBLEM 3 (Vasconcelos [24; p. 93]). *Exhibit all positive integers as λ-dimensions of commutative rings.*

The aim of the present paper is to solve this problem. More precisely, we shall prove the following.

THEOREM A. *Let* $(R,\underline{m})$ *be a local commutative noetherian Gorenstein ring (for example, take* R *to be regular!), and let* I(k) *be the injective hull of the residue field* $k = R/\underline{m}$. *Let*

$$S_R = R \propto I(k) \tag{2}$$

be the "trivial" extension of the ring R *with the* R*-module* I(k). [Thus S_R is the set of pairs (r,i), $r \in R$, $i \in I(k)$, with pairwise addition and multiplication given by $(r,i).(r',i') = (r.r', r.i' + i.r')$.] *Then*

$$\lambda\text{-dim } S_R = \dim R, \tag{3}$$

where dim R *is the Krull dimension of* R.

Thus if we take, for example, $R = \mathbb{C}[[T_1,\ldots,T_n]]$ (formal power series), we obtain $\lambda\text{-dim } S_R = n$, and the problem of Vasconcelos is solved. Note that in this special case $I(\mathbb{C})$ can be interpreted (cf., for example, [8; Exercise 32(b), p. 173]) as $\mathbb{C}[X_1,\ldots,X_n]$ where R operates by $T_i.P = \partial P/\partial X_i$ for $P \in \mathbb{C}[X_1,\ldots,X_n]$, so that S_R is indeed a very explicit ring.

Previously, in view of the few examples of λ-dimension known (only 0, 1 and 2), it was not clear that λ-dimension was an interesting notion. With the examples of Theorem A in our hands, we have more confidence, since the λ-dimension of each of the rings (2) is equal to the more familiar Krull dimension. In §4 we shall prove (Theorem A′) that Theorem A remains true if we only suppose that $R_{\underline{p}}$ is Gorenstein for all primes $\underline{p} \neq \underline{m}$.

Here is the plan of the present paper.

Aiming at a study of the λ-dimension of trivial extensions in general, we start in §1 with the even more general case of a ring S with an ideal J and we relate finiteness properties of S/J to similar properties of S in some cases ("generalized Nakayama lemmata"). In §2 we study the special case of a trivial extension $S = R \propto M$, where M is an R-R-bimodule and $J = 0 \propto M$ (so that $J^2 = 0$). In §3 we give a fairly complete treatment of the case where R is a local commutative noetherian ring and M is an *artinian* R-module. As a result we get "half" Theorem A, that is an inequality ≤ in (3). The remaining inequality ≥ follows from a general estimate in §4. The theory of §4 gives rise to some problems about a duality dimension for modules over commutative rings (§5). In §6 the λ-dimension of Papick's ring is determined (here trivial extensions also come up!). Finally in §7 several open problems and generalizations are mentioned.

I wish to thank Christer Lech, Clas Löfwall and Peter Schenzel for stimulating discussions.

1. *On the higher Nakayama lemmata*

Let S be a ring and let J be a two-sided ideal in S. Assuming that the ring S/J has good finiteness properties, we want to deduce similar assertions about S. To be more precise, we introduce the following.

Definition 3. An S-module U is said to have a t-*finite presentation* if there is an exact sequence of S-modules

$$F_t \longrightarrow F_{t-1} \longrightarrow \cdots \longrightarrow F_1 \longrightarrow F_0 \longrightarrow U \longrightarrow 0$$

where the F_i are finitely generated and S-free.

Thus λ-dim S ≤ n if and only if each n-finitely presented

module is $(n+1)$-finitely presented.

THEOREM 1. *Let us suppose that* S/J *is a left coherent ring and that the following condition is satisfied.*

If V *is a left* S*-module with* $V = J.V$, *then* V *is finitely generated.* (*)

[This is true if, for example, J is nilpotent and then it follows that $V = 0$.] *Then an* S*-module* U *has a* t*-finite presentation over* S *if and only if*

$\operatorname{Tor}_i^S(S/J,U)$ *has a* $(t-i)$*-finite presentation over* S/J *for all* $i = 0,\ldots,t$. $(**)_t$

Remark 1. Since S/J is coherent, possession of a 1-finite presentation over S/J is equivalent to possession of an n-finite presentation for all $n \geq 1$.

Proof of Theorem 1. The proof is by induction on t. If $t = 0$, the assertion is this: U is a finitely generated S-module if and only if U/JU is a finitely generated S/J-module. This result is true even if S/J is not coherent. Of course, we only have to prove the " $\Leftarrow$ " implication. If U_F is the S-submodule of U generated by the lifting of a finite set of S/J-generators for U/JU, it follows that $U_F + JU = U$. But this implies that $J.(U/U_F) = U/U_F$, so that (*) gives that U/U_F is finitely generated. But then U is also finitely generated, being an extension of two finitely generated S-modules.

Assume now that $t \geq 1$, and let us prove that if U has a t-finite S-presentation, then $(**)_t$ is satisfied. We have an exact sequence

$$0 \longrightarrow U' \longrightarrow F_0 \longrightarrow U \longrightarrow 0 \tag{4}$$

where F_0 is a free, finitely generated S-module and U' has a finite $(t-1)$-presentation. If we apply the functor $\operatorname{Tor}_*^S(S/J,\,.\,)$ to (4), we obtain an exact sequence of left S/J-modules

$$0 \to \operatorname{Tor}_1^S(S/J,U) \to S/J\otimes_S U' \to S/J\otimes_S F_0 \to S/J\otimes_S U \to 0 \tag{5}$$

and an isomorphism

$$\operatorname{Tor}_i^S(S/J,U) \xrightarrow{\sim} \operatorname{Tor}_{i-1}^S(S/J,U'), \quad i \geq 2. \tag{6}$$

It follows from the induction hypothesis that $\operatorname{Tor}_i^S(S/J,U')$ is $(t-1-i)$-presented for $0 \leq i \leq t-1$. Now (5) implies that $\operatorname{Tor}_1^S(S/J,U)$

is (t-1)-presented (we use the fact that S/J is coherent), and that $S/J\otimes_S U$ is t-presented. Combining these observations with (6), we obtain $(**)_t$. That $(**)_t$ implies that U is t-finitely presented now follows essentially by reversing the preceding reasoning, and Theorem 1 is proved. □

COROLLARY. *Under the hypotheses of Theorem* 1, *the following assertions are equivalent* ($t \geq 1$):

(a) $\lambda\text{-dim } S \leq t$;

(b) if $\mathrm{Tor}_i^S(S/J,U)$ *is* S/J-*finitely presented for* $i \leq t-1$ *and finitely generated for* $i = t$, *it follows that it is also finitely presented for* $i = t$ *and finitely generated for* $i = t+1$. □

Example (cf. our paper [21] in these proceedings). Let

$$k \longrightarrow H_1 \longrightarrow H_2 \longrightarrow H_3 \longrightarrow k$$

be an extension of graded Hopf algebras. Put $S = H_2$ and let J be the two-sided ideal in H_2 generated by the augmentation ideal of H_1. Now

$$\mathrm{Tor}_i^S(S/J,U) = \mathrm{Tor}_i^{H_2}(k\otimes_{H_1} H_2,U) = \mathrm{Tor}_i^{H_1}(k,U).$$

Furthermore, the condition (*) of Theorem 1 is satisfied for positively graded modules. Using the natural variant of the λ-dimension, defined by means of the category of positively graded H_2-modules, we therefore easily obtain that *if* H_3 is coherent, then

$$\lambda\text{-dim } H_2 \leq 1 + \mathrm{gldim}\ H_1.$$

Here gldim means the global homological dimension, defined by means of positively graded modules. This result is best possible (cf. [21]).

More examples are given in the next section.

2. *Trivial extensions*

Consider now the following special case of §1. Let R be a ring, let M be an R-R-bimodule, and let

$$S = R \propto M$$

be the set of pairs (r,m) with pairwise addition and multiplication given by

$$(r,m).(r',m') = (r.r', r.m' + m.r').$$

This is called the "trivial" extension of R by M. Here $J = 0 \propto M$

is a two-sided ideal in S, $J^2 = 0$, and $S/J = R$, so that the theory of §1 can be applied, if R is left coherent. Let us first note that, in general, giving a left $R \propto M$-module is the same as giving a pair (U,f), consisting of a left R-module U and a left R-module map $M \otimes_R U \xrightarrow{f} U$, such that

$$M \otimes_R M \otimes_R U \xrightarrow{id_M \otimes_R f} M \otimes U \xrightarrow{f} U \tag{7}$$

is a complex.

Applying the theory of §1, we now obtain the following.

THEOREM 2. *Assume that* R *is left coherent, and that* M *is an* R-R-*bimodule. A left* $R \propto M$-*module* (U,f) *is* t-*finitely presented if and only if*

$$\mathrm{Tor}_i^{R \propto M}((R,0),(U,f))$$

is a (t-i)-*finitely presented* R-*module for* $0 \le i \le t$. □

Remark 1. If t = 0, the condition in Theorem 2 is that Coker f is a finitely generated R-module (no coherence condition on R is needed). If t = 1, we tensor the exact sequence of $R \propto M$-modules

$$0 \longrightarrow (M,0) \longrightarrow R \propto M \longrightarrow (R,0) \longrightarrow 0$$

with (U,f) and we obtain an exact sequence

$$0 \longrightarrow \mathrm{Tor}_1^{R \propto M}((R,0),(U,f)) \longrightarrow M \otimes_R \mathrm{Coker}\ f \xrightarrow{\bar{f}} U \longrightarrow \mathrm{Coker}\ f \longrightarrow 0, \tag{8}$$

where $\bar{f}$ is the natural map induced by f (use the fact that (7) is a complex). Thus (U,f) is 1-finitely presented, that is finitely presented in the usual terminology, if and only if Coker f and Ker $\bar{f}$ are finitely generated!

Remark 2. With the information we now have, it is easy to give a necessary and sufficient condition for $R \propto M$ to be left coherent in general (cf. [11; p. viii, ℓℓ. 8-16] and note in particular that if $R \propto M$ is left coherent, then so also is R).

While we are at it, let us deduce a simple consequence of Theorem 2 if M is (say) right R-flat.

COROLLARY 1. *Let* M *be right* R-*flat. Then* (U,f) *is* t-*finitely presented if and only if* $H_i((MfU)_*)$ *is* (t-i)-*finitely presented for* $0 \le i \le t$. *Here* $(MfU)_*$ *is the complex*

$$\cdots \longrightarrow M^{\otimes n}\otimes_R U \xrightarrow{\mathrm{id}_{M^{\otimes(n-1)}}\otimes_R f} M^{\otimes(n-1)}\otimes_R U \longrightarrow \cdots \xrightarrow{f} U.$$

Proof. Under the hypotheses made, it is easy to see (see [14],[18] or give a direct proof) that

$$\mathrm{Tor}_i^{R \propto M}((R,0),(U,f)) = H_i((MfU)_*).$$

This proves Corollary 1. □

It is possible to give a more explicit formula for the groups of Corollary 1. First recall (cf., for example, [18; Lemma 1, p. 385]) that we have an exact sequence of $R \propto M$-modules

$$0 \longrightarrow (\mathrm{Im}\ f,0) \longrightarrow (U,f) \longrightarrow (\mathrm{Coker}\ f,0) \longrightarrow 0.$$

This gives an exact sequence of complexes (recall that M is flat)

$$0 \longrightarrow (M0\mathrm{Im}\ f)_* \longrightarrow (MfU)_* \longrightarrow (M0\mathrm{Coker}\ f)_* \longrightarrow 0. \qquad (9)$$

The long exact sequence of homology associated to (9) breaks down into exact sequences

$$0 \longrightarrow H_i((MfU)_*) \longrightarrow M^{\otimes i}\otimes_R \mathrm{Coker}\ f \longrightarrow M^{\otimes(i-1)}\otimes_R \mathrm{Im}\ f \longrightarrow 0$$

for $i \geq 1$, and this gives the following useful formula:

$$H_i((MfU)_*) = M^{\otimes(i-1)}\otimes_R \mathrm{Ker}\left\{M\otimes_R \mathrm{Coker}\ f \xrightarrow{\bar{f}} \mathrm{Im}\ f\right\} \text{ for } i \geq 1, \qquad (10)$$

where $\bar{f}$ is the evident map induced by f (cf. Remark 1 above). Using (10) it is now possible to deduce the following result, which for simplicity we only formulate when R is noetherian.

COROLLARY 2. *Let* R *be a left noetherian ring and let* M *be an* R-R-*bimodule, which is right* R-*flat. Then* λ-dim $R \propto M \leq n$ $(n \geq 1)$ *if and only if, for all finitely generated left* R-*modules* C *and for all finitely generated left* R-*submodules* F of $M\otimes_R C$ *such that* $M^{\otimes i}\otimes_R F$ *is finitely generated for* $1 \leq i \leq n-1$, *we have that* $M^{\otimes n}\otimes_R F$ *is finitely generated too.*

Proof. We know that λ-dim $R \propto M \leq n$ if and only if all n-finitely presented $R \propto M$-modules are (n+1)-finitely presented. Therefore, using (10), we see that the condition in Corollary 2 is sufficient. Conversely, suppose that λ-dim $R \propto M \leq n$, and let C be any finitely generated left R-module and F a finitely generated R-submodule of $M\otimes_R C$. We construct a suitable $R \propto M$-module (U,f) as follows. Put $U = C \oplus (M\otimes_R C)/F$ and define $M\otimes_R U \xrightarrow{f} U$ to be the natural epimorphism $M\otimes_R C \longrightarrow (M\otimes_R C)/F$ on $M\otimes_R C$ and to be zero elsewhere. Then $\operatorname{Ker} \bar{f} = F$ and it follows from (10) and Corollary 1 that the condition in Corollary 2 is also necessary. □

Corollary 2 has the following immediate consequence.

COROLLARY 3. *Let* R *be a commutative noetherian ring, and let* M *be a free* R-*module. Then*

$$\lambda\text{-dim } R \propto M = \begin{cases} 0 & \text{if } M \text{ has finite rank,} \\ 2 & \text{if } M \text{ has infinite rank.} \end{cases}$$ □

Remark 1. We can express Corollary 3 by saying that for R noetherian

$$\lambda\text{-dim } R[X_1,\dots,X_n,\dots]/(X_1,\dots,X_n,\dots)^2 = 2$$

if we have infinitely many variables.

Remark 2. It is easy to see, using essentially the preceding theory (be inspired by [18] or [14]!), that if $S = k \propto M$, where k is a field and M is an infinite-dimensional k-vector space, then flat $\dim_S \prod_I S$ is infinite (I big). On the other hand we have just seen that λ-dim S = 2. This gives a counter-example to a surmise that for any non-noetherian ring S

$$\text{flat } \dim_S \prod_I S + 1 = \lambda\text{-dim } S \quad (I \text{ big}). \tag{11}$$

It is *known* that (11) is true if λ-dim S = 1 (left λ-dimension corresponds to right flat dimension).

3. *Trivial extensions of noetherian rings with artinian modules*

Our aim in this section is to prove the following result.

THEOREM 3. *Let* R *be a local commutative noetherian ring, let* M *be an artinian* R-*module, and let* $S = R \propto M$ *be the trivial extension of* R *by* M. *An* S-*module* (U,f) *is* n-*finitely presented* ($n \geq 0$) *if and only if the following three conditions are satisfied:*

(a) $\operatorname{Coker}(M\otimes_R U \xrightarrow{f} U)$ *is a finitely generated* R-*module;*

(b) $\operatorname{Ker} \bar{f} = \operatorname{Ker}(M\otimes_R \operatorname{Coker} f \xrightarrow{\bar{f}} U)$ *is an* R-*module of finite length* (*if* $n \geq 1$) (here $\bar{f}$ is the natural map induced by f; cf. §2);

(c) $\mathrm{Tor}_i^R(M,\mathrm{Coker}\ f)$ *is an* R-*module of finite length for* $1 \le i \le n-1$ (*if* $n \ge 2$).

Proof. We use part of the theory of §1 and §2. We have already calculated (cf. Remark 1 following Theorem 2 in §2) that

$$\mathrm{Tor}_i^{R \propto M}((R,0),(U,f)) = \begin{cases} \mathrm{Coker}\ f & \text{if } i = 0, \\ \mathrm{Ker}\ \bar{f} & \text{if } i = 1. \end{cases}$$

In view of Theorem 2, this proves Theorem 3 if $n \le 1$. Indeed, if Coker f is finitely generated, we obtain that $M\otimes_R \mathrm{Coker}\ f$ is artinian (M is artinian) and its submodule $\mathrm{Ker}\ \bar{f}$ is therefore finitely generated if and only if it is of finite length. To handle the case where $n \ge 2$, write [assuming (a) and (b) are satisfied] (U,f) as a quotient of a finitely generated free $R \propto M$-module as follows (cf. [18; p. 387]). Take an exact sequence $P \xrightarrow{\pi'} \mathrm{Coker}\ f \longrightarrow 0$, where P is finitely generated and R-free, and factorize π' as $P \xrightarrow{\pi} U \longrightarrow \mathrm{Coker}\ f$. Consider the R-module $K_0 = (M\otimes_R P) \oplus P$. This module is mapped *onto* U by the map $p = (f(\mathrm{Id}_M \otimes_R \pi), \pi)$. We define a natural map $M\otimes_R K_0 \xrightarrow{f_0} K_0$ by taking the identity map on $M\otimes_R P$ and the zero map elsewhere. Then (K_0, f_0) is a finitely generated free $R \propto M$-module and we have a natural epimorphism of $R \propto M$-modules $(K_0,f_0) \longrightarrow (U,f)$, whose kernel we shall denote by (K,ψ). We therefore have an exact sequence of $R \propto M$-modules

$$0 \longrightarrow (K,\psi) \longrightarrow (K_0,f_0) \longrightarrow (U,f) \longrightarrow 0,$$

which gives rise to an exact and commutative diagram

$$
\begin{array}{ccccccc}
 & & 0 & & & & \\
 & & \downarrow & & & & \\
M\otimes_R \mathrm{Coker}\ \psi & \xrightarrow{\bar{\psi}} & K & \longrightarrow & \mathrm{Coker}\ \psi & \longrightarrow & 0 \\
\downarrow {\scriptstyle \mathrm{Id}_M\otimes_R \tau} & & \downarrow & & \downarrow {\scriptstyle \tau} & & \\
M\otimes_R P & \xrightarrow{\bar{f}_0} & (M\otimes_R P)\oplus P & \longrightarrow & P & \longrightarrow & 0 \\
\downarrow {\scriptstyle \mathrm{Id}_M\otimes_R \pi'} & & \downarrow {\scriptstyle p} & & \downarrow {\scriptstyle \pi'} & & \\
M\otimes_R \mathrm{Coker}\ f & \xrightarrow{\bar{f}} & U & \longrightarrow & \mathrm{Coker}\ f & \longrightarrow & 0 \\
\downarrow & & \downarrow & & \downarrow & & \\
0 & & 0 & & 0 & &
\end{array} \quad . \qquad (12)
$$

Here $\bar{f}$, $\bar{f}_0$ and $\bar{\psi}$ are the natural maps induced by f, f_0 and ψ [note that $\bar{f}_0$ turns out to be the natural inclusion in (12)] and τ is defined by the diagram. Apply the snake lemma to (12)! This gives an exact sequence

$$0 \longrightarrow \mathrm{Ker}\ \bar{f} \longrightarrow \mathrm{Coker}\ \psi \xrightarrow{\tau} P \longrightarrow \mathrm{Coker}\ f \longrightarrow 0. \qquad (13)$$

Put $H = \mathrm{Im}\ \tau$, break up (13) into two short exact sequences and apply the functor $\mathrm{Tor}_*^R(M, .)$! We easily obtain two exact sequences

$$M\otimes_R \mathrm{Ker}\ \bar{f} \longrightarrow \mathrm{Ker}(\mathrm{Id}_M\otimes_R \tau) \longrightarrow \mathrm{Tor}_1^R(M,\mathrm{Coker}\ f) \longrightarrow 0 \qquad (14)$$

and, for $i \geq 1$,

$$\mathrm{Tor}_i^R(M,\mathrm{Ker}\ \bar{f}) \longrightarrow \mathrm{Tor}_i^R(M,\mathrm{Coker}\ \psi) \longrightarrow \mathrm{Tor}_{i+1}^R(M,\mathrm{Coker}\ f)$$
$$\longrightarrow \mathrm{Tor}_{i-1}^R(M,\mathrm{Ker}\ \bar{f}) \longrightarrow . \qquad (15)$$

[Note that $\mathrm{Tor}_{i+1}^R(M,\mathrm{Coker}\, f) \xrightarrow{\sim} \mathrm{Tor}_i^R(M,H)$ if $i \geq 1$.]

Now, since (U,f) is n-finitely presented if and only if (K,ψ) is (n-1)-finitely presented, we can easily deduce Theorem 3 by induction, using (13), (14), (15), using the fact that $\mathrm{Ker}\ \bar{\psi} = \mathrm{Ker}(\mathrm{Id}_M\otimes_R \tau)$ [which follows from (12), since $\bar{f}_0$ is a monomorphism] and using the following general lemma.

LEMMA. *Let* $(R,\underline{m})$ *be a local noetherian ring, let* M *be an artinian* R*-module, and let* L *be an* R*-module of finite length. Then* $\mathrm{Tor}_i^R(M,L)$ *is of finite length for all* $i \geq 0$.

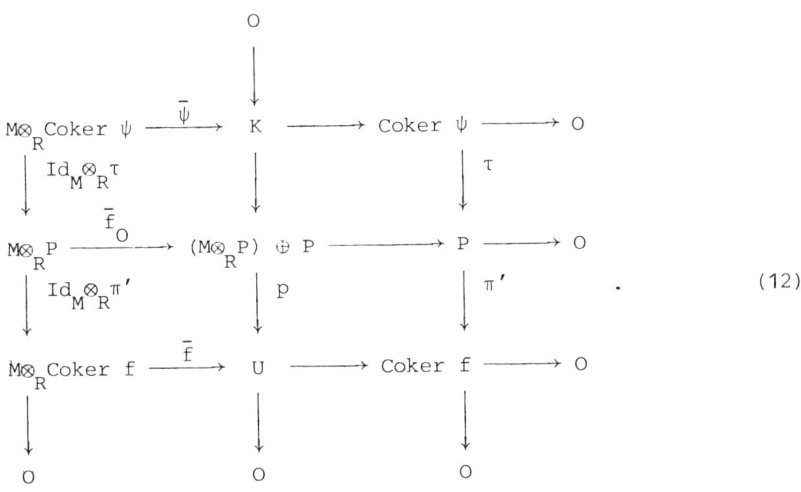

Proof. Let $I(k)$ be the injective envelope of $k = R/\underline{m}$, and let $P_*(L)$ be an R-free resolution of L. We have

$$\mathrm{Hom}_R(\mathrm{Tor}_i^R(M,L),I(k)) = H^i(\mathrm{Hom}_R(M\otimes_R P_*(L),I(k)))$$
$$= H^i(\mathrm{Hom}_R(P_*(L),\mathrm{Hom}_R(M,I(k))))$$
$$= \mathrm{Ext}_R^i(L,\mathrm{Hom}_R(M,I(k))) = \mathrm{Ext}_{\hat{R}}^i(L,\mathrm{Hom}_R(M,I(k))),$$

where $\hat{R}$ is the completion of R, $\mathrm{Hom}_R(M,I(k))$ is a finitely generated $\hat{R}$-module and L is an R- [and $\hat{R}$-] module of finite length. It follows that $\mathrm{Hom}_R(\mathrm{Tor}_i^R(M,L),I(k))$ is an $\hat{R}$-module of finite length, so that $\mathrm{Tor}_i^R(M,L)$ is an R-module of finite length. This proves the lemma and Theorem 3 is completely proved. □

COROLLARY 1. *Let* $(R,\underline{m})$ *be a local commutative noetherian ring and let* M *be an artinian* R-*module. Then* λ-dim R α M ≤ σ *if and only if the following condition is satisfied.*

For all finitely generated R-*modules* V *such that* $\mathrm{Tor}_i^R(M,V)$ *is of finite length for* $1 \le i \le \sigma-1$, *it follows that* $\mathrm{Tor}_\sigma^R(M,V)$ *is of finite length too.* $\quad (C)_\sigma$

Proof. Corollary 1 is clearly true if σ = 0. Assume now that σ ≥ 1. It follows from Theorem 3 that λ-dim R α M ≤ σ if and only if, for every R α M-module (U,f) with Coker f finitely generated, Ker $\bar{f}$ of finite length, and $\mathrm{Tor}_i^R(M,\mathrm{Coker}\ f)$ of finite length for $1 \le i \le \sigma - 1$ [this is an empty condition if σ = 1!], we have that $\mathrm{Tor}_\sigma^R(M,\mathrm{Coker}\ f)$ has finite length. Therefore $(C)_\sigma$ implies that λ-dim R α M ≤ σ.

Assume now conversely that λ-dim R α M ≤ σ, and let V be any finitely generated R-module such that $\mathrm{Tor}_i^R(M,V)$ is of finite length for $1 \le i \le \sigma - 1$. Construct an R α M-module (U,f) by taking $U = (M\otimes_R V) \oplus V$ and by defining $M\otimes_R U \xrightarrow{f} U$ to be the identity on $M\otimes_R V$ and zero elsewhere. Then Coker f = V, Ker $\bar{f}$ = 0, so that (U,f) is σ-finitely presented, and therefore (σ + 1)-finitely presented, since λ-dim R α M ≤ σ. But this gives that $\mathrm{Tor}_\sigma^R(M,V)$ is of finite length, so that $(C)_\sigma$ is verified and Corollary 1 is completely proved. □

COROLLARY 2. *Let* $(R,\underline{m})$ *be a local commutative noetherian ring, let* N *be a finitely generated* R-*module and let* $M = \mathrm{Hom}_R(N,I(k))$. *Then* λ-dim R α M ≤ σ *if and only if, for all finitely generated*

R-modules V *such that* $\mathrm{Ext}^i_R(V,N)$ *is of finite length for* $1 \le i \le \sigma - 1$, *we have that* $\mathrm{Ext}^{\sigma}_R(V,N)$ *is of finite length.*

Proof. It is known that the M of Corollary 2 is artinian. Corollary 2 now follows from Theorem 3 and the natural isomorphism

$$\mathrm{Hom}_R(\mathrm{Tor}^R_i(\mathrm{Hom}_R(N,I(k)),V),I(k)) \simeq \mathrm{Ext}^i_R(V,N)\otimes_R \hat{R}. \quad \square$$

COROLLARY 3. *Let* $(R,\underline{m})$ *be a local commutative noetherian ring such that* $R_{\underline{p}}$ *is a Gorenstein ring for all prime ideals* $\underline{p} \neq \underline{m}$. *Then*

$\lambda\text{-dim } R \,\alpha\, I(k) \le \dim R$.

Proof. Take N = R in Corollary 2. In view of Corollary 2 it is now *sufficient* to prove that for a ring R as in Corollary 3, $\mathrm{Ext}^{\dim R}_R(V,R)$ has finite length for *all* finitely generated R-modules. But this Ext-group, being a finitely generated R-module, has finite length if and only if its localizations at all non-maximal primes are zero. But

$$\mathrm{Ext}^{\dim R}_R(V,R)_{\underline{p}} = \mathrm{Ext}^{\dim R}_{R_{\underline{p}}}(V_{\underline{p}},R_{\underline{p}}) = 0$$

if $\underline{p} \subsetneq \underline{m}$, for then $\dim R_{\underline{p}} < \dim R$ and $\mathrm{injdim}\, R_{\underline{p}} = \dim R_{\underline{p}}$, since $R_{\underline{p}}$ is supposed to be Gorenstein [5]. $\square$

Remark. In the next section we shall prove a general result which implies that we always have equality in Corollary 3.

4. *An estimate for the λ-dimension*

The aim of this section is to prove the following theorem.

THEOREM 4. *Let* $(R,\underline{m})$ *be any local commutative noetherian ring. Then*

$$\lambda\text{-dim } R \,\alpha\, I(k) \ge \dim R. \tag{16}$$

Proof. Assume that (16) is false, that is that $\lambda\text{-dim } R \,\alpha\, I(k) < \dim R$. But this last inequality implies that $\dim R \ge 2$: otherwise we would have $\lambda\text{-dim } R \,\alpha\, I(k) = 0$, that is I(k) is noetherian, which is only possible if dim R = 0, in which case we have equality in (16). Thus $n = \dim R \ge 2$ and $\lambda\text{-dim } R \,\alpha\, I(k) \le n-1$. Corollary 2 (for N = R) of Theorem 3 in §3 now implies that *if* a finitely generated R-module V is such that $\mathrm{Ext}^i_R(V,R)$ has finite length for $1 \le i \le n-2$ (this is an empty condition if n = 2), then

$\mathrm{Ext}_R^{n-1}(V,R)$ has finite length too. We shall construct a module V that contradicts this. According to Proposition 5.2 in [4], there exists in R (recall that dim R = n) a prime ideal $\underline{p}$ of height n-1 such that $R_{\underline{p}}$ is a Cohen-Macaulay ring. I claim that $V = R/\underline{p}$ now gives the desired contradiction. Indeed,

$$\mathrm{Ext}_R^i(R/\underline{p},R)_{\underline{p}} = \mathrm{Ext}_{R_{\underline{p}}}^i(R_{\underline{p}}/\underline{p}R_{\underline{p}},R_{\underline{p}}) = 0$$

for $i < n-1$ and is $\neq 0$ for $i = n-1$, since $R_{\underline{p}}$ is Cohen-Macaulay of dimension n-1. If $\underline{p}'$ is any other prime $\neq \underline{m}$, we have $\underline{p} \not\subset \underline{p}'$, so that $(R/\underline{p})_{\underline{p}'} = 0$, and therefore $\mathrm{Ext}_R^i(R/\underline{p},R)_{\underline{p}'} = 0$ for all i. Since a finitely generated R-module is of finite length if and only if its localizations at all primes $\neq \underline{m}$ are zero, it follows that $\mathrm{Ext}_R^i(R/\underline{p},R)$ is of finite length for $0 \leq i \leq n-2$, and *not* of finite length for $i = n-1$. Therefore $V = R/\underline{p}$ does indeed give the desired contradiction and Theorem 4 is proved. □

Combining Theorem 4 with Corollary 3 of Theorem 3 in §3, we obtain the following generalization of Theorem A that was announced in the Introduction.

THEOREM A′. *Let* $(R,\underline{m})$ *be a local commutative noetherian ring such that* $R_{\underline{p}}$ *is a Gorenstein ring for all prime ideals* $\underline{p} \neq \underline{m}$. *Then*

$\lambda\text{-dim } R \propto I(k) = \dim R$. □

When do we have equality in (16)? The following problem seems natural.

PROBLEM 4. *Let* $(R,\underline{m})$ *be a local ring such that*

$\lambda\text{-dim } R \propto I(k) = \dim R$.

Does it follow that $R_{\underline{p}}$ *is a Gorenstein ring for all primes* $\underline{p} \neq \underline{m}$?

Here follows a partial result in this direction.

THEOREM 5. *Let* $(R,\underline{m})$ *be a local ring such that* $\lambda\text{-dim } R \propto I(k) \leq 2$. *Then* $R_{\underline{p}}$ *is a Gorenstein ring for all primes* $\underline{p} \neq \underline{m}$ *(and it follows that* $\dim R = \lambda\text{-dim } R \propto I(k) \leq 2$*).*

Proof. If M is a finitely generated R-module, we denote by M^* the dual $\mathrm{Hom}_R(M,R)$ of M and by M^{**} the bidual $(M^*)^*$. We have a natural biduality map $M \xrightarrow{\sigma} M^{**}$ whose kernel and cokernel can be calculated as follows (cf. [3; Chapter 2] or [23; Chapitre 3]).

Let

$$P_1 \xrightarrow{\phi} P_0 \longrightarrow M \longrightarrow 0 \qquad (17)$$

be the beginning of a finitely generated projective resolution of M. Dualize (17)! We obtain an exact sequence, which defines a module D(M):

$$0 \longrightarrow M^* \longrightarrow P_0^* \xrightarrow{\phi^*} P_1^* \longrightarrow D(M) \longrightarrow 0. \qquad (18)$$

(Of course D(M) depends on the resolution (17).) Put V equal to the image of ϕ^* in (18), break up (18) into two short exact sequences, and apply the functor $\mathrm{Ext}_R^*(\ .\ ,R)$! This gives (more details are available in *loc. cit.*) an exact sequence

$$0 \longrightarrow \mathrm{Ext}_R^1(D(M),R) \longrightarrow M \xrightarrow{\sigma} M^{**} \longrightarrow \mathrm{Ext}_R^2(D(M),R) \longrightarrow 0. \qquad (19)$$

Let us recall that M is said to be *torsionless* if σ is a monomorphism. The torsionless modules are exactly the submodules of finitely generated projective modules. If M is torsionless, then (19) can be written

$$0 \longrightarrow M \longrightarrow M^{**} \longrightarrow \mathrm{Ext}_R^1(V,R) \longrightarrow 0$$

where V, being a submodule of P_1^* (a first syzygy of D(M)) is also torsionless. Repeating the argument above for the torsionless module V, choosing a finitely generated projective module P_2 that maps onto M^* and using this map and (18) to get

$$P_2 \longrightarrow P_0^* \longrightarrow V \longrightarrow 0$$

as the beginning of a projective resolution of V, we easily obtain an exact sequence (cf. [12] or the papers cited above)

$$0 \longrightarrow V \longrightarrow V^{**} \longrightarrow \mathrm{Ext}_R^1(M,R) \longrightarrow 0, \qquad (20)$$

since M turns out to be a first syzygy of D(V). Note that in (20) M can be any torsionless R-module. All this is quite general (it even works in a non-commutative setting). Let us now return to the case of Theorem 5, where we suppose that λ-dim R α I(k) ≤ 2. If M is *any* torsionless module, and if V is constructed as above, it follows from (20) that $\mathrm{Ext}_R^1(D(V),R) = 0$, and so is of finite length. Thus the condition on the λ-dimension implies that $\mathrm{Ext}_R^2(D(V),R)$ is of finite length, and since $\mathrm{Ext}_R^2(D(V),R) \simeq \mathrm{Ext}_R^1(M,R)$ we obtain that

$\mathrm{Ext}_R^1(M,R)$ is of finite length for all torsionless R-modules M. Taking in particular for M any prime ideal $\underline{p}$, we obtain that $\mathrm{Ext}_R^1(\underline{p},R) \simeq \mathrm{Ext}_R^2(R/\underline{p},R)$ is of finite length. Therefore, if $\underline{p} \neq \underline{m}$, we obtain that

$$\mathrm{Ext}^2_{R_{\underline{p}}}(R_{\underline{p}}/\underline{p}R_{\underline{p}},R_{\underline{p}}) = 0,$$

and $\dim R_{\underline{p}} \leq 1$ (recall that $\dim R \leq 2$ according to Theorem 4). But according to [5; Lemma 3.5], this implies that $R_{\underline{p}}$ is Gorenstein and Theorem 5 is proved. □

Remark. Theorem 5 generalizes the results about coherence on pp. 31-32 of [11].

Here is a strange consequence of Theorem 5.

COROLLARY. *Let* $(R,\underline{m})$ *be a local ring of Krull dimension* 1. *Then either* λ-dim R α I(k) *is* 1, *or it is* ≥ 3. □

In the next section we shall develop a theory that, in particular, will show that the λ-dimension *can* be 3 in the preceding Corollary.

5. *A duality dimension for local rings*

In [20] several examples of local rings $(R,\underline{m})$ are given for which there is a universal constant n(R) such that if V is a finitely generated R-module with $\mathrm{Ext}_R^i(V,R) = 0$, $1 \leq i \leq n(R)$, then $\mathrm{Ext}_R^i(V,R) = 0$ for $i \geq 1$. In particular, if $\underline{m}^2 = 0$, then $n(R) = 2$ works; if R is Gorenstein it is evident that $n(R) = \dim R$ works, etc. This leads us to the following definition.

Definition. Let $(R,\underline{m})$ be a local ring. Put

$$\alpha(R) = \inf\left\{ n \;\middle|\; \begin{array}{l}\text{for each finitely generated R-module V with}\\ \mathrm{Ext}_R^i(V,R) = 0,\ 1 \leq i \leq n, \text{ we have that } \mathrm{Ext}_R^{n+1}(V,R) = 0\end{array}\right\}.$$

If no such n exists, we put $\alpha(R) = \infty$.

In what follows we shall call $\alpha(R)$ the *duality dimension* of R. Clearly $\alpha(R) = 0$ if and only if R is artinian and Gorenstein, and we shall see that $\alpha(R) = 1$ can never occur if R is artinian.

PROBLEM 5. *Is it true that* $\alpha(R) < \infty$ *for all local commutative noetherian rings?*

One has $\alpha(R) = 0$ or 2 if $\underline{m}^2 = 0$, the first case only occurring if the embedding dimension of R is ≤ 1, and one can prove (cf. Theorem 7 below) that for "most" local rings $(R,\underline{m})$ with $\underline{m}^3 = 0$, we have that $\alpha(R) < \infty$.

The following result connects λ-dimension with α-dimension and it will make the Corollary of Theorem 5 in §4 more precise.

THEOREM 6. *Let* $(A,\underline{n})$ *be a local artinian ring and let* $R = A[[T]]$ *be the local ring (of Krull dimension* 1*) of formal power series over* A. *Let* $I(k)$ *be the* R*-injective hull of the residue field* k *of* R. *Let* $B = R_{\underline{n}.(T)}$ *be the localization of* R *at the minimal prime* $\underline{n}.(T)$ *of* R. *Then*

$$\lambda\text{-dim } R \;\alpha\; I(k) = \alpha(B) + 1.$$

Remark. Note that the natural map $A \rightarrow B$ is a faithfully flat local ring homomorphism, whose fibre $k \otimes_A B$ is a field. Therefore A and B have the same Krull dimension, same Hilbert series, same multiplicity, and we *probably* also have $\alpha(A) = \alpha(B)$. This is at least true if $\underline{n}^2 = 0$ (this follows from Theorem 7(b) below). In any case, $\alpha(A) \leq \alpha(B)$ (as is proved below). Constructions, similar to $A \rightarrow B$, are studied in section 3 of [22].

Proof of Theorem 6. Let us first prove that

$$\lambda\text{-dim } R \;\alpha\; I(k) \leq \alpha(B) + 1. \tag{21}$$

We may assume that $\alpha(B) < \infty$. Let V be a finitely generated R-module such that $\mathrm{Ext}_R^i(V,R)$ is of finite length for $1 \leq i \leq \alpha(B)$. Localizing at the only prime ideal $\underline{p} = \underline{n}.(T)$ different from the maximal ideal $\underline{m} = \underline{n} + (T)$ in R, we obtain (recall that $B = R_{\underline{p}}$)

$$\mathrm{Ext}_B^i(V_{\underline{p}},B) = 0, \quad 1 \leq i \leq \alpha(B),$$

so that $\mathrm{Ext}_B^i(V_{\underline{p}},B) = 0$, $i \geq 1$, and since $\underline{p}$ is the only prime $\neq \underline{m}$ in R, we also get that $\mathrm{Ext}_R^i(V,R)$ is of finite length for all $i \geq 1$. Thus the inequality (21) is proved. We now prove that

$$\alpha(B) + 1 \leq \lambda\text{-dim } R \;\alpha\; I(k). \tag{22}$$

Let the right hand side of (22) be λ. We can assume that $\lambda < \infty$. Let M be a finitely generated B-module such that $\mathrm{Ext}_B^i(M,B) = 0$, $1 \leq i \leq \lambda - 1$ (this is to be considered as an empty condition if

$\lambda = 1$). Since $B = R_{\underline{p}}$ there exists a finitely generated R-module N such that $N_{\underline{p}} = M$. Thus

$$\mathrm{Ext}^i_R(N,R)_{\underline{p}} = \mathrm{Ext}^i_{R_{\underline{p}}}(N_{\underline{p}},R_{\underline{p}}) = \mathrm{Ext}^i_B(M,B). \tag{23}$$

Therefore our assumption on M gives, using (23), that $\mathrm{Ext}^i_R(N,R)$ is of finite length for $1 \le i \le \lambda - 1$, and therefore the higher $\mathrm{Ext}^i_R(N,R)$ are also of finite length; it follows that [use (23)] $\mathrm{Ext}^i_B(M,B) = 0$, $i \ge 1$. Therefore (22) is proved and the proof of Theorem 6 is completed. □

Remark 1. Theorem 6 can be easily generalized to the following. *Let* $(R,\underline{m})$ *be any local commutative noetherian ring of Krull dimension* 1, *having only one minimal prime* $\underline{p}$. *Then* λ-dim R α I(k) = dim R + $\alpha(R_{\underline{p}})$.

Remark 2. Let us prove that $\alpha(A) \le \alpha(B)$ if $A \to B$ is a faithfully flat ring homomorphism. If N is a finitely generated A-module, it follows, since B is A-flat and A,B are noetherian, that we have

$$\mathrm{Ext}^i_B(B\otimes_A N,B) = \mathrm{Ext}^i_A(N,B) = \mathrm{Ext}^i_A(N,A)\otimes_A B. \tag{24}$$

Thus if $\mathrm{Ext}^i_A(N,A) = 0$, $1 \le i \le \alpha(B)$, it follows from (24) that $\mathrm{Ext}^i_B(B\otimes_A N,B) = 0$, $1 \le i \le \alpha(B)$, and since $B\otimes_A N$ is finitely generated as a B-module, we obtain that $\mathrm{Ext}^i_B(B\otimes_A N,B) = 0$, $i \ge 1$. Going backwards, using (24) and the fact that B is *faithfully* A-flat, we obtain that $\mathrm{Ext}^i_A(N,A) = 0$, $i \ge 1$, and we have proved that $\alpha(A) \le \alpha(B)$. □

How should one calculate $\alpha(B)$ for a given local noetherian ring? Here are some partial results, some new and some old.

THEOREM 7. *Let* $(R,\underline{m})$ *be a local commutative noetherian ring.*

(a) *If* R *is artinian and not Gorenstein, then* $\alpha(R) \ge 2$ (*in the Gorenstein case we have* $\alpha(R) = 0$).

(b) *If* $\underline{m}^2 = 0$, *we have* $\alpha(R) = 2$, *whenever* R *is not Gorenstein, that is if and only if* $\dim_k \underline{m} \ge 2$.

(c) *If* $\underline{m}^3 = 0$, *we have* $\alpha(R) < \infty$ *for "most"* $(R,\underline{m})$. *More precisely, if* $\alpha(R) = \infty$, *then the "Bass series"*

$$I^R(Z) = \sum_{i\geq 0} \mu_i(R).Z^i,$$

where $\mu_i(R) = \dim_k \mathrm{Ext}_R^i(k,R)$, *would be a rational function (even a geometric series from* $\mu_1(R)$ *on), and this is false for most* $(R,\underline{m})$; *in particular this is false for those* $(R,\underline{m})$, *studied in* [9], *having transcendental* $I^R(Z)$. ([9] uses [2] and [15].)

(d) *Let* x *be a non-zerodivisor in the maximal ideal of* R *and let* $S = R/xR$. *Then*

$$\alpha(S) \leq \alpha(R) \leq \alpha(S) + 2.$$

In particular, $\alpha(R)$ *and* $\alpha(S)$ *are finite at the same time.*

Proof. (a) Let $D(k)$ be the "transpose" of the R-module $k = R/\underline{m}$ [cf. formula (18) in §4]. Recall the exact sequence (19) of §4

$$0 \longrightarrow \mathrm{Ext}_R^1(D(k),R) \longrightarrow k \longrightarrow k^{**} \longrightarrow \mathrm{Ext}_R^2(D(k),R) \longrightarrow 0. \quad (25)$$

Since soc R (the socle of R) is $\neq 0$, the map $k \to k^{**}$ is a monomorphism, and therefore $\mathrm{Ext}_R^1(D(k),R) = 0$. Thus if $\alpha(R) \leq 1$, we would have $\mathrm{Ext}_R^2(D(k),R) = 0$. But it follows easily from (25) that this is only possible if soc(R) is a one-dimensional vector space, that is if R is a Gorenstein ring (I thank Clas Löfwall for this proof).

(b) We shall prove the stronger but known (cf. [16],[20]) result that if (under the hypotheses made) $\mathrm{Ext}_R^2(M,R) = 0$, then M is free. Let $P(M) \longrightarrow M$ be the projective envelope of M, and consider the exact sequence

$$0 \longrightarrow K \longrightarrow P(M) \longrightarrow M \longrightarrow 0.$$

If M is not free, then $K \neq 0$. But since $P(M) \longrightarrow M$ is a projective envelope, we have $K \subset \underline{m}.P(M)$, so that $\underline{m}.K = 0$ ($\underline{m}^2 = 0$). Therefore $K \simeq \bigoplus_1^s k$ $(s \geq 1)$ and

$$0 = \mathrm{Ext}_R^2(M,R) \simeq \mathrm{Ext}_R^1(K,R) \simeq \bigoplus_1^s \mathrm{Ext}_R^1(k,R),$$

so that $\mathrm{Ext}_R^1(k,R) = 0$; this implies that R is Gorenstein, since R is artinian [5].

(c) To prove (c) we prove the following more precise result.

PROPOSITION 1. *Let* $(R,\underline{m})$ *be a local commutative noetherian ring with* $\underline{m}^3 = 0$, *and let* γ *be an integer* > 2 *such that there exists a finitely generated* R-*module* M *with* $\mathrm{Ext}_R^i(M,R) = 0$ *for* $1 \leq i \leq \gamma$, *but is such that* $\mathrm{Ext}_R^i(M,R) \neq 0$ *for some* $i > \gamma$. *Then* $\mu_{i+1}(R)/\mu_i(R)$ *is constant for* $1 \le i \le \gamma-2$.

Proof. Under the hypotheses made, R is *not* Gorenstein, and therefore all $\mu_i(R) \neq 0$. Let, as before, P(M) be the projective envelope of M and consider the exact sequence

$$0 \longrightarrow K \longrightarrow P(M) \longrightarrow M \longrightarrow 0.$$

We have $K \subset \underline{m}.P(M)$ and therefore $\underline{m}^2.K = 0$. But according to the hypotheses made, we have $\mathrm{Ext}_R^i(K,R) = 0$ for $1 \le i \le \gamma-1$, and the exact sequence

$$0 \longrightarrow \underline{m}.K \longrightarrow K \longrightarrow K/\underline{m}.K \longrightarrow 0$$

gives rise to isomorphisms

$$\mathrm{Ext}_R^i(\underline{m}.K,R) \simeq \mathrm{Ext}_R^{i+1}(K/\underline{m}.K,R) \text{ for } 1 \le i \le \gamma-2.$$

Thus since $\underline{m}.(\underline{m}.K) = 0$, we obtain (for a k-vector space V, we define $|V| = \dim_k V$)

$$|\underline{m}.K|.\mu_i(R) = |K/\underline{m}.K|.\mu_{i+1}(R) \text{ for } 1 \le i \le \gamma-2. \tag{26}$$

If one of $|\underline{m}.K|$ or $|K/\underline{m}.K|$ is zero then according to (26) the other one would be so too, so that we would have $K=0$, which contradicts the fact that M is not free. Therefore (26) implies that $\mu_{i+1}(R)/\mu_i(R)$ is a non-zero constant for $1 \le i \le \gamma-2$. This proves Proposition 1 and thereby also assertion (c) of Theorem 7. □

(d) First we observe that we have a natural change of rings spectral sequence (M is any finitely generated S-module)

$$E_2^{p,q} = \mathrm{Ext}_S^p(M,\mathrm{Ext}_R^q(S,R)) \Rightarrow \mathrm{Ext}_R^n(M,R),$$

which degenerates into an isomorphism [we use the fact that $\mathrm{Ext}_R^i(S,R)$ is 0 for $i \neq 1$ and S for $i = 1$]

$$\mathrm{Ext}_S^{n-1}(M,S) \simeq \mathrm{Ext}_R^n(M,R) \text{ for all } n. \tag{27}$$

Let us now prove the inequality $\alpha(S) \le \alpha(R)$ of (d)! Suppose therefore further that M is such that $\mathrm{Ext}_S^i(M,S) = 0$, $1 \le i \le \alpha(R)$. Now (27) implies that $\mathrm{Ext}_R^i(M,R) = 0$, $2 \le i \le \alpha(R) + 1$. But if K is a first R-syzygy of any R-module M, we have $\mathrm{Ext}_R^j(K,R) \simeq \mathrm{Ext}_R^{j+1}(M,R)$ for all $i \ge 1$. In our particular case we obtain that $\mathrm{Ext}_R^j(K,R) = 0$, $1 \le j \le \alpha(R)$, which by the definition of $\alpha(R)$ implies that all $\mathrm{Ext}_R^j(K,R)$ are 0 for $j \ge 1$; thus $\mathrm{Ext}_R^{j+1}(M,R) = 0$, $j \ge 1$, and so (use (27)) $\mathrm{Ext}_S^j(M,S) = 0$, $j \ge 1$; thus $\alpha(S) \le \alpha(R)$ is proved.

Finally, to prove that $\alpha(R) \leq \alpha(S) + 2$, we suppose that M is a finitely generated R-module and that $\text{Ext}_R^i(M,R) = 0$, $1 \leq i \leq \alpha(S) + 2$. Then for a finitely generated first R-syzygy K of M we have $\text{Ext}_R^i(K,R) = 0$, $1 \leq i \leq \alpha(S) + 1$. Now x is a non-zerodivisor on K and the exact sequence

$$0 \longrightarrow K \xrightarrow{.x} K \longrightarrow K/xK \longrightarrow 0$$

gives that $\text{Ext}_R^i(K/xK,R) = 0$, $2 \leq i \leq \alpha(S) + 1$. But K/xK is a finitely generated S-module, and from (27) it now follows that $\text{Ext}_S^j(K/xK,S) = 0$ for $1 \leq j \leq \alpha(S)$, so that $\text{Ext}_S^j(K/xK,A) = 0$, for $j \geq 1$, by the definition of $\alpha(S)$. Therefore (use (27) again) $\text{Ext}_R^n(K/xK,R) = 0$ for $n \geq 2$, so that

$$\text{Ext}_R^i(K,R) \xrightarrow{.x} \text{Ext}_R^i(K,R)$$

is onto for $i \geq 1$. According to the Nakayama lemma, this implies that $\text{Ext}_R^i(K,R) = 0$ for $i \geq 1$, so that $\text{Ext}_R^i(M,R) = 0$ for $i \geq 1$, and therefore $\alpha(R) \leq \alpha(S) + 2$. Thus Theorem 7 is completely proved. □

Remark. The idea of the proof of the last inequality of (d) comes from [20; Proposition 2.8], which is a variant of a result of Menzin [16; Proposition 7].

COROLLARY. *Let* $(A,\underline{n})$ *be a local ring with* $\underline{n}^2 = 0$, *and of embedding dimension* ν. *Put* $R = A[[T]]$. *Then* $\lambda\text{-dim } R\ \alpha\ I(k) = 3$ *if* $\nu \geq 2$, *and for* $\nu \leq 1$, *the* λ*-dimension is* 1.

Proof. A and $B = R_{\underline{n}.(T)}$ have the same embedding dimension, and both rings have the squares of their maximal ideals equal to 0. Now apply Theorem 6. Theorem 7(b) shows that $\alpha(B) = \alpha(A) = (2 \text{ or } 0)$. This gives the result. □

6. *On the rings between a noetherian domain and its field of fractions*

In [19] Papick proved the following result.

Let R *be a commutative noetherian domain with field of fractions* K. *The following two conditions are equivalent:*

(a) *each ring* S *with* $R \subset S \subset K$ *is coherent;*

(b) $\dim R \leq 1$.

To prove that (a) implies (b), Papick reduces everything to the case when R is a local 2-dimensional Cohen-Macaulay domain. Let

(x,y) be an R-sequence in the maximal ideal $\underline{m}$ of R. Then the subring S of K generated by R and $\{x^n/y^{n-1}\}_{n\geq 2}$ is proved in [19] to be non-coherent. Our aim here is to study briefly λ-dim S. Let us introduce variables $\{X_n\}_{n\geq 2}$, and let S' be the ring

$$S' = R[X_2,\ldots,X_n,\ldots]/(X_2y-x^2, X_3y-xX_2,\ldots,X_{n+1}y-xX_n,\ldots). \quad (28)$$

Map S' onto S by the identity map on R and by $X_n \longmapsto x^n/y^{n-1}$, $n \geq 2$. It is easy to see that this map is an isomorphism. Indeed, the elements of S' have representations similar to the representation (1) in [19]. Taking a non-zero element in the kernel of $S' \longrightarrow S$ having smallest possible n_t easily gives a contradiction. Thus we have $S' \simeq S$ and we use (28) as a representation for S. From this we see that y is a non-zerodivisor in S and that S/yS is a coherent ring! Let us, just for simplicity, take the case when $R = k[[X,Y]]$ and $(x,y) = (X,Y)$. Then

$$S/YS = k[X,X_2,X_3,\ldots,X_n,\ldots]/(X^2,XX_2,XX_3,\ldots,XX_n,\ldots).$$

But this last quotient is just the trivial extension $k[X_2,\ldots,X_n,\ldots] \ltimes k$, which is easily seen to be coherent (cf. Remark 2 following Theorem 2 in §2). We can now apply for (S,Y) the same arguments as those used for (B,T) on the top of p. 93 of [24]. This gives λ-dim S ≤ 2, and here we have equality, since S is not coherent.

PROBLEM 6. *Let* R *be a noetherian local domain, with field of fractions* K, *and integral closure* $\bar{R}$. *When is it true that* $\bar{R}$, *or, more generally,* all *rings between* R *and* K, *have* λ*-dimension* $\leq n$ (*for some* $n \geq 2$)?

7. *Some open questions and generalizations*

In the long paper [3] (cf. also [23]) three classes of modules over a (not necessarily commutative) noetherian ring R were studied, namely

(a) the n-torsionfree modules M, defined by $\mathrm{Ext}_R^i(D(M),R) = 0$ for $1 \leq i \leq n$ (here D(M) is the "transpose" of M; cf. §4);

(b) the n-reflexive modules M, defined by requiring that a

map $M \longrightarrow (D_n)^2(M)$ induces an isomorphism after application of $\mathrm{Ext}^1_R(\,.\,,R)$ (cf. [3; p.3]);

(c) those modules M that are n-th modules of syzygies, that is for which there are exact sequences

$$0 \longrightarrow M \longrightarrow P_{n-1} \longrightarrow \dots \longrightarrow P_0 \longrightarrow N \longrightarrow 0$$

with P_i finitely generated and projective.

In the commutative case (to which we shall restrict ourselves now), all these notions coincide (for all n) *if and only if* R *is a Gorenstein ring.* If R is not Gorenstein several problems about these notions remain unanswered. Since every finitely generated module is of the form D(M), with M finitely generated, it follows that the "duality dimension" $\alpha(R)$ of §5 can be interpreted as

$$\alpha(R) = \inf\left\{ t \;\middle|\; \begin{array}{l}\text{every finitely generated t-torsion free module} \\ \text{is (t+1)-torsion free}\end{array} \right\}.$$

In an analogous way, we introduce the *reflexive dimension* $\beta(R)$ and the *syzygetic dimension* $\gamma(R)$ as follows:

$$\beta(R) = \inf\{\, t \mid \text{every f.g. t-reflexive module is (t+1)-reflexive}\};$$

$$\gamma(R) = \inf\{\, t \mid \text{every f.g. t-th syzygy is a (t+1)-st syzygy}\}.$$

PROBLEM 7. *What are the relations between the preceding three integers* $\alpha(R)$, $\beta(R)$ *and* $\gamma(R)$? *In particular, are they all finite for a commutative noetherian local ring* $(R,\underline{m})$?

The preceding problem is closely related to the problem of deciding whether $\lambda\text{-dim}\; R \propto I(k) < \infty$. Indeed, it is easy to see that, for example,

$$\lambda\text{-dim}\; R \propto I(k) \le \sup_{\underline{p}}\{\alpha(R_{\underline{p}}) + 1\}$$

where the sup is taken over those prime ideals that lie immediately below the maximal ideal $\underline{m}$ of the local ring R.

Even more general problems come up when one studies $\lambda\text{-dim}\; R \propto M$, where M is an artinian module or a module in one of the categories studied by Zink [25].

References

1. H. Åberg, "Coherence of amalgamations", J. Algebra, to appear.

2. D.J. Anick, "A counterexample to a conjecture of Serre", Ann. of Math., 115 (1982), 1-33.

3. M. Auslander and M. Bridger, Stable module theory, Memoirs of the American Mathematical Society 94 (American Mathematical Society, Providence, 1969).

4. H. Bass, "Injective dimension in noetherian rings", Trans. Amer. Math. Soc., 102 (1962), 18-29.

5. H. Bass, "On the ubiquity of Gorenstein rings", Math. Z., 82 (1963), 8-28.

6. R. Bieri and R. Strebel, "Soluble groups with coherent group rings", Homological group theory: proceedings of a symposium held at Durham 1977, London Mathematical Society Lecture Notes 36 (Ed. C.T.C. Wall, Cambridge University Press, Cambridge, 1979), pp. 235-240.

7. N. Bourbaki, Algèbre commutative, Chapitres 1-2 (Hermann, Paris, 1961).

8. N. Bourbaki, Algèbre, Chapitre 10 (Algèbre homologique) (Masson, Paris, 1980).

9. R. Bøgvad, "Gorenstein rings with transcendental Poincaré series", Math. Scand., to appear.

10. E.D. Davis, "Overrings of commutative rings. I. Noetherian overrings", Trans. Amer. Math. Soc., 104 (1962), 52-61.

11. R. Fossum, P. Griffith and I. Reiten, Trivial extensions of abelian categories, Lecture Notes in Mathematics 456 (Springer, Berlin, Heidelberg, New York, 1975).

12. J.P. Jans, "Duality in noetherian rings", Proc. Amer. Math. Soc., 12 (1961), 829-835.

13. I. Kaplansky, Commutative rings (University of Chicago Press, Chicago, 1974).

14. C. Löfwall, "The global homological dimensions of trivial extensions of rings", J. Algebra, 39 (1976), 287-307.

15. C. Löfwall and J.-E. Roos, "Cohomologie des algèbres de Lie graduées et séries de Poincaré-Betti non rationnelles", C.R. Acad. Sci. Paris Sér. A, 290 (1980), 733-736.

16. M.S. Menzin, "The condition $\mathrm{Ext}^1(M,R) = 0$ for modules over local Artin algebras $(R,\underline{m})$, with $\underline{m}^2 = 0$", Proc. Amer. Math. Soc., 43 (1974), 47-52.

17. M. Nagata, Local rings (Wiley-Interscience, New York, London, 1962).

18. I. Palmér and J.-E. Roos, "Explicit formulae for the global homological dimensions of trivial extensions of rings", J. Algebra, 27 (1973), 380-413.

19. I.J. Papick, "A remark on coherent overrings", Canad. Math. Bull., 21 (1978), 373-375.

20. M. Ramras, "Betti numbers and reflexive modules", Proceedings of a conference on ring theory, Park City, Utah, 1971 (ed. R. Gordon, Academic Press, New York, London, 1972), pp. 297-308.

21. J.-E. Roos, "On the use of graded Lie algebras in the theory of local rings", Commutative algebra: Durham 1981, London Mathematical Society Lecture Notes 72 (ed. R.Y. Sharp, Cambridge University Press, Cambridge, 1982), pp. 204-230.

22. P. Samuel, "Anneaux factoriels", Bull. Soc. Math. France, 89 (1961), 155-173.

23. P. Samuel, "Séminaire d'algèbre commutative, Anneaux de Gorenstein et torsion en algèbre commutative (exposés de M. Auslander) (Secrétariat Mathématique, 11 rue Pierre Curie, Paris 5ème, 1967).

24. W.V. Vasconcelos, The rings of dimension 2, Lecture Notes in Pure and Applied Mathematics 22 (Marcel Dekker, New York, Basel, 1976).

25. T. Zink, "Endlichkeitsbedingungen für Moduln über einem Noetherschen Ring", Math. Nachr., 64 (1974), 239-252.

Department of Mathematics,
University of Stockholm,
Box 6701,
S-113 85 Stockholm, Sweden.

ON THE USE OF GRADED LIE ALGEBRAS IN THE THEORY OF LOCAL RINGS

JAN-ERIK ROOS

Introduction

Many lectures at this Symposium have been about finite free resolutions or about finite complexes of modules over commutative rings.

I shall study a different aspect of commutative algebra. I shall try to say something systematic about infinite resolutions. The problems (and also the methods) that we encounter there seem to be very different. But there *are* relations with finite resolutions as we shall see later in this paper.

I begin by formulating two well-known questions.

I. QUESTION OF SERRE AND KAPLANSKY (1950's and 1960's; see [37], [17] and the historical discussion in [35]). *Let* $(R,\underline{m})$ *be a local commutative noetherian ring, with maximal ideal* $\underline{m}$ *and residue field* $k = R/\underline{m}$. *Consider a minimal free resolution*

$$\cdots \longrightarrow F_r \longrightarrow \cdots \longrightarrow F_1 \longrightarrow F_0 \longrightarrow k \longrightarrow 0$$

of the R-*module* k. (If R is *not* regular, this resolution never stops.) *Does the series*

$$P_R(Z) = \sum_{i\geq 0} (\text{rank}_R F_i).Z^i \qquad (1)$$

(which converges in a neighbourhood of 0*) represent a rational function of* Z?

Before I discuss the recent progress on this question, I must also mention the following.

II. ANOTHER QUESTION OF SERRE (1950's; see [37] and the historical discussion in [35]). *Let* X *be a finite, simply-connected* CW-*complex and let* ΩX *be the loop space of* X. *Is*

$$\sum_{i\geq 0} \dim_{\mathbb{Q}} H_i(\Omega X,\mathbb{Q}).Z^i$$

a rational function of Z?

In 1978 I had proved [34] that a positive answer to Question I for local rings $(R,\underline{m})$ with $\underline{m}^3 = 0$ (and $R/\underline{m} = \mathbb{Q}$) was equivalent to a positive answer to Question II for CW-complexes X with $\dim X \leq 4$, and indeed even a more precise result was given (cf. Theorem B of [34]). But in 1980 David Anick [2] found a counterexample X, with $\dim X \leq 4$, to Question II, which therefore by the result just mentioned also gives a counterexample to Question I. Thus here we have a situation where two questions of Serre (one of them also asked by Kaplansky) *do* have negative answers! But my point is that we should not leave these questions too early, and this for at least three reasons (from now on I shall almost exclusively stick to Question I, in view of the theme of this Symposium).

(i) There seem to be interesting classes of local rings for which we do have a positive answer to Question I.

(ii) It is quite possible that slight reformulations of Questions I and II give rise to questions that do have positive answers. Indeed, saying that $P_R(Z)$ is rational is equivalent to saying that the numbers $\mathrm{rank}_R F_i$ in (1) satisfy a *linear* recursive relation. If we replace linear recursiveness by some more general recursive concept, it is quite possible that the answer to the corresponding Question I is positive. Indeed, recent results about the homology of finitely presented groups [13] give support to this surmise.

(iii) Interesting mathematical tools have been developed while we have been trying to answer Questions I and II, and these tools can be used for other problems as well. For example, a case in point is the Gulliksen-Avramov-André-Löfwall-Jacobsson... analysis of local flat ring homomorphisms (cf. §7 below).

One of the aims of this paper is to report briefly both about the earlier progress on Question I and about the recent constructions of counterexamples. We shall also prove some new results related to these questions. Thus in §1 we explain old attempts to reduce Question I to questions about finite resolutions, leading in particular to the theory of so-called Golod rings. In §2 we study the Yoneda Ext-algebra of a local ring. This Ext-algebra is a Hopf algebra, which is equal to the enveloping algebra of a graded Lie algebra. We prove some new results about the finitistic global dimension and the λ-dimension of Hopf algebras, and we apply these results to Ext-algebras.

In §3 we recall the theory of Golod maps, introduced by Levin, and we study several examples. A class AG of local rings that *might* have rational series $P_R(Z)$ is introduced in §4. The class AG consists of those local rings (having completions) R which are Attached to a regular ring $\tilde{R}$ by a sequence of Golod maps, as in

$$\tilde{R} \longrightarrow R_1 \longrightarrow R_2 \longrightarrow \cdots \longrightarrow R_s = R \qquad (2)$$

(the terminology is due to J. Backelin). We prove that if R is as in (2), then $\mathrm{Ext}_R^*(k,k)$ has finitistic global dimension $\leq s$ and λ-dimension $\leq s$. In §5 we recall the graded Lie algebra constructions of counterexamples $(R,\underline{m})$ to rationality of $P_R(Z)$, and we use the theory of §4 to show that *all* these counterexamples are outside the class AG.

In §6 we mention some results about the "Bass" series

$$I^R(Z) = \sum_{i\geq 0} \dim_k \mathrm{Ext}_R^i(k,R).Z^i.$$

In particular we explain how the analogue for Hopf algebras (which are groups in the category of coalgebras!) of the theory of ends of groups would solve the problem of rationally relating $P_R(Z)$ to $I^R(Z)$ for a local ring $(R,\underline{m})$ having $\underline{m}^3 = 0$ (these are results of Bøgvad).

In §7 a brief discussion of local flat ring homomorphisms is given. This theory is closely related to a study of the centre of $\mathrm{Ext}_R^*(k,k)$ (§8). Finally, §9 is devoted to general remarks and open problems.

In retrospect, one can say that many constructions and results in this paper are related to graded Lie algebras (or, equivalently, to a certain class of graded Hopf algebras), and this explains the title of this paper. We have made some effort to make the overlap between the present paper and our recent survey [35] as small as possible.

1. *Early attempts to reduce Question* I *to problems about finite free resolutions*

Recall that if $(R,\underline{m})$ is a local commutative noetherian ring with residue field $k = R/\underline{m}$ as in the Introduction, and if the F_i are (as above) the different modules of a minimal R-free resolution of k, then

$$\operatorname{rank}_R F_i = \dim_k \operatorname{Tor}_i^R(k,k).$$

If $(\hat{R},\hat{\underline{m}})$ is the completion of R, we clearly have $k = \hat{R}/\hat{\underline{m}}$ and $\operatorname{Tor}_i^R(k,k) \simeq \operatorname{Tor}_i^{\hat{R}}(k,k)$. Thus if we wish to study, say, the series $P_R(Z)$ in (1), we might as well assume from the beginning that $(R,\underline{m})$ is a *complete* local ring. Then it is known (Cohen) that we may write R as a quotient of a "minimal" regular local ring $(\tilde{R},\tilde{\underline{m}})$, that is that we have an exact sequence

$$0 \longrightarrow \underline{a} \longrightarrow \tilde{R} \longrightarrow R \longrightarrow 0 \tag{3}$$

where $\underline{a}$ is an ideal in $\tilde{R}$ satisfying $\underline{a} \subset \tilde{\underline{m}}^2$. It follows that the embedding dimensions $\dim_k \tilde{\underline{m}}/\tilde{\underline{m}}^2$ of $\tilde{R}$ and $\dim_k \underline{m}/\underline{m}^2$ of R are equal. From now on we shall only study local rings R that are quotients of regular rings as in (3). Then we have a change of rings spectral sequence associated to (3),

$$E^2_{p,q} = \operatorname{Tor}_p^R(\operatorname{Tor}_q^{\tilde{R}}(R,k),k) \Rightarrow \operatorname{Tor}_n^{\tilde{R}}(k,k), \tag{4}$$

but this spectral sequence *never degenerates* unless $\tilde{R} = R$.

Both k and R have as $\tilde{R}$-modules *finite* free minimal resolutions X_* and Y_*, respectively. Let us first concentrate on X_*, which has a very explicit form; it is even a differential graded algebra, whose underlying algebra structure over $\tilde{R}$ is an exterior algebra [17]. It follows that $K_* = R \otimes_{\tilde{R}} X_*$ is a finite complex of free R-modules, which is even a differential graded commutative algebra (the so-called Koszul algebra), whose homology $H_*(K_*) = \operatorname{Tor}_*^{\tilde{R}}(R,k)$ is also an algebra. It is exactly when this algebra has "trivial multiplication" that the spectral sequence (4) comes as close as possible to degeneration, without being trivial. The corresponding local rings R for which this holds true are called *Golod* rings, and they can be characterized in many different ways [17, 26]. Let me be slightly more precise:

(i) "trivial multiplication" means here not only that $H_+(K).H_+(K) = 0$ (+ denotes the elements of positive degree), but also that "the higher Massey products of $H_*(K)$ vanish";

(ii) that the spectral sequence (4) comes as close as possible to degeneration means that the differentials

$$d^r_{p,q} : E^r_{p,q} \longrightarrow E^r_{p-r,q+r-1}$$

are zero for $r \geq 2$ *and* $q > 0$ (in the situation (3) we always have $E^\infty_{p,q} = 0$ for $q > 0$).

An equivalent reformulation of (ii) that does not mention the spectral sequence (4) is that R should be such that, in a representation (3), the natural map

$$\mathrm{Tor}^{\tilde{R}}_*(\underline{m},k) \longrightarrow \mathrm{Tor}^R_*(\underline{m},k)$$

is a *monomorphism* [25,26]. It then follows more generally [26] that if

$$\cdots \longrightarrow F_i \xrightarrow{d_i} F_{i-1} \longrightarrow \cdots \xrightarrow{d_1} F_0 \longrightarrow k \longrightarrow 0$$

is a minimal R-free resolution of k ($F_0 = R$), then all the maps

$$\mathrm{Tor}^{\tilde{R}}_*(\mathrm{Im}\, d_i,k) \longrightarrow \mathrm{Tor}^R_*(\mathrm{Im}\, d_i,k)$$

are monomorphisms for $i \geq 1$. (Note that $\mathrm{Im}\, d_1 = \underline{m}$.)

A *third* equivalent reformulation of (ii) that is of special interest for Question I will now be formulated. First we introduce a general notation. If M is a module over a local ring R such that the $\mathrm{Tor}^R_i(M,k)$ are finite-dimensional vector spaces over k (this is true if, for example, M is finitely generated), then we introduce $P^M_R(Z)$ by

$$P^M_R(Z) = \sum_{i\geq 0} \dim_k \mathrm{Tor}^R_i(M,k).Z^i. \tag{5}$$

Note that $P^k_R(Z)$ is the $P_R(Z)$ of (1) (it is time to mention that these series are called *Poincaré-Betti series*). Now by formal manipulations with the spectral sequence (4) it is easy to see [25,26] that we *always* have (<< means a coefficientwise inequality of formal power series)

$$P_R(Z) << \frac{P_{\tilde{R}}(Z)}{1 - Z.(P^R_{\tilde{R}}(Z) - 1)} \tag{6}$$

and that we have equality in (6) *if and only if* $(R,\underline{m})$ *is a Golod ring.* In particular, since $P_{\widetilde{R}}(Z)$ and $P^R_{\widetilde{R}}(Z)$ are polynomials, it follows that *for Golod rings, Question* I *does have a positive answer.*

Examples of Golod rings (we denote by $(\widetilde{R},\underline{\widetilde{m}})$ any regular local ring). (a) Any $R = \widetilde{R}/\underline{\widetilde{m}}^n$ $(n \geq 1)$ is a Golod ring.

(b) Any local ring of the form $R = \widetilde{R}/x.I$, where $x \in \underline{\widetilde{m}}$ and $I \subset \widetilde{R}$ is an ideal such that $x.I \subset \underline{\widetilde{m}}^2$, is a Golod ring.

(c) Let T be an $r \times s$-matrix $(2 \leq r \leq s)$ with entries in $\underline{\widetilde{m}}$, and let $\underline{a}$ be the ideal in $\widetilde{R}$ generated by all the $r \times r$-minors of T. Then $\widetilde{R}/\underline{a}$ is a Golod ring if prof $\underline{a} = s - r + 1$ [7,26]. We return to these examples later.

The spectral sequence (4) is in general difficult to analyse. Levin [24] has proved (and used) that (4) is a spectral sequence of differential $H_*(K_*) = \mathrm{Tor}^{\widetilde{R}}_*(R,k)$-modules. But we now turn to an even more sophisticated way of relating the rings $\widetilde{R}$, R and the ring homomorphism $\widetilde{R} \longrightarrow R$ of (3) to the multiplicative structure of $H_*(K_*) = \mathrm{Tor}^{\widetilde{R}}_*(R,k)$.

THEOREM A (Avramov [5]). *Let* $\widetilde{R} \longrightarrow R$ *be a map of local rings as in* (3) *(with* R *regular,...). Then there is a spectral sequence of bigraded Hopf algebras*

$$E^2_{p,q} = \mathrm{Tor}_p^{\mathrm{Tor}^{\widetilde{R}}_*(R,k)}(k,k)_q \Rightarrow E^\infty = \mathrm{gr}(\mathrm{Tor}^R_*(k,k)//\mathrm{Tor}^{\widetilde{R}}_*(k,k)) \quad (7)$$

where the differentials d^r *are zero for* $r > \frac{1}{2}(\dim_k \underline{m}/\underline{m}^2 - \mathrm{prof}\ R)$. □

Several remarks are in order here.

Remark 1. In the E^2 term of (7), we take the Tor_p's of the (graded) $\mathrm{Tor}^{\widetilde{R}}_*(R,k)$-modules k and k. These Tor_p's are naturally graded and this grading gives the extra index q.

Remark 2. For any local ring $(R,\underline{m})$, $\mathrm{Tor}^R_*(k,k)$ is both an algebra *and* a coalgebra (the dual algebra is - more details are given in §2 - $\mathrm{Ext}^*_R(k,k)$ with the Yoneda product) and these two structures are related so that $\mathrm{Tor}^R_*(k,k)$ becomes a graded Hopf algebra [24,17]. There is also a graded version of this result, so that $E^2_{*,*}$ becomes a bigraded Hopf algebra.

Remark 3. The natural map (a monomorphism)

$$\mathrm{Tor}^{\widetilde{R}}_*(k,k) \longrightarrow \mathrm{Tor}^R_*(k,k) \quad (8)$$

is not only a map of algebras, but also a map of coalgebras, and thus a map of Hopf algebras. The quotient // in the E^∞-term on the right in (7) is the Hopf algebra cokernel of (8).

Remark 4. The differentials in (7) are derivations and are compatible with the Hopf algebra structure. These and the other assertions about the differentials are made explicit and are proved in [5].

Remark 5. The spectral sequence (7) is the "local algebra version" of the Eilenberg-Moore spectral sequence in algebraic topology [5].

Arguments with differential graded Tor, inspired by arguments in algebraic topology, give the following important corollary [7].

COROLLARY TO THEOREM A. *Let* $\tilde{R} \longrightarrow R$ *be as in* (3) *and assume that the finite minimal* $\tilde{R}$*-module resolution* Y_* *of* R *admits a multiplicative structure [associative, commutative (graded)], compatible with the differential in the usual way. Then the spectral sequence* (8) *degenerates.* □

Remark 7. Note that there are several explicit well-known situations in commutative algebra where $\tilde{R} \longrightarrow R$ satisfies the assumptions of the preceding Corollary (cf. [7], where results of Buchsbaum and Eisenbud, Gover,... are treated in an even more general "relative" setting; see Remark 9 below). Therefore in these cases, we have, for example, that $P_R(Z)/P_{\tilde{R}}(Z)$ only depends on the finite-dimensional graded algebra $\mathrm{Tor}_*^{\tilde{R}}(R,k)$. Reversing the preceding arguments - thereby using methods originally invented for attacking Question I - one can prove that in some cases Y_* *does not have a nice multiplicative structure* (cf. [8], where Avramov gives counterexamples to conjectures [14] of Buchsbaum and Eisenbud).

Remark 8. If $(R,\underline{m})$ is a Golod ring or a complete intersection, one can show that the spectral sequence (7) degenerates [5].

Remark 9. There are *relative versions* of all the preceding results. Relative Golod maps $R_1 \longrightarrow R_2$ (R_1 not necessarily regular) are treated in §3 below. The relative Avramov spectral sequence (generalizing (7)) was first treated by Avramov in [7], for so-called *small* maps (that is $\mathrm{Tor}_*^{R_1}(k,k) \longrightarrow \mathrm{Tor}_*^{R_2}(k,k)$ is a monomorphism), but Avramov has recently found a more complicated version [9] for general maps. However, the multiplicative structure involved is more complicated than first thought [9]. Löfwall [30] and Jacobsson [18] have found that *central extensions of Hopf algebras* come up here in a natural way.

2. *Finitistic global dimension, λ-dimension and extensions of Hopf algebras; the Yoneda* Ext-*algebra of a Golod ring*

For any ring R and R-modules L, M, N, the natural pairing (composition of maps)

$$\operatorname{Hom}_R(M,N) \times \operatorname{Hom}_R(L,M) \longrightarrow \operatorname{Hom}_R(L,N)$$

extends naturally to a pairing of Ext,

$$\operatorname{Ext}_R^i(M,N) \times \operatorname{Ext}_R^j(L,M) \longrightarrow \operatorname{Ext}_R^{i+j}(L,N).$$

This is the so-called Yoneda product. In particular, if $(R,\underline{m})$ is a local ring, $\operatorname{Ext}_R^*(k,k)$ becomes a graded associative algebra over $k = R/\underline{m}$, and $\operatorname{Ext}_R^*(L,k)$ becomes a graded left $\operatorname{Ext}_R^*(k,k)$-module, and similarly $\operatorname{Ext}_R^*(k,N)$ is a graded right $\operatorname{Ext}_R^*(k,k)$-module. If

$$0 \longrightarrow \underline{a} \longrightarrow \tilde{R} \longrightarrow R \longrightarrow 0$$

is a representation of R, with $\tilde{R}$ regular and $\underline{a} \subset \tilde{\underline{m}}^2$, it follows that the induced *algebra* map (which is dual to the map of Tor's in §1)

$$\operatorname{Ext}_R^*(k,k) \longrightarrow \operatorname{Ext}_{\tilde{R}}^*(k,k) \tag{9}$$

is *onto*. But (9) is also a map of coalgebras, even of Hopf algebras [24,17], and if we let K denote the coalgebra kernel of (9), then K has an induced Hopf algebra structure. It follows that we have an "extension" of Hopf algebras

$$k \longrightarrow K \longrightarrow \operatorname{Ext}_R^*(k,k) \longrightarrow \operatorname{Ext}_{\tilde{R}}^*(k,k) \longrightarrow k. \tag{10}$$

Moreover, in these cases we even have that the Hopf algebras $\operatorname{Ext}_R^*(k,k)$ and $\operatorname{Ext}_{\tilde{R}}^*(k,k)$ are the enveloping algebras of graded Lie algebras $\underline{\underline{g}}$ and $\tilde{\underline{\underline{g}}}$, respectively, and that K is the enveloping algebra of the graded Lie algebra kernel $\underline{\underline{k}}$ of a natural Lie algebra map $\underline{\underline{g}} \longrightarrow \tilde{\underline{\underline{g}}}$. Therefore, underlying (10) is an extension of graded Lie algebras

$$0 \longrightarrow \underline{\underline{k}} \longrightarrow \underline{\underline{g}} \longrightarrow \tilde{\underline{\underline{g}}} \longrightarrow 0. \tag{11}$$

Now it was shown by Avramov [7] and Löfwall [28] that, in those cases where R is a Golod ring, it follows that the K of (10) is a *free* graded algebra (it is even the free algebra over the graded vector space $\{\mathrm{Ext}^{i-1}_{\widetilde{R}}(R,k)\}_{i\geq 2}$), that is it has global homological dimension 1. It follows from (10) (or (11)) that the Ext-algebra $\mathrm{Ext}^{*}_{R}(k,k)$ of a Golod ring $(R,\underline{m})$ has several interesting properties, closely related to those of a free algebra. Indeed, it is known that $\mathrm{Ext}^{*}_{\widetilde{R}}(k,k)$ for $\widetilde{R}$ a regular ring is an exterior algebra. Now we have the following general result.

THEOREM 1. *Let* H *be a graded connected (cocommutative) Hopf algebra that is an extension of a finitely generated exterior algebra* E *by means of a free algebra* F: *thus we have*

$$k \longrightarrow F \longrightarrow H \longrightarrow E \longrightarrow k.$$

Then

(a) *the finitistic global dimension of* H, *that is*

$$\mathrm{f.gldim}\, H = \sup\left\{ \mathrm{hd}_H M \;\middle|\; \begin{array}{l} M \text{ a (left) positively graded } H\text{-module} \\ \\ \text{of finite homological dimension (hd)} \end{array} \right\},$$

is ≤ 1;

(b) H *is graded coherent, that is all finitely generated left ideals are finitely presented* (only graded ideals are studied).

Proof. We start with case (a), which will follow from the following even more general result that we shall need later on in §4.

THEOREM 2. *Let*

$$k \longrightarrow H_1 \longrightarrow H_2 \longrightarrow H_3 \longrightarrow k$$

be an extension of graded connected Hopf algebras. Then we have the following inequality for the corresponding finitistic global dimensions:

$$\mathrm{f.gldim}\, H_2 \leq \mathrm{f.gldim}\, H_1 + \mathrm{f.gldim}\, H_3. \qquad (12)$$

Proof. We may of course assume that $t_1 = \mathrm{f.gldim}\, H_1 < \infty$ and that $t_3 = \mathrm{f.gldim}\, H_3 < \infty$. Let M be any positively graded left H_2-module *of finite homological dimension*. Let N be a t_1-th syzygy of M, so that we have an exact sequence of positively graded left H_2-modules

$$0 \longrightarrow N \longrightarrow F_{t_1-1} \longrightarrow \cdots \longrightarrow F_1 \longrightarrow F_0 \longrightarrow M \longrightarrow 0, \quad (13)$$

where the F_i are H_2-free (or projective - this is equivalent [20]).

Now to the ring map $H_2 \longrightarrow H_3$ there is associated a change of rings spectral sequence (T is *any* left H_2-module)

$$E^2_{p,q} = \mathrm{Tor}_p^{H_3}(k, \mathrm{Tor}_q^{H_2}(H_3, T)) \Rightarrow \mathrm{Tor}_n^{H_2}(k, T). \qquad (14)$$

I claim that the spectral sequence (14) degenerates when T is the module N of (13). Indeed, $\mathrm{hd}_{H_2} M < \infty$, and since H_1 is a sub Hopf algebra of H_2, it follows by a well-known result [32; Theorem 4.4] that H_2 is free as an H_1-module. Therefore we also obtain that $\mathrm{hd}_{H_1} M < \infty$, and since f.gldim $H_1 = t_1$, we have $\mathrm{hd}_{H_1} M \le t_1$, so that (13) gives that N is H_1-free. Therefore

$$\mathrm{Tor}_q^{H_2}(H_3, N) \simeq \mathrm{Tor}_q^{H_2}(k \otimes_{H_1} H_2, N) \simeq \mathrm{Tor}_q^{H_1}(k, N) = 0$$

for $q > 0$, so that (14) *does* degenerate into an isomorphism

$$\mathrm{Tor}_n^{H_2}(k, N) \simeq \mathrm{Tor}_n^{H_3}(k, H_3 \otimes_{H_2} N). \qquad (15)$$

By hypothesis, the left hand side of (15) is zero for large n. But then the right hand side of (15) is also zero for large n, so that the left H_3-module $H_3 \otimes_{H_2} N$ has finite homological dimension (it is sufficient to test homological dimension with $\mathrm{Tor}_*^H(k, \,.\,)$, for a connected H; see, for example, [20; Appendix]), and this homological dimension must be $\le t_3$, since f.gldim $H_3^{\cdot} = t_3$. Thus, by (15) again, we have $\mathrm{hd}_{H_2} N \le t_3$, and this combined with (13) gives that

$$\mathrm{hd}_{H_2} M \le t_1 + t_3.$$

Theorem 2 is now proved. □

Now Theorem 1(a) *follows*, since f.gldim E = 0 if E is an exterior algebra (an exterior algebra being an iterated extension of $k[X]/(X^2)$'s), and since f.gldim F = gldim F = 1 if F is a free algebra.

Now we shall prove Theorem 1(b) *about coherence*. We shall prove a more general result about λ-dimension that we shall need in its *graded* form.

Definition. Let H be a positively graded connected k-algebra.

We say that H has (left, graded) λ-*dimension* $\leq n$ (λ-dim $H \leq n$) if the following holds. Each exact sequence of (positively graded) left H-modules

$$F_n \longrightarrow \cdots \longrightarrow F_0 \longrightarrow M \longrightarrow 0, \tag{16}$$

where the F_i are finitely generated free H-modules, can be continued one step to the left to an exact sequence

$$F_{n+1} \longrightarrow F_n \longrightarrow \cdots \longrightarrow F_0 \longrightarrow M \longrightarrow 0$$

where F_{n+1} is a finitely generated free H-module.

Since λ-dimension has been studied in detail in [36] of these proceedings, we shall just recall here that if we have an exact sequence like (16), then we say that M is (graded) n-finitely presented. It follows easily from [36] that this is equivalent to saying that $\mathrm{Tor}_i^H(k,M)$ is finite-dimensional as a vector space over k, for $0 \leq i \leq n$. *Note also that* λ-dim $H \leq 0$ (≤ 1) *if and only if* H *is noetherian (coherent).*

THEOREM 3. *Let*

$$k \longrightarrow H_1 \longrightarrow H_2 \longrightarrow H_3 \longrightarrow k$$

be an exact sequence of Hopf algebras with H_1 *free. Then*

$$\lambda\text{-dim } H_2 \leq 1 + \lambda\text{-dim } H_3.$$

Proof. We consider again the change of rings spectral sequence (14) above, where T is any left H_2-module. Put $\lambda = \lambda$-dim H_3 (we suppose that it is finite). Now let M be any H_2-module such that $\mathrm{Tor}_i^{H_2}(k,M)$ is finite-dimensional for $i \leq \lambda + 1$. We want to prove that $\mathrm{Tor}_{\lambda+2}^{H_2}(k,M)$ is finite-dimensional too. Present M as a quotient of a finitely generated free H_2-module F, say

$$0 \longrightarrow N \longrightarrow F \longrightarrow M \longrightarrow 0. \tag{17}$$

If we apply the functor $\mathrm{Tor}_*^{H_2}(H_3, .)$ to the exact sequence (17) we obtain the isomorphism

$$\mathrm{Tor}_i^{H_2}(H_3,N) \simeq \mathrm{Tor}_{i+1}^{H_2}(H_3,M) \text{ for } i \geq 1. \tag{18}$$

But $\mathrm{Tor}_{i+1}^{H_2}(H_3,M) \simeq \mathrm{Tor}_{i+1}^{H_2}(k \otimes_{H_1} H_2, M) \simeq \mathrm{Tor}_{i+1}^{H_1}(k,M) = 0$ for $i \geq 1$, since gldim $H_1 = 1$. Combining this with (18) we obtain

$\mathrm{Tor}_i^{H_2}(H_3,N) = 0$ for $i \geq 1$, and therefore the spectral sequence (14) for $T = N$ degenerates into an isomorphism

$$\mathrm{Tor}_n^{H_3}(k, H_3 \otimes_{H_2} N) \simeq \mathrm{Tor}_n^{H_2}(k,N), \quad n \geq 0. \tag{19}$$

Now the fact that $\mathrm{Tor}_i^{H_2}(k,M)$ is finite-dimensional for $i \leq \lambda + 1$ implies (use (17)) that $\mathrm{Tor}_s^{H_2}(k,N)$ is finite-dimensional for $s \leq \lambda$, and therefore, by (19) the H_3-module $H_3 \otimes_{H_2} N$ is λ-finitely presented. But since λ-dim $H_3 = \lambda$, this implies that $H_3 \otimes_{H_2} N$ also is $(\lambda+1)$-finitely presented over H_3. Now by (19) again, this gives that $\mathrm{Tor}_{\lambda+1}^{H_2}(k,N)$, that is $\mathrm{Tor}_{\lambda+2}^{H_2}(k,M)$, is finite-dimensional, and Theorem 3 is proved, and thereby also the remaining part (b) of Theorem 1.

COROLLARY 1. *Any extension of a finitely generated exterior algebra by a free algebra is coherent.* □

Remark 1. Theorem 3 is best possible in the following sense. Let $H = k\langle T_1,T_2\rangle \otimes_k k\langle T_3,T_4\rangle$ be the tensor product of two free associative algebras, each in two variables. Then H is an extension of $k\langle T_1,T_2\rangle$ with $k\langle T_3,T_4\rangle$ and, according to Theorem 3,

$$\lambda\text{-dim } H \leq 1 + 1 = 2. \tag{20}$$

But H is *not* coherent [35], and therefore λ-dim $H \geq 2$. Thus we have equality in (20).

Remark 2. If we have an extension as in Theorem 3, where we only assume that H_1 has finite global dimension, it also follows by a similar argument that λ-dim $H_2 \leq$ gldim H_1 + λ-dim H_3.

Remark 3. One would *like* to have the following generalization of our previous Theorem 3.

Let

$$k \longrightarrow H_1 \longrightarrow H_2 \longrightarrow H_3 \longrightarrow k$$

be an extension of Hopf algebras. Then

$$\lambda\text{-dim } H_2 \leq \lambda\text{-dim } H_1 + \lambda\text{-dim } H_3. \tag{21}$$

We do not even know if this is true when the λ-dimensions on the right of (21) are 0, that is *we do not know whether the Hopf algebra extension of two noetherian Hopf algebras is noetherian.*

The following result will be used several times in §4 and §5.

THEOREM 4. *Let* A *be a sub Hopf algebra of a Hopf algebra* B *(graded, connected over* k*). Then*

(a) λ-dim A ≤ λ-dim B;

(b) f.gldim A ≤ f.gldim B.

Proof. Assume that λ-dim B = λ < ∞. Let M be a left A-module. Then since B is A-free, we have as before an isomorphism

$$\mathrm{Tor}_i^A(k,M) \simeq \mathrm{Tor}_i^B(k,B\otimes_A M), \quad i \geq 0. \qquad (22)$$

Assume that M is λ-finitely presented over A, that is that the left member of (22) is finite-dimensional for i ≤ λ. Then (22) implies that $\mathrm{Tor}_i^B(k,B\otimes_A M)$ is finite-dimensional for i ≤ λ, so that $\mathrm{Tor}_{\lambda+1}^B(k,B\otimes_A M) \simeq \mathrm{Tor}_{\lambda+1}^A(k,M)$ is also finite-dimensional. Therefore λ-dim A ≤ λ and (a) is proved.

The proof of (b) is similar. Let f.gldim B = μ < ∞. If M is any A-module of finite homological dimension, it follows from (22) that $\mathrm{Tor}_i^B(k,B\otimes_A M)$ is zero for i >> 0, and thus is zero for i > μ, since f.gldim B = μ < ∞. Thus (22) gives that $\mathrm{hd}_A M \leq \mu$ and (b) is proved. □

COROLLARY 2 (of Theorem 3 and Theorem 2). *Let* (R,$\underline{m}$) *be a Golod ring. Then* $\mathrm{Ext}_R^*(k,k)$ *is coherent and has finitistic global dimension equal to* 1. □

Part of this was proved in [33], where we also had some more precise results.

PROBLEM. *Does the preceding Corollary* 2 *have a converse?*

3. *Golod maps*

In [25] Gerson Levin introduced a relative version of the Golod rings of §1.

Definition. A surjective local ring map $(R_0,\underline{m}_0) \longrightarrow (R_1,\underline{m}_1)$ of the form

$$0 \longrightarrow \underline{a} \longrightarrow R_0 \longrightarrow R_1 \longrightarrow 0 \qquad (23)$$

with $\underline{a} \subset \underline{m}_0^2$ is called a *Golod map* if the two natural maps

$$\mathrm{Ext}_{R_1}^*(k,k) \longrightarrow \mathrm{Ext}_{R_0}^*(k,k) \text{ and } \mathrm{Ext}_{R_1}^*(\underline{m}_1,k) \longrightarrow \mathrm{Ext}_{R_0}^*(\underline{m}_1,k)$$

are onto.

If R_0 is regular, then the ring map $R_0 \longrightarrow R_1$ in (23) is Golod if and only if R_1 is a Golod ring in the sense of §1. All the characterizations of Golod rings in §1 have analogues for Golod maps. We briefly state some results.

(a) For any surjective ring map $R_0 \longrightarrow R_1$ we have a (coefficientwise) inequality of Poincaré-Betti series

$$P_{R_1}(Z) << \frac{P_{R_0}(Z)}{1 - Z.(P_{R_0}^{R_1}(Z) - 1)}, \tag{24}$$

with equality in (24) if and only if $R_0 \longrightarrow R_1$ is a Golod map.

(b) For any surjective local ring map $R_0 \longrightarrow R_1$, the change of rings spectral sequence

$$E^2_{p,q} = \operatorname{Tor}_p^{R_1}(k, \operatorname{Tor}_q^{R_0}(R_1,k)) \Rightarrow \operatorname{Tor}_n^{R_0}(k,k)$$

satisfies $d^r_{p,q} = 0$, $r \geq 2$, $q > 0$, and $E^\infty_{p,q} = 0$, $q > 0$, if and only if $R_0 \longrightarrow R_1$ is a Golod map.

(c) *If* $R_0 \longrightarrow R_1$ is a Golod map, then the kernel of the surjective *Hopf algebra map* $\operatorname{Ext}^*_{R_1}(k,k) \longrightarrow \operatorname{Ext}^*_{R_0}(k,k)$ is the free algebra on the graded vector space $\{\operatorname{Ext}^{i-1}_{R_0}(R_1,k)\}_{i\geq 2}$.

Remark. In case $R_0 \longrightarrow R_1$ is Golod, we have by (a) that

$$P_{R_1}(Z) = P_{R_0}(Z).[1 - Z.(P_{R_0}^{R_1}(Z) - 1)]^{-1}, \tag{25}$$

but, unlike the case when R_0 is regular, rationality of $P_{R_0}(Z)$ does not necessarily imply rationality of $P_{R_1}(Z)$ in general, since $P_{R_0}^{R_1}(Z)$ might be an infinite, perhaps non-rational, series if R_0 is non-regular.

A very interesting case of a Golod map, where we *can* calculate the relative series in (25), is given by the following theorem of Levin [25] (the Artin-Rees lemma is used in an essential way).

THEOREM (Levin [25]). *Let* $(R,\underline{m})$ *be any local commutative noetherian ring. Then there exists a number* ν_0 *such that*

$R \longrightarrow R/\underline{m}^n$ *is a Golod map for all* $n \geq \nu_0$. □

In this special case we can even choose ν_0 so large (cf. [25]) that all the maps $\mathrm{Ext}_R^*(R/\underline{m}^n,k) \longrightarrow \mathrm{Ext}_R^*(R/\underline{m}^{n+1},k)$ are zero for $n \geq \nu_0$ and $* > 0$. In this case, the long exact sequences obtained by applying $\mathrm{Ext}_R^*(\,.\,,k)$ to

$$0 \longrightarrow \underline{m}^n/\underline{m}^{n+1} \longrightarrow R/\underline{m}^{n+1} \longrightarrow R/\underline{m}^n \longrightarrow 0$$

decompose into short exact sequences if $n \geq \nu_0$, giving the formula

$$P_R^{R/\underline{m}^n}(Z) - 1 = (-1)^n . Z^{-(n-1)} . P_R(Z) . \Big\{ \sum_{s \geq n} \dim_k(\underline{m}^s/\underline{m}^{s+1}) . (-Z)^s \Big\}, \quad n \geq \nu_0,$$

and this, together with formula (25) for $R \longrightarrow R/\underline{m}^n$, now gives, after some simplifications,

$$P_{R/\underline{m}^n}(Z)^{-1} - P_R(Z)^{-1} = -(-Z)^{-(n-2)} . \sum_{s \geq n} \dim_k(\underline{m}^s/\underline{m}^{s+1}) . (-Z)^s,$$

for $n \geq \nu_0$. But the last series is rational, since the Hilbert series

$$\sum_{i \geq 0} \dim_k(\underline{m}^i/\underline{m}^{i+1}) . Z^i$$

is rational. Therefore $P_R(Z)$ is rational if and only if $P_{R/\underline{m}^n}(Z)$ is rational (for some and then for all $n \geq \nu_0$) and the rationality questions are reduced to the artinian cases.

Further examples of Golod maps (cf. [26]). (a) If $(R,\underline{m})$ is local and if $x \in \underline{m}^2$ is a non-zerodivisor, then $R \longrightarrow R/(x)$ is a Golod map.

(b) More generally, if $(R,\underline{m})$ is local, $x \in \underline{m}$ is a non-zerodivisor, and I is an ideal in R such that $x.I \subset \underline{m}^2$, it follows that $R \longrightarrow R/xI$ is a Golod map.

More examples are given in [26] and [7]. *Note that it is not true in general that the composites of Golod maps are again Golod maps.* The most illuminating counterexample is probably that of a complete intersection R, which is obtained by dividing out an $\tilde{R}$-sequence $(t_1,\ldots,t_\nu)$ in $\underline{\tilde{m}}^2$ of a regular local ring $\tilde{R}$: we have

$$\tilde{R} \longrightarrow \tilde{R}/(t_1) \longrightarrow \tilde{R}/(t_1,t_2) \longrightarrow \ldots \longrightarrow \tilde{R}/(t_1,\ldots,t_\nu) = R. \quad (26)$$

It follows from (a) above that each map in (26) is a Golod map.

But if composites of Golod maps were Golod, it would follow that R were a Golod ring. But this is impossible for $\nu > 1$. This known result could be deduced in an over-sophisticated way, using Ext-algebras, and the theory of §2...

4. *A class of local rings which might have rational Poincaré-Betti series*

Let $\underline{\underline{AG}}$ be the class consisting of those local rings (having completions) R which can be reached by a finite sequence of Golod maps from a regular local ring $\tilde{R}$, as in

$$\tilde{R} \longrightarrow R_1 \longrightarrow R_2 \longrightarrow \cdots \longrightarrow R_s = R.$$

This class $\underline{\underline{AG}}$ contains the Golod rings and the complete intersections. It also contains all quotients of a regular ring $\tilde{R}$ by monomials in an $\tilde{R}$-sequence [16,11,18] (by iterated use of Example (b) of a Golod map). It *might* also contain every quotient of $\tilde{R}$ by a determinantal ideal.

It is known that if R can be reached by ≤ 2 Golod maps from a regular local ring (or even from a complete intersection), then $P_R(Z)$ is rational [27]. *So far, no local ring* $R \in \underline{\underline{AG}}$ *is known for which* $P_R(Z)$ *is not rational.* Part (i) of the following theorem will be used in §5 to prove that the known counterexamples to rationality of $P_R(Z)$ do *not* belong to $\underline{\underline{AG}}$.

THEOREM 5. *Let* $(R,\underline{m})$ *be a local ring whose completion can be reached by a sequence of* s *Golod maps from a regular local ring* $(\tilde{R},\underline{\tilde{m}})$, *say*

$$\tilde{R} \longrightarrow R_1 \longrightarrow \cdots \longrightarrow R_s = \hat{R}. \tag{27}$$

Then

(i) $\operatorname{f.gldim} \operatorname{Ext}^*_R(k,k) \leq s$;

(ii) $\lambda\text{-}\dim \operatorname{Ext}^*_R(k,k) \leq s$.

Proof. We can assume that $R = \hat{R}$. Let us prove Theorem 5 by induction on s. The case in which $s = 1$ is Corollary 2 (of Theorem 3 and Theorem 2) in §2. If the Theorem is proved for R_{s-1} $(s \geq 2)$, it follows from (27) and property (c) of Golod maps in §3 that we have an exact sequence of Hopf algebras

$$k \longrightarrow F_s \longrightarrow \operatorname{Ext}^*_{R_s}(k,k) \longrightarrow \operatorname{Ext}^*_{R_{s-1}}(k,k) \longrightarrow k$$

where F_s is a free algebra. Now it follows from Theorems 2 and 3 that

$$\text{f.gldim } \operatorname{Ext}^*_{R_s}(k,k) \le \text{f.gldim } \operatorname{Ext}^*_{R_{s-1}}(k,k) + 1$$

and

$$\lambda\text{-dim } \operatorname{Ext}^*_{R_s}(k,k) \le \lambda\text{-dim } \operatorname{Ext}^*_{R_{s-1}}(k,k) + 1$$

respectively, and Theorem 5 follows. □

Remark 1. We even have a stronger result than (i) in Theorem 5: *the Hopf algebra kernel of*

$$\operatorname{Ext}^*_{R_s}(k,k) \longrightarrow \operatorname{Ext}^*_{\tilde{R}}(k,k)$$

has gldim $\le$ s.

COROLLARY. *Let* $(R,\underline{m})$ *be a local ring whose completion can be reached by* s *Golod maps from a regular local ring. Let* H *be a sub Hopf algebra of* $\operatorname{Ext}^*_R(k,k)$ *of finite global dimension. Then* gldim H $\le$ s.

Proof. This follows from Theorem 5 and Theorem 4. □

Remark 2. In §5 we shall show that the examples of $(R,\underline{m})$'s with $P_R(Z)$ not rational that have been constructed up to now are all such that $\operatorname{Ext}^*_R(k,k)$ contains sub Hopf algebras of arbitrarily high finite global dimension. In view of the Corollary above, these $(R,\underline{m})$'s cannot belong to $\underline{\underline{AG}}$.

Remark 3. There *are* rings R that can be reached from a regular ring by *three* Golod maps, and that have rather strange properties. Here is an example $(R,\underline{m})$, where $\underline{m}^3 = 0$, which we studied in [34] (I thank Calle Jacobsson, who told me why R is in $\underline{\underline{AG}}$!).

$$R = k[X_1,\ldots,X_5]/(X_1^2,\ldots,X_5^2,X_1(X_2+\ldots+X_5),X_2X_3,X_4X_5,X_1X_2X_4).$$

Since

$$R' = k[[X_1,\ldots,X_5]]/(X_2^2,\ldots,X_5^2,X_2X_3,X_4X_5)$$

only has monomial relations, it is, as we remarked above, in $\underline{\underline{AG}}$. Now observe that X_1 is a non-zerodivisor in R' and that $R = R'/X_1.I$, where I is the ideal in R' generated by $X_1, X_2+\ldots+X_5, X_2X_4$. It therefore follows from Example (b) of Golod maps in §3 that $R' \longrightarrow R'/X_1I = R$ *is* a Golod map, so that $R \in \underline{\underline{AG}}$. In this case

$\mathrm{Ext}_R^*(k,k)$ *cannot* be generated as an algebra by a finite number of elements (when char $k \neq 2$), λ-dim $\mathrm{Ext}_R^*(k,k) = 3$ and gldim $\mathrm{Ext}_R^*(k,k) = 3$. Furthermore (cf. [34]),

$$P_R(Z) = (1 - Z)/(1 - 6Z + 11Z^2 - 8Z^3),$$

and this expression is *not* of the form

$$(1 + Z)^{\text{embedding dim}(R)}/\text{polynomial in } Z,$$

which we always obtain for rings that can be reached by ≤ 2 Golod maps from a regular ring (or from a complete intersection). Thus new phenomena occur when one studies rings in $\underline{\underline{AG}}$ with ≥ 3 Golod maps, and perhaps one should be careful with conjectures here.

5. *All the known counterexamples to rationality of* $P_R(Z)$ *are outside the class* $\underline{\underline{AG}}$

We do not describe in detail here the ingenious constructions found earlier by Anick [2], using certain quotients of semi-tensor products of free algebras, that give counterexamples to Question II of the Introduction, and thereby also counterexamples to Question I. Instead, we shall briefly discuss a construction by Löfwall and myself [31] that is perhaps slightly easier to explain and to put into use to prove the assertion in the title of this section.

We start with some general remarks. It follows from the Theorem of Levin and the discussion following it in §3 that the rationality questions for Poincaré-Betti series of local rings can be reduced to the artinian case. The first non-trivial case is that of $(R,\underline{m})$ with $\underline{m}^3 = 0$, and from now on we shall stick to that case. Consider $\mathrm{Ext}_R^*(k,k)$ and its subalgebra A generated by $\mathrm{Ext}_R^1(k,k)$. Here A turns out to be the enveloping algebra of a graded Lie algebra, generated by elements in degree 1, and having relations in degree 2, so that A is a Hopf algebra of a very special type (we shall say that A is (1,2)-presented). Conversely, all (1,2)-presented Hopf algebras come from a suitable local ring $(R,\underline{m})$ in the way we have just described. Indeed, given such a Hopf algebra A, we obtain an R by taking the "diagonal part" of the cohomology algebra of the Hopf algebra A, and by dividing out the cube of the augmentation ideal of the cohomology algebra. More details can be found in [28,29], where it is also proved that, in the equicharacteristic case, R and A determine each other. Furthermore we have the following formula, which was first discovered by Löfwall [28,29] (see also [34] for an alternative treatment of this formula):

$$P_R(Z)^{-1} = (1 + Z^{-1}).A(Z)^{-1} - Z^{-1}.H_R(-Z), \tag{28}$$

where $A(Z) = \sum_{i\geq 0} (\dim_k A^i).Z^i$ is the Hilbert series of the graded (non-commutative) algebra A, and $H_R(Z) = \sum_{i\geq 0} \dim_k(\underline{m}^i/\underline{m}^{i+1}).Z^i$ (in our special case this is a polynomial of degree 2) is the Hilbert series of the local ring $(R,\underline{m})$. Therefore, in view of (28), we get an R with transcendental $P_R(Z)$ if we can construct a (1,2)-presented Hopf algebra A with transcendental A(Z). Here is a general construction of such A that Löfwall and I made in [31] (inspired by [2], of course).

We want a (1,2)-presented *graded* Lie algebra $\underline{\underline{g}}$ such that, for the enveloping algebra $U(\underline{\underline{g}})$, the Hilbert series $U(\underline{\underline{g}})(Z)$ is transcendental. It follows from the Poincaré-Birkhoff-Witt theorem that

$$U(\underline{\underline{g}})(Z) = \prod_{i\geq 1} \frac{(1+Z^{2i-1})^{|\underline{\underline{g}}_{2i-1}|}}{(1-Z^{2i})^{|\underline{\underline{g}}_{2i}|}}, \tag{29}$$

where $|\ .\ | = \dim_k$. The point now is that it is easy to achieve a product (29) which is transcendental if we abandon the condition that $\underline{\underline{g}}$ is (1,2)-presented. Take for example any graded connected k-algebra N, consider the augmentation ideal N^+, and equip this graded vector space with the trivial (abelian) Lie algebra structure. Then in most cases [3,4] $U(N^+)(Z)$ is transcendental, the simplest case being that where N = k<T> and deg T = 1 (the free associative algebra). We now show that if the associative graded algebra N is (1,2)-presented, then we can embed the abelian Lie algebra N^+ *as an ideal of a* (1,2)-*presented graded Lie algebra* $\underline{\underline{g}}$ such that the Lie algebra quotient $\underline{\underline{g}}/N^+ = \Phi$ is a "nice" Lie algebra, that is such that $U(\Phi)(Z)$ is a rational function. The exact sequence of graded Lie algebras

$$0 \longrightarrow N^+ \longrightarrow \underline{\underline{g}} \longrightarrow \Phi \longrightarrow 0 \tag{30}$$

combined with the general formula (29) then gives that $U(\underline{\underline{g}})(Z) = U(\Phi)(Z).U(N^+)(Z)$ and, since $U(\Phi)(Z)$ is rational, it would follow that $U(\underline{\underline{g}})(Z)$ is transcendental at the same time as $U(N^+)(Z)$. Here is a description of how the construction is made (it uses the multiplicative structure of N; see [31; pp. 735-736]). Let us take as Φ the product of the free graded Lie algebras on the vector spaces $N^1 \oplus W$ and $\tilde{N}^1 \oplus k.a$, respectively, where $W = N^2$, considered as being in degree 1 (if $x \in N^2$, then x_W is the corresponding element in W, considered as being in degree 1). Furthermore, $\tilde{N}^1$ is just another copy of N^1, and a is an extra generator in degree 1. We define a Φ-module structure on N^+ by $W * N^+ = a * N^+ = 0$ and $n * n^+ = n.n^+$ if $n \in N^1$, $n^+ \in N^+$, and $\tilde{n} * n^+ = -(-1)^{\deg n^+} .n^+.\tilde{n}$ if $\tilde{n} \in \tilde{N}^1$. We define a 2-cocycle of degree 0 in the graded vector space $H^2(\Phi, N^+)$ by $\xi(x_W, a) = x$ (more details are available in [31]). This Φ-module structure on N^+ and the corresponding (non-trivial) 2-cocycle ξ now define an extension (30) and thereby a graded Lie algebra $\underline{\underline{g}}$. One shows that $\underline{\underline{g}}$ is (1,2)-presented if N is, either by using the Hochschild-Serre spectral sequence of the extension (30), or by an estimate with Hilbert series. For the corresponding local ring R_N, we have that the Hopf algebra $\mathrm{Ext}^*_{R_N}(k,k)$ contains the sub Hopf algebra $U(\underline{\underline{g}})$, where $\underline{\underline{g}}$ is given by (30). But the Hopf algebra $U(\underline{\underline{g}})$ contains the sub Hopf algebra $U(N^+)$, and since N^+ is abelian we have

$$U(N^+) = \mathrm{Pol}(N^+_{even}) \otimes E(N^+_{odd}),$$

where Pol means the ordinary polynomial algebra, and E means the exterior algebra. If N is infinite-dimensional as a vector space, then since N is generated by its elements of degree 1, it follows that N^+_{even} is infinite-dimensional. Thus $\mathrm{Pol}(N^+_{even})$ contains a polynomial sub Hopf algebra of arbitrarily high finite global homological dimension. In view of Remark 2 following Corollary 2 of Th[illegible] in §4, we now obtain the following.

THEOREM 6. *Any local ring* R_N, *obtained by the preceding construction from an infinite-dimensional* (1,2)-*presented associative algebra* N, *is outside the class* $\underline{\underline{\mathrm{AG}}}$ *of* §4. □

Example. Taking $N = k\langle T\rangle$, we obtain

$$R = R_N = k[X_1,\dots,X_5]/(X_1^2,X_2^2,X_3^2,X_5^2,X_1X_2,X_3X_5,X_1X_3+X_2X_4+X_4X_5,\underline{n}^3) \quad (31)$$

where $\underline{n} = (X_1,\dots,X_5)$ and

$$P_R(Z)^{-1} = (1+Z^{-1})(1-2Z)^2.\prod_{i=1}^{\infty}(1-Z^{2i})/(1+Z^{2i-1}) - Z^{-1}(1-5Z+8Z^2). \quad (32)$$

Note that the ring (31) does not look very different from the ring in Remark 3 of §4, but their Poincaré-Betti series are *very* different!

Remark 1. In all the preceding theory we have assumed that char $k \neq 2$. If char $k = 2$ only minor changes have to be made.

Remark 2. The exact sequence of Hopf algebras corresponding to (30) is

$$k \longrightarrow U(N^+) \longrightarrow U(\underline{\underline{g}}) \longrightarrow U(\Phi) \longrightarrow k.$$

Since gldim $U(\Phi) = 2$ ($U(\Phi)$ is the tensor product of two free algebras) it follows that λ-dim $U(\Phi) = 2$ (see Remark 1 following Corollary 1 of Theorem 3 in §2). Furthermore λ-dim $U(N^+) = 1$ if N is infinite-dimensional (write N^+ as an inductive limit of finite-dimensional abelian sub Lie algebras). Thus if the inequality (21) were true, we would have λ-dim $U(\underline{\underline{g}}) \leq 3$, where we probably would have equality, and then we might also have λ-dim $\mathrm{Ext}^*_{R_N}(k,k) = 3$.

PROBLEM. *In general, if* $(R,\underline{m})$ *is local and such that* λ-dim $\mathrm{Ext}^*_R(k,k) \leq 2$, *does it follow that* $P_R(Z)$ *is rational?*

6. *The Bass series of a local ring*

The Bass numbers of a local ring $(R,\underline{m})$ are the numbers $\mu_i(R) = \dim_k \mathrm{Ext}^i_R(k,R)$. They were introduced and studied by Bass in [12]. In general, they are difficult to calculate. The "Bass" series

$$I^R(Z) = \sum_{i\geq 0} \mu_i(R).Z^i \quad (33)$$

is known to be rational for a Golod ring [33] (an explicit formula is given in [10]), but Bøgvad proved in [15] that (33) is transcendental for the ring (31), for example. The following conjecture seems reasonable, however.

CONJECTURE. *Let* $(R,\underline{m})$ *be a local ring with* $\underline{m}^3 = 0$. *Then*

$I^R(Z)$ *and* $P_R(Z)$ *are rationally related.*

Remark 1. In [15], Bøgvad proves that the above conjecture is false if one only assumes that $\underline{m}^4 = 0$.

Remark 2. If a suitable form of the theory of ends of groups were true for a finitely presented Hopf algebra (which is a group in the category of coalgebras) H, that is if $\mathrm{Ext}_H^1(k,H) \neq 0$ implies either that H is a free Hopf algebra on one variable, or that H is a non-trivial co-product of smaller Hopf algebras or that H is an extension of an exterior algebra by one of the preceding examples, then the conjecture would follow. This follows easily from [15], but the results about ends seem hard to prove (they might be found in a forthcoming paper by Bøgvad).

Remark 3. In [21,22] Lescot has determined the Bass series of some fibre products, and in [23] Lescot reduces the determination of the Bass series to the artinian case by means of very nice results about a certain change of rings spectral sequence.

7. *Results by André and others about local flat ring homomorphisms*

Let $A \xrightarrow{\phi} B$ be a local flat ring homomorphism (assume that A and B have the same residue field for simplicity) and let $\bar{B} = k \otimes_A B$ be the fibre of ϕ. Generalizing earlier results by Gulliksen [17] and Avramov [6], André proved in [1], using methods originally intended to be used in the study of rationality questions for Poincaré-Betti series, that

$$P_A(Z).P_{\bar{B}}(Z) = P_B(Z).\prod_{j=1}^{N}\left(\frac{1+Z^{2j-1}}{1-Z^{2j}}\right)^{\delta_{2j}}$$

and he and others have conjectured that $\delta_{2j} = 0$, $j \geq 2$.

Now Avramov [9] has constructed a more general spectral sequence than that of (7) in §1. Applying this result to our ϕ we obtain a spectral sequence of Hopf algebras

$$E^2_{p,q} = \mathrm{Tor}_p^{\mathrm{Tor}_*^A(k,B)}(k,k)_q \Rightarrow E, \qquad (34)$$

where E is a certain Hopf algebra that sits in the middle of an exact sequence of Hopf algebras

$$k \longrightarrow \mathrm{Tor}_*^B(k,k)//\phi_* \longrightarrow E \longrightarrow U(V(\widetilde{A|B})) \longrightarrow k,$$

whose dual is a *central* extension of Hopf algebras, which in general is nonsplit. (See [9] and the corrections by Löfwall and Jacobsson.) In our special case of a flat homomorphism, we have that $\mathrm{Tor}_*^A(k,B) = \bar{B}$, so that (34) degenerates. This leads one to study the centre of the algebra $\mathrm{Ext}_{\bar{B}}^*(k,k)$ and Jacobsson proved in [18] that the so-called special variables of André correspond to central elements of the Lie algebra $\bar{\underline{\underline{g}}}$ of the Yoneda Ext-algebra of the fibre $\bar{B}$ (recall that $\mathrm{Ext}_{\bar{B}}^*(k,k) = U(\bar{\underline{\underline{g}}})$). Jacobsson put forward the conjecture (which by the result mentioned implies the truth of the André conjecture) that for any local commutative ring $(R,\underline{m})$, the centre of the Lie algebra of the Yoneda Ext-algebra $\mathrm{Ext}_R^*(k,k)$ is finite-dimensional and concentrated in degrees one and two. Using extensions of Hopf algebras associated to Golod maps, he was then able to prove his conjecture when $R \in \underline{\underline{AG}}$ [18]. It follows, for example, that the André conjecture holds true if the fibre of ϕ is a quotient of a regular ring $\tilde{R}$ by a set of monomials in an $\tilde{R}$-sequence.

8. *The centre of the* Ext-*algebra of a local ring*

We have already seen in the preceding section (in connection with the conjecture of Jacobsson) that the centre of $\mathrm{Ext}_R^*(k,k)$, for $(R,\underline{m})$ a local commutative noetherian ring, might be very small. Löfwall (unpublished) has determined the centre when $\underline{m}^3 = 0$, with the exception of the case when $\mathrm{gldim}\,\mathrm{Ext}_R^*(k,k) = 2$. If the analogues for Hopf algebras of some results of Bieri for groups were true, we would have the centre in general for $\underline{m}^3 = 0$. In particular, if we had the results about ends of Hopf algebras mentioned in Remark 2 of §6, we would also have the Bieri results for Hopf algebras, and thereby the proof of the André conjecture, when the fibre has $\underline{m}^3 = 0$. Bøgvad also has an alternative approach to these problems....

9. *Final remarks*

Many questions and many conjectures have been put forward in the preceding sections, and there is no point in repeating them here.

Instead, I would like to end with a beautiful application of the so-called Larfeldt-Lech construction [19], which was suggested by Gerson Levin in discussions with David Eisenbud, during a recent visit to Stockholm. Recall that the Larfeldt-Lech construction associates in particular to each *artinian* local ring $(R,\underline{m})$ a local 1-dimensional *domain* $(D,\underline{n})$ such that the completion $\hat{D}$ of D is isomorphic to R[[T]]. If we choose as $(R,\underline{m})$ any local ring with $\underline{m}^3 = 0$, having transcendental $P_R(Z)$, it follows that

$$P_D(Z) = P_{\hat{D}}(Z) = P_{R[[T]]}(Z) = (1+Z)P_R(Z)$$

is *also* transcendental. Clearly the local domain D is very strange, and in particular, it is very far from being excellent in the sense of Grothendieck. Several natural questions arise here....

References

1. M. André, "Le caractère additif des déviations des anneaux locaux", preprint, École Polytechnique Féderale de Lausanne, 1982.

2. D. Anick, "A counterexample to a conjecture of Serre", Ann. of Math. (2), 115 (1982), 1-33.

3. D. Anick, "The smallest singularity of a Hilbert series", Math. Scand., to appear.

4. D. Anick and C. Löfwall, "Some results on Hilbert and Poincaré series", preprint, University of Stockholm, 1982.

5. L.L. Avramov, "On the Hopf algebra of a local ring", Izv. Akad. Nauk SSSR Ser. Mat., 38 (1974), 253-277; Math. USSR-Izv., 8 (1974), 259-284.

6. L.L. Avramov, "Homology of local flat extensions and complete intersection defects", Math. Ann., 228 (1977), 27-37.

7. L.L. Avramov, "Small homomorphisms of local rings", J. Algebra, 50 (1978), 400-453.

8. L.L. Avramov, "Obstructions to the existence of multiplicative structures on minimal free resolutions", Amer. J. Math., 103 (1981), 1-31.

9. L.L. Avramov, "The homology of a tensor product diagram", Reports No. 2, Department of Mathematics, University of Stockholm, 1982.

10. L.L. Avramov and J. Lescot, "Bass numbers and Golod rings", Math. Scand., to appear.

11. J. Backelin, "Monomial ideal residue class rings and iterated Golod maps", Math. Scand., to appear.

12. H. Bass, "On the ubiquity of Gorenstein rings", Math. Z., 82 (1963), 8-28.

13. G. Baumslag, G. Dyer and C.F. Miller, "On the integral homology of finitely presented groups", Bull. Amer. Math. Soc., 4 (1981), 321-324.

14. D. Buchsbaum and D. Eisenbud, "Algebra structures for finite free resolutions, and some structure theorems for ideals of codimension 3", Amer. J. Math., 99 (1977), 447-485.

15. R. Bøgvad, "Gorenstein rings with transcendental Poincaré series", Math. Scand., to appear.

16. T. Gulliksen and F. Ghione, "Some reduction formulas for the Poincaré series of modules", Atti Acad. Naz. Lincei Rend. Cl. Sci. Fis. Mat. Natur. (8), 58 (1975), 82-91.

17. T. Gulliksen and G. Levin, Homology of local rings, Queen's University Papers on Pure and Applied Mathematics 20 (Queen's University, Kingston, Ontario, 1969).

18. C. Jacobsson, "On local flat homomorphisms and the Yoneda Ext-algebra of the fibre", Les méthodes d'algèbre homotopique en topologie, Journées S.M.F., Juin 1-5, 1982, Marseille-Luminy, Astérisque, to appear.

19. T. Larfeldt and C. Lech, "Analytic ramifications and flat couples of local rings", Acta Math., 146 (1981), 201-208.

20. J.M. Lemaire, Algèbres connexes et homologie des espaces de lacets, Lecture Notes in Mathematics 422 (Springer, Berlin, Heidelberg, New York, 1974).

21. J. Lescot, "La série de Bass d'un produit fibré d'anneaux locaux", C. R. Acad. Sci. Paris Sér. A, 293 (1981), 569-571.

22. J. Lescot, La série de Bass d'un produit fibré, Prépublication No. 9, Département de Mathématiques et de Mécanique, Université de Caen, 1982.

23. J. Lescot, "Sur la série de Poincaré d'un module", Reports No. 19, Department of Mathematics, University of Stockholm, 1982.

24. G. Levin, "Two conjectures in the homology of local rings", J. Algebra, 30 (1974), 56-74.

25. G. Levin, "Local rings and Golod homomorphisms", J. Algebra, 37 (1975), 266-289.

26. G. Levin, "Lectures on Golod homomorphisms", Reports No. 15, Department of Mathematics, University of Stockholm, 1976.

27. G. Levin, "Finitely generated Ext-algebras", Math. Scand., 49 (1981), 161-180.

28. C. Löfwall, "On the subalgebra generated by the one-dimensional elements of the Yoneda Ext-algebra", Reports No. 5, Department of Mathematics, University of Stockholm, 1976.

29. C. Löfwall, "Une algèbre nilpotente dont la série de Poincaré-Betti est non rationelle", C. R. Acad. Sci. Paris Sér. A, 288 (1979), 327-330.

30. C. Löfwall, "A change of rings theorem for local rings", Les méthodes d'algèbre homotopique en topologie, Journées S.M.F., Juin 1-5, 1982, Marseille-Luminy, Astérisque, to appear.

31. C. Löfwall and J.-E. Roos, "Cohomologie des algèbres de Lie graduées et séries de Poincaré-Betti non rationelles", C. R. Acad. Sci. Paris Sér. A, 290 (1980), 733-736.

32. J. Milnor and J.C. Moore, "On the structure of Hopf algebras", Ann. of Math. (2), 81 (1965), 210-264.

33. J.-E. Roos, "Sur l'algèbre Ext de Yoneda d'un anneau local de Golod", C. R. Acad. Sci. Paris Sér. A, 286 (1978), 9-12.

34. J.-E. Roos, "Relations between the Poincaré-Betti series of loop spaces and of local rings", Lecture Notes in Mathematics 740 (Springer, Berlin, Heidelberg, New York, 1979), pp.285-322.

35. J.-E. Roos, "Homology of loop spaces and of local rings", Proceedings of the 18th Scandinavian Congress of Mathematicians, Århus, 1980, Progress in Mathematics 11 (ed. E. Balslev, Birkhäuser, Boston, Basel, Stuttgart, 1981), pp. 441-468.

36. J.-E. Roos, "Finiteness conditions in commutative algebra and solution of a problem of Vasconcelos", Commutative algebra: Durham 1981, London Mathematical Society Lecture Notes 72 (ed. R.Y. Sharp, Cambridge University Press, Cambridge 1982), pp. 179-203.

37. J.-P. Serre, Algèbre locale: multiplicités, Lecture Notes in Mathematics 11 (Springer, Berlin, Heidelberg, New York, 1965).

Department of Mathematics,
University of Stockholm,
Box 6701,
S-113 85 Stockholm, Sweden.

Stop press (added 10 September, 1982)

Jörgen Backelin has recently proved, by means of beautiful arguments, that if k is a field, if $k[[X_1,\dots,X_n]]$ is the *commutative* ring of formal power series over k in variables $X_1,\dots,X_n$ and if $R = k[[X_1,\dots,X_n]]/\underline{a}$, where $\underline{a}$ is *any* ideal, generated by *any* set of *monomials* in the X_i, then the Poincaré-Betti series $P_R(Z)$ is a rational function of the form

$$(1 + Z)^{\text{embedding dim}(R)}/\text{polynomial in } Z$$

(even slightly more general rings R can be treated). Note that these R are all in the class $\underline{\underline{AG}}$ of §4. Backelin also proves for these R that the Yoneda Ext-algebra $\mathrm{Ext}_R^*(k,k)$ is finitely presented. He also has a multigraded version of the previous results, and an essential ingredient in his proof is indeed the idea of using a multigrading systematically. Backelin's results will appear in a note to be submitted to C. R. Acad. Sci. Paris (see also his thesis at the University of Stockholm, which will be publicly discussed on 21 January, 1983).

Jan-Erik Roos

REDUCTIONS, LOCAL COHOMOLOGY AND HILBERT FUNCTIONS OF LOCAL RINGS

JUDITH D.SALLY

Throughout this exposition, $(R,\underline{m})$ is a d-dimensional local Cohen-Macaulay ring of multiplicity e. The goal is to compute the Hilbert function of $(R,\underline{m})$. This is, of course, the Hilbert function H_R of the associated graded ring $\operatorname{gr} R = R/\underline{m} \oplus \underline{m}/\underline{m}^2 \oplus \ldots$. The chief problem is that, in general, the ring $\operatorname{gr} R$ has few, if any, good properties so that what is known about Hilbert functions for nice graded rings is often not applicable. So the game is twofold: try to get information about the Hilbert function in spite of bad properties of $\operatorname{gr} R$ and try to find properties of $(R,\underline{m})$ which lead to reasonable associated graded rings $\operatorname{gr} R$.

One way to recognize some Cohen-Macaulay local rings with good associated graded rings is to use the fact that v, the embedding dimension of R, satisfies

$$d \le v \le e+d-1.$$

We write $v = e+d-1-h$ with $0 \le h \le e-1$. For certain values of h, namely $h = 0,(1),e-2,e-1$, $\operatorname{gr} R$ is known to be a good ring (cf.[S_1]; (1) indicates that "most of the time" the value $h = 1$ gives a good associated graded ring for $d > 1$ - cf.[S_2]). "Good" means, at least, that there are homogeneous non-zerodivisors in $\operatorname{gr} \underline{m}$. (The existence of non-zerodivisors in $\operatorname{gr} \underline{m}$ means that one can reduce dimension so that the structure of $\operatorname{gr} R$ and the computation of the Hilbert function is made simpler.) The fact that Cohen-Macaulay local rings with small values of h share some of the good properties of regular local rings and hypersurfaces makes h a number worth further examination. If $h = 0$, for example, we get, for any d, the class of regular local rings ($e = 1$) and for $h = 0$, $e > 1$, we get a class of local rings whose structure is well understood because, among other things, their associated graded rings are Cohen-Macaulay and they have linear resolutions (cf. [S_3] and [Sch]). This class includes the class of rational surface singularities - a nice one to think about is $R = k[[(X,Y)^s]]$, where k is a field, X,Y are indeterminates, and s is any positive integer.

Another number attached to R which gives information about gr R and H_R is the reduction number of R. (For convenience, throughout $R/\underline{m}$ will be assumed to be an infinite field.) The ideal in R generated by a set $x_1,\ldots,x_d$ of preimages of a degree one system of parameters $\bar{x}_1,\ldots,\bar{x}_d$ in gr R is called a *minimal reduction* of $\underline{m}$. (The notation $\bar{a}$ means that $\bar{a}$ is the initial form in gr R of the element a in R.) If $x_1,\ldots,x_d$ is a minimal reduction of $\underline{m}$ then there is an s such that $\underline{m}^{s+1} = (x_1,\ldots,x_d)\underline{m}^s$. The *reduction number for* $\underline{m}$ *with respect to* $\underline{x} = (x_1,\ldots,x_d)$ is the least integer $r_{\underline{x}}$ such that $\underline{m}^{r_{\underline{x}}+1} = \underline{x}\,\underline{m}^{r_{\underline{x}}}$. The *reduction number* of R is the least integer r such that there is a minimal reduction $\underline{x} = (x_1,\ldots,x_d)$ of $\underline{m}$ with $\underline{m}^{r+1} = \underline{x}\,\underline{m}^r$. (Here the rather annoying - more than important - question of whether $r = r_{\underline{x}}$ intrudes. We shall return to this point later.) Thus modules of the form $\underline{m}^{i+1}/\underline{x}\,\underline{m}^i$ naturally come into play. They arise also in another way. The first place to look for non-zerodivisors in $\mathrm{gr}\,\underline{m} = \underline{m}/\underline{m}^2 \oplus \underline{m}^2/\underline{m}^3 \oplus \ldots$ is at a degree one system of parameters $\bar{x}_1,\ldots,\bar{x}_d$. Now $\bar{x}_1$ is a non-zerodivisor in gr R if and only if $(x_1) \cap \underline{m}^{i+1} = x_1\underline{m}^i$ for all $i \geq 1$. Similarly, $\bar{x}_1,\ldots,\bar{x}_j$ is a regular sequence in gr R if and only if $(x_1,\ldots,x_j) \cap \underline{m}^{i+1} = (x_1,\ldots,x_j)\underline{m}^i$ for all $i \geq 1$ (cf. [V-V],[S_4]). But even if one cannot find elements satisfying such equalities for all i, it is true that certain carefully chosen minimal reductions will satisfy such equalities for all large i. So one is forced to look at modules $\underline{m}^{i+1}/\underline{x}\,\underline{m}^i$ and $\underline{m}^{i+1}/(\underline{x}) \cap \underline{m}^{i+1}$.

Now modules of the form $\underline{m}^{i+1}/\underline{x}\,\underline{m}^i$ arise when one looks at certain local cohomology modules. The purpose of this exposition is to look at a cohomological interpretation of these numbers and modules to try to gain more insight into the problem of computing Hilbert functions.

The additional notation will be as follows. First, $B = R \oplus \underline{m} \oplus \underline{m}^2 \oplus \ldots$ and B_+ is the ideal of B which is the direct sum of components of positive degree. Note that height B_+ = grade B_+ = 1. Secondly, $G = \operatorname{gr} R = R/\underline{m} \oplus \underline{m}/\underline{m}^2 \oplus \underline{m}^2/\underline{m}^3 \oplus \dots$ and $G_+ = \operatorname{gr} \underline{m}$, the direct sum of components of positive degree.

Set $(X, \mathcal{O}_X)$ = Proj B, the blow-up of R at $\underline{m}$. The geometrical interpretation of minimal reduction is that the corresponding open sets give a minimal cover of Proj B. For if $\underline{x} = x_1, \dots, x_d$ is any minimal reduction of $\underline{m}$, then $X = \bigcup_{i=1}^{d} D_+(\bar{x}_i)$ because the only homogeneous ideal of B containing $\bar{x}_1, \dots, \bar{x}_d$ is B_+. If E is any finitely generated graded B-module and $\tilde{E}$ is the corresponding sheaf, the local cohomology modules $H^{\cdot}_{B_+}(E)$ can be expressed in terms of the Grothendieck cohomology of the sheaves $\tilde{E}(n)$. Set $H^{\cdot}(X, \tilde{E}(*)) = \bigoplus_{n \in \mathbb{Z}} H^{\cdot}(X, \tilde{E}(n))$. Then, by [EGA; (2.1.5)], there exist degree 0 canonical isomorphisms, functorial in E, of graded B-modules

$$H^p(X, \tilde{E}(*)) \xrightarrow{\cong} H^{p+1}_{B_+}(E) \text{ for } p \geq 1$$

and an exact sequence of degree 0 homomorphisms, functorial in E,

$$0 \to H^0_{B_+}(E) \to E \to H^0(X, \tilde{E}(*)) \to H^1_{B_+}(E) \to 0.$$

Now, for $i \geq 0$, the local cohomology module $H^i_{B_+}(E)$ can be computed as a direct limit of cohomology modules of Koszul complexes on sequences $\underline{x}^k = x_1^k, \dots, x_d^k$: cf. [Hz-K]. Consequently, we have for the d-th local cohomology,

$$H^d_{B_+}(B)_n = \varinjlim_k \underline{m}^{dk+n}/\underline{x}^k \underline{m}^{(d-1)k+n}$$

and

$$H^d_{G_+}(G)_n = \varinjlim_k \underline{m}^{dk+n}/\underline{x}^k \underline{m}^{(d-1)k+n} + \underline{m}^{dk+n+1},$$

where the maps in the direct systems are multiplication by $x_1 \dots x_d$. (The idea will be to try to assume as little as possible about the vanishing of $H^i_{B_+}(B)$ for $i < d$ and say as much as we can about R by exploiting the fact that $H^d_{B_+}(B)$ looks computable.) It is known (the result of Serre, cf. [EGA; Proposition 2.2.2]) that the modules $H^i(X, \tilde{E}(n))$ are finitely generated R-modules and that there exists N such that $H^i(X, \tilde{E}(n)) = 0$ for all $n \geq N$ and all $i > 0$.

For E = B or G we get the following additional information after first choosing a special minimal reduction for $\underline{m}$. By prime avoidance there exists a minimal reduction $x_1,\ldots,x_d$ for $\underline{m}$ and an integer i_o such that

$$(x_{j_1},\ldots,x_{j_\ell}) \cap \underline{m}^{i+1} = (x_{j_1},\ldots,x_{j_\ell})\underline{m}^i \text{ for all } i \geq i_o,\ 1 \leq \ell \leq d.$$

We call such a reduction *a smooth minimal reduction for* $\underline{m}$.

PROPOSITION 1. *Let* $\underline{x}$ *be a smooth minimal reduction for* $\underline{m}$. *Let* $\tilde{r}$ *be the reduction number for* $\underline{x}$. *Fix* $n \in \mathbb{Z}$. *If* $s = \max(i_o,\tilde{r},-d-n+\tilde{r}+1)$, *then for* $k \geq s$ *the maps*

$$\underline{m}^{dk+n}/\underline{x}^k\underline{m}^{(d-1)k+n} \xrightarrow{x_1\cdots x_d} \underline{m}^{d(k+1)+n}/\underline{x}^{k+1}\underline{m}^{(d-1)(k+1)+n}$$

are isomorphisms.

The proof of Proposition 1 is a tedious manipulation with regular sequences.

PROPOSITION 2. *Let* $\underline{x}$ *be a minimal reduction for* $\underline{m}$ *and let* $r_{\underline{x}}$ *be the reduction number of* $\underline{m}$ *with respect to* $\underline{x}$.

(i) $$H^d_{B_+}(B)_{r_{\underline{x}}-d} \cong H^d_{G_+}(G)_{r_{\underline{x}}-d} \text{ and } H^d_{B_+}(B)_n = H^d_{G_+}(G)_n = 0$$

for $n \geq r_{\underline{x}}-d+1$.

(ii) $$\lambda\left(H^d_{B_+}(B)_{n+1}\right) \leq \lambda\left(H^d_{B_+}(B)_n\right) \text{ for all } n \in \mathbb{Z}, \text{ where } \lambda$$

denotes length as an R-*module.*

(iii) *If* $d > 1$, *then* $H^1_{B_+}(B)_{-1} \cong R$.

(iv) *Assume that* $\operatorname{depth}_{G_+} G \geq d-1$ *and* $d > 1$. *Then* $H^1_{B_+}(B)_n = 0$ *for all* $n \geq \max(0,r_{\underline{x}}-d)$, $H^i_{B_+}(B) = 0$ *for* $1 < i < d-1$ *and* $H^{d-1}_{B_+}(B)_n = 0$ *for* $n \geq \max(0,r_{\underline{x}}-d)$.

Remark. The same notation has been used for $x_1,\ldots,x_d$ in R and $x_1,\ldots,x_d$ as degree 1 elements in B.

Proof. (i) $\underline{m}^{kd+r_{\underline{x}}-d+j} = (\underline{x})^{kd-d+j}\underline{m}^{r_{\underline{x}}}$ for $j \geq 1$ and each monomial $x_1^{\alpha_1}\ldots x_d^{\alpha_d}$ generating $(\underline{x})^{(k-1)d+j}$ must have some $\alpha_i \geq k$, and so $\underline{m}^{kd+r_{\underline{x}}-d+j} = \underline{x}^k\underline{m}^{(d-1)k+r_{\underline{x}}+j}$. Thus

$$\underline{m}^{dk+r_{\underline{x}}-d}/\underline{x}^k\underline{m}^{(d-1)k+r_{\underline{x}}-d} + \underline{m}^{dk+r_{\underline{x}}-d+1} = \underline{m}^{dk+r_{\underline{x}}-d}/\underline{x}^k\underline{m}^{(d-1)k+r_{\underline{x}}-d}$$

and $\underline{m}^{dk+r_{\underline{x}}-d+j}/\underline{x}^k\underline{m}^{(d-1)k+r_{\underline{x}}-d+j} = 0$ for $j \geq 1$.

(ii) $0 \to B(-1) \xrightarrow{x_1} B \to B/x_1B \to 0$ is exact and so we get that

$$\dots \to H^d_{B_+}(B)_{n-1} \to H^d_{B_+}(B)_n \to H^d_{B_+}(B/x_1B)_n \to 0$$

is exact; but $H^d_{B_+}(B/x_1B) = 0$, and so the desired conclusion follows.

(iii) $0 \to B \to \bigoplus_{n\in\mathbb{Z}} H^0(X,\tilde{B}(n)) \to H^1_{B_+}(B) \to 0$ is exact.

Since R is C.-M. of dim $d \geq 2$, $H^0(X,\tilde{B}) = \bigcup_{k\geq 0}(\underline{m}^k:\underline{m}^k) = R$, and so $H^1_{B_+}(B)_0 = 0$. From the exact sequence

$$0 \to B(-1) \xrightarrow{x_1} B \to B/x_1B \to 0,$$

we get that

$$0 \to H^0_{B_+}(B/x_1B)_0 \xrightarrow{x_1} H^1_{B_+}(B)_{-1} \to H^1_{B_+}(B)_0 \to \dots \text{ is exact,}$$

and, as $H^1_{B_+}(B)_0 = 0$ and $H^0_{B_+}(B/x_1B)_0 \cong x_1R$, it follows that $H^1_{B_+}(B)_{-1} \cong R$.

(iv) Since depth $G \geq d-1$, $(B/x_1B)_n \cong B(R/x_1R)_n$ for all $n > 0$ where

$$B(R/x_1R) = R/x_1R \oplus \underline{m}/x_1R \oplus (\underline{m}/x_1R)^2 \oplus \dots$$

and $r_{\underline{x}}$ is the reduction number for $\underline{m}/x_1R$ with respect to the images of $x_2,\dots,x_d$ in R/x_1R. We have the exact sequence

$$\dots \to H^{i-1}(B)_{n+1} \to H^{i-1}(B/x_1B)_{n+1} \to H^i(B)_n \xrightarrow{x_1} H^i(B)_{n+1}$$
$$\to H^i(B/x_1B)_{n+1} \to \dots.$$

Set $i = 1$ and $n \geq \max(0,r_{\underline{x}}-d)$. Since $n \geq 0$, $H^0(B/x_1B)_{n+1} = 0$. The fact that every element of $H^1(B)_n$ is annihilated by a power of x_1 means that $H^1(B)_n = 0$ for all $n \geq \max(0,r_{\underline{x}}-d)$. A similar argument shows that $H^{d-1}(B)_n = 0$ for all $n \geq \max(0,r_{\underline{x}}-d)$. Now use the exact sequence

$$0 \to \underline{m}B \to B \to G \to 0$$

to get the exact sequence

$$\dots \to H^{i-1}(G)_n \to H^i(\underline{m}B)_n \to H^i(B)_n \to H^i(G)_n$$
$$\to H^{i+1}(\underline{m}B)_n \to \dots .$$

If i is any integer such that $1 < i < d-1$, then $H^i(G) = H^{i-1}(G) = 0$ and (since $i > 1$) $H^i(\underline{m}B)_n \cong H^i(B)_{n+1}$. Thus $H^i(B)_{n+1} \cong H^i(B)$ for all $n \in \mathbb{Z}$. But since $H^i(B)_n = 0$ for all large n, it follows that $H^i(B) = 0$ for $1 < i < d-1$. □

Remark. (iv) puts strong hypotheses on G. (iv) was included in Proposition 2 to illustrate the type of computation that can be done with slightly less than the full Cohen-Macaulay hypothesis on G.

Proposition 2 shows some ways in which r fits into the picture cohomologically. The question about whether r is independent of $\underline{x}$, a situation which seems unlikely to me except in some generic sense, could be phrased in terms of the non-vanishing of $H^d(B)_{r_{\underline{x}}-d}$. If depth $G \geq d-1$ it is true that $H^d_{B_+}(B)_{r_{\underline{x}}-d} \neq 0$. However, one can easily show directly that if $\operatorname{depth}_{G_+} G \geq d-1$ then the reduction number is independent of $\underline{x}$. One underlying reason is the behavior of the Hilbert polynomial P_R where $P_R(n) = H_R(n)$ for n large.

PROPOSITION 3. *If* $\operatorname{depth}_{G_+} G \geq d-1$, *then*

$$H_R(r_{\underline{x}}-d) \neq P_R(r_{\underline{x}}-d)$$

and

$$H_R(n) = P_R(n) \text{ for all } n \geq r_{\underline{x}}-d+1$$

where $\underline{x}$ *is any minimal reduction for* $\underline{m}$ *and* $r_{\underline{x}}$ *is the reduction number of* $\underline{m}$ *with respect to* $\underline{x}$.

Proof. Application of [EGA; Proposition 2.1.5] gives

$$H_R(n)-P_R(n) = \sum_{i\geq 0}(-1)^i \dim_{R/\underline{m}} H^i_{G_+}(G)_n \quad \text{for all } n \in \mathbb{Z}$$

(cf.[Sch]). Since $\operatorname{depth}_{G_+} G \geq d-1$,

$$H_R(n)-P_R(n) = \pm \dim_{R/\underline{m}} H^{d-1}_{G_+}(G)_n \mp \dim_{R/\underline{m}} H^d_{G_+}(G)_n.$$

By induction on d, we shall show that $H^{d-1}_{G_+}(G)_n = 0$ for all $n \geq r_{\underline{x}}-d$ and $H^d_{G_+}(G)_{r_{\underline{x}}-d} \neq 0$. Since $H^d_{G_+}(G)_n = 0$ for $n \geq r_{\underline{x}}-d+1$ by Proposition 2(i), this will complete the proof. Let $d = 1$ and let x be a minimal reduction for $\underline{m}$. $H^0_{G_+}(G)_n = \{\bar{z} \in \underline{m}^n/\underline{m}^{n+1} \mid z\underline{m}^{\ell} \subseteq \underline{m}^{n+\ell+1}$, some $\ell\}$. If $n \geq r_x-1$, say $n = r_x-1+j$, then $zx^\ell \in \underline{m}^{\ell+r_x+j} = x^\ell \underline{m}^{r_x+j}$ gives the contradiction $z \in \underline{m}^{n+1}$. It is also clear that $H^1_{G_+}(G)_{r_x-1} \cong \underline{m}^{r_x}/x\underline{m}^{r_x-1} \neq 0$.

Suppose that $d > 1$. Given any minimal reduction $x_1,\ldots,x_d$ of $\underline{m}$ we may assume that $\bar{x}_1$ is a non-zerodivisor in G. We have the exact sequence

$$0 \longrightarrow G(-1) \xrightarrow{\bar{x}_1} G \longrightarrow G/\bar{x}_1 G \longrightarrow 0$$

from which we get the exact sequence

$$0 \longrightarrow H^{d-2}_{G_+}(G/\bar{x}_1G)_n \longrightarrow H^{d-1}_{G_+}(G)_{n-1} \xrightarrow{\bar{x}_1} H^{d-1}_{G_+}(G)_n \longrightarrow H^{d-1}_{G_+}(G/\bar{x}_1G)_n$$
$$\longrightarrow H^d_{G_+}(G)_{n-1} \xrightarrow{\bar{x}_1} H^d_{G_+}(G)_n \longrightarrow 0. \qquad (*)$$

Since depth $G \geq d-1$, $r_{\underline{x}}$ is the reduction number for $\underline{m}/x_1R$ with respect to the images of $x_2,\ldots,x_d$ and $G/\bar{x}_1G \cong G(R/x_1R)$. Set $n = r_{\underline{x}}-(d-1)+j$ with $j \geq 1$. By induction, $H^{d-2}_{G_+}(G/\bar{x}_1G)_{r_{\underline{x}}-(d-1)+j} = 0$ and by Proposition 2(i), $H^{d-1}(G/\bar{x}_1G)_{r_{\underline{x}}-(d-1)+j} = 0$; hence

$$0 \longrightarrow H^{d-1}_{G_+}(G)_{r_{\underline{x}}-d+j} \xrightarrow{\bar{x}_1} H^{d-1}_{G_+}(G)_{r_{\underline{x}}-d+j+1} \longrightarrow 0$$

is exact. Thus, since every element of $H^{d-1}_{G_+}(G)_{r_{\underline{x}}-d+j}$ is annihilated by a power of $\bar{x}_1$, $H^{d-1}_{G_+}(G)_{r_{\underline{x}}-d+j} = 0$ for $j \geq 1$. Now set $n = r_{\underline{x}}-d+1$ in (*). We get $H^{d-1}_{G_+}(G)_{r_{\underline{x}}-d} = 0$ and

$$0 \longrightarrow H^{d-1}_{G_+}(G/\bar{x}_1G)_{r_{\underline{x}}-d+1} \longrightarrow H^d_{G_+}(G)_{r_{\underline{x}}-d}$$

is exact. Since, by induction, $H^{d-1}_{G_+}(G/\bar{x}_1G)_{r_{\underline{x}}-d+1} \neq 0$, it must be true that $H^d_{G_+}(G)_{r_{\underline{x}}-d} \neq 0$. □

Remark. It should be noted that $\sup\{r_{\underline{x}} : \underline{x}$ is a minimal reduction for $\underline{m}\}$ is a positive integer. This follows from the work of G. Hermann [Her]; cf. also [Sei], where it is shown that the degree of nilpotency of an ideal I in a polynomial ring in n variables over a field is bounded by a number depending only on n and the degrees of the generators of I.

Now let us see how h fits into the picture. Recall that we have

embedding dim R = v = e+d-1-h.

PROPOSITION 4. (i) $\lambda\left(H^d_{B_+}(B)_{2-d}\right) \geq h.$

(ii) $\lambda\left(H^d_{B_+}(B)_{1-d}\right) \geq e-1-h.$

(iii) $\lambda\left(H^d_{B_+}(B)_{-j-d}\right) \geq e$ *for all* $j \geq 0$.

Proof. Let $\underline{x}$ be any minimal reduction for $\underline{m}$. The exact sequence

$$0 \to \underline{x}R/\underline{xm} \to \underline{m}/\underline{xm} \to R/\underline{x}R \to R/\underline{m} \to 0$$

yields $\lambda(\underline{m}/\underline{xm}) = e+d-1$. The exact sequence

$$0 \to \underline{m}^2/\underline{xm} \to \underline{m}/\underline{xm} \to \underline{m}/\underline{m}^2 \to 0$$

yields $h = \lambda(\underline{m}^2/\underline{xm})$. Now $H^d_{B_+}(B)_{2-d} = \varinjlim_k \underline{m}^{kd+2-d}/\underline{x}^k\underline{m}^{k(d-1)+2-d}$.

It is easy to check that $\underline{m}^2/\underline{xm}$ embeds into $H^d_{B_+}(B)_{2-d}$ under multiplication by $x_1^{k-1}\dots x_d^{k-1}$.

Parts (ii) and (iii) follow even more directly. □

Thus we have the following.

COROLLARY. (i) *The following conditions are equivalent:*

(1) $h = 0$;

(2) $r = 1$;

(3) $H^d_{B_+}(B)_{2-d} = 0$.

(ii) R *is regular if and only if* $H^d_{B_+}(B)_{1-d} = 0$.

This seems to exhibit graphically why the rings with small h share properties with regular local rings - namely the cohomology of the blow-up is "very close" to that of the blow-up of a regular ring. In particular we see why 2-dimensional rational singularities must have embedding dimension v = e+2-1 = e+1 and that there is no restriction on the embedding dimension of such a singularity if d > 2. For if R is the local ring of a rational surface singularity and $Z \to \operatorname{Spec} R$ is a desingularization, then $H^1(Z,\mathcal{O}_Z) = 0$. But $Z \to \operatorname{Spec} R$ can be factored into a product of blow-ups, and so we have $Z \to X \to \operatorname{Spec} R$ and $H^1(X,\mathcal{O}_X) = 0$: cf. [L].

Thus we have $H^2_{B_+}(B)_0 = 0$ and so $h = 0$ and $v = e+d-1 = e+1$. In view of this it seems not unreasonable to ask if the local ring of a rational singularity in dimension d satisfies $\underline{m}^d = \underline{xm}^{d-1}$.

We conclude this exposition with an amusing and quick computation exhibiting the relationship between the coefficients of the Hilbert polynomial p_R which gives the length of $R/\underline{m}^n$ for large n and the lengths of certain $H^d_{B_+}(B)_n$ in the special case where dim R = 2. We write

$$p_R(n) = e\binom{n}{2} + a_1 n + a_2.$$

PROPOSITION 5. *With the notation as above,*

(i) $\lambda\left(H^2_{B_+}(B)_0\right) = a_2 \qquad (\Rightarrow h \leq a_2)$,

(ii) $\lambda\left(H^2_{B_+}(B)_1\right) = a_1+a_2-1 \qquad (\Rightarrow a_1 \leq 1)$,

(iii) $\lambda\left(H^2_{B_+}(B)_{-1}\right) = e-a_1+a_2$.

Proof. For any $\underline{m}$-primary ideal $\underline{q}$ we have

$$\lambda(\underline{q}^2/\underline{yq}) = e(\underline{q}) + \lambda(R/\underline{q}) - \lambda(\underline{q}/\underline{q}^2), \qquad \text{(i*)}$$

where $\underline{y}$ is a minimal reduction of $\underline{q}$. For $\lambda(\underline{q}/\underline{yq}) = \lambda(R/\underline{yq})-\lambda(R/\underline{q}) = \lambda(R/\underline{y}R)+\lambda(\underline{y}R/\underline{yq})-\lambda(R/\underline{q})$ and $\lambda(\underline{y}R/\underline{yq}) = 2\lambda(R/\underline{q})$: cf. [V]. Similarly,

$$\lambda(\underline{mq}^2/\underline{yqm}) = e(\underline{q})+2\lambda(R/\underline{qm})-1-\lambda(R/\underline{q}^2\underline{m}). \qquad \text{(ii*)}$$

For $\lambda(\underline{yq}/\underline{yqm}) = 2\lambda(\underline{q}/\underline{qm})-1$. We will now set $\underline{q} = \underline{m}^n$ for n large. From (i*) we get

$$\lambda\left(H^d_{B_+}(B)_0\right) = \lambda(\underline{m}^{2n}/\underline{x}^n\underline{m}^n) = n^2e + 2p_R(n)-p_R(2n) = a_2.$$

From (ii*) we get

$$\lambda\left(H^d_{B_+}(B)_1\right) = \lambda\left(\underline{m}^{2n+1}/\underline{x}^n\underline{m}^{n+1}\right) = n^2e+2p_R(n+1)-1-p_R(2n+1)$$
$$= a_1 + a_2 - 1.$$

Finally,

$$\lambda\left(H^d_{B_+}(B)_{-1}\right)=\lambda\left(\underline{m}^{2n-1}/\underline{x}\,\underline{m}^{n-1}\right)=\lambda\left(R/\underline{x}^n\underline{m}^n\right)-\lambda\left(\underline{x}^n\underline{m}^{n-1}/\underline{x}^n\underline{m}^n\right)-\lambda\left(R/\underline{m}^{2n-1}\right)$$
$$=\lambda\left(R/\underline{x}^nR\right)+2\lambda\left(R/\underline{m}^n\right)-2\lambda\left(\underline{m}^{n-1}/\underline{m}^n\right)-\lambda\left(R/\underline{m}^{2n-1}\right)$$
$$= n^2e + 2p_R(n-1) - p_R(2n-1)$$
$$= e - a_1 + a_2. \ \square$$

Remark 1. Lipman has a proof, using intersection theory, of most of the proposition above in his paper [L]. He assumes that R is 2-dimensional and normal, but only for the purpose of obtaining $H^0(X,\mathcal{O}_X) = R$, and so the weaker Cohen-Macaulay hypothesis works for his proof as well.

Remark 2. If $P_R(n)$ is the polynomial giving $\lambda(\underline{m}^n/\underline{m}^{n+1})$ for large n, then $P_R(n) = en+a_1$ and so a_1 is the Euler characteristic:

$$a_1 = \lambda(H^0(X,\tilde{G})) - \lambda(H^1(X,\tilde{G})) = 1 + \lambda(H^1(G)_0) - \lambda(H^2(G)_0).$$

The fact that $\lambda(H^1(G)_0) - \lambda(H^2(G)_0) = \lambda(H^2(B)_1) - \lambda(H^2(B)_0)$ follows from the exact sequence

$$0 \to H^1_{B_+}(\underline{m}B)_0 \to H^1_{B_+}(B)_0 \to H^1_{G_+}(G)_0$$
$$\to H^2_{B_+}(\underline{m}B)_0 \to H^2_{B_+}(B)_0 \to H^2_{G_+}(G)_0 \to 0.$$

The fact that R is Cohen-Macaulay of dimension 2 implies that $H^1_{B_+}(B)_0 = H^1_{B_+}(B)_1 = 0$, for the maps $R \to H^0(X,\mathcal{O}_X)$ and $\underline{m} \to H^0(X,\mathcal{O}_X(1))$ are isomorphisms. The arithmetic genus $1-a_1$ is then $-\lambda(H^1(G)_0) + \lambda(H^2(G)_0) = -\lambda(H^2(B)_1) + \lambda(H^2(B)_0)$.

Remark 3. At the Symposium, D. Kirby spoke on his very recent work on Koszul complexes and coefficients of Hilbert polynomials. The computations in Proposition 5 above are special cases in his much more general set up. See [K-M_1] and [K-M_2].

References

[EGA] A. Grothendieck and J. Dieudonné, Éléments de géométrie algébrique III, Publications Mathématiques 11 (Institut des Hautes Études Scientifiques, Paris, 1961).

[Her] G. Hermann, "Die Frage der endlich vielen Schritte in der Theorie der Polynomideale", Math.Ann., 95 (1926), 736-788.

[Hz-K] J. Herzog and E. Kunz, Der kanonische Modul eines Cohen-Macaulay-Rings, Lecture Notes in Mathematics 238 (Springer, Berlin, 1971).

[H-R] M. Hochster and J.L. Roberts, "Rings of invariants of reductive groups acting on regular rings are Cohen-Macaulay", Adv. in Math., 13 (1974), 115-175.

[K-M_1] D. Kirby and H.A. Mehran, "Hilbert functions and the Koszul complex", J.London Math.Soc. (2), 24 (1981), 459-466.

[K-M_2] D.Kirby and H.A. Mehran, "A note on the coefficients of the Hilbert-Samuel polynomial for a Cohen-Macaulay module", J.London Math.Soc. (2), 25 (1982), 449-457.

[L] J. Lipman, "Rational singularities, with applications to algebraic surfaces and unique factorization", Publications Mathématiques 36 (Institut des Hautes Études Scientifiques, Paris, 1969), pp. 195-279.

[S_1] J.D. Sally, "On the associated graded ring of a local Cohen-Macaulay ring", J.Math.Kyoto Univ., 17 (1977), 19-21.

[S_2] J.D. Sally, "Cohen-Macaulay local rings of embedding dimension e + d - 2", in preparation.

[S_3] J.D. Sally, "Cohen-Macaulay local rings of maximal embedding dimension", J. Algebra, 56 (1979), 168-183.

[S_4] J.D. Sally, "Super-regular sequences", Pacific J.Math., 84 (1979), 465-481.

[Sch] P. Schenzel, "Über die freien Auflösungen extremaler Cohen-Macaulay-Ringe", J. Algebra,64 (1980), 93-101.

[Sei] A. Seidenberg, "Constructions in algebra", Trans.Amer.Math. Soc., 197 (1974), 273-313.

[V-V] P. Valabrega and G. Valla, "Form rings and regular sequences", Nagoya Math. J., 72 (1978), 93-101.

[V] G. Valla, "On form rings which are Cohen-Macaulay", J.Algebra, 58 (1979), 247-250.

Department of Mathematics,
Northwestern University,
Evanston,
Illinois 60201, U.S.A.

FURTHER PROBLEMS

FURTHER PROBLEMS

SUBMITTED BY SYMPOSIUM PARTICIPANTS

Introduction

Participants at the Symposium were invited to submit open problems in commutative algebra for inclusion in a Problem Section in these proceedings, and what follows represents the response to that invitation. In fact, the response was rather limited: perhaps the fact that much of the Symposium was concerned with various well-known open problems in the subject provides an explanation for this. Indeed, many of the foregoing articles contain statements of and discussions about various problems and conjectures, and thus it seems appropriate to entitle this section 'Further problems'.

The material below is arranged into three sections: Section 1 was submitted by Chr. Lech and is concerned with problems on inequalities for couples of local rings; Section 2 was submitted by L.J. Ratliff, Jr. and is concerned with problems on asymptotic prime divisors; and Section 3 is an amalgamation of the remaining problems suggested by various participants.

1. *Problems on inequalities for couples of local rings (submitted by Chr. Lech and received on 27 May, 1982)*

Simple geometric considerations suggest that a (Noetherian) local ring Q should not be less complicated (or singular) than any of its localizations $Q_{\underline{p}}$. Various aspects of this vague idea can be expressed in terms of certain formal power series that can be associated with a local ring, viz. the Hilbert series, the deviation series, and the μ-series. The resulting statements take the form of coefficientwise inequalities between power series, one belonging to $Q_{\underline{p}}$ and the other to Q. Remarkably enough, these statements turn out to have equivalent counterparts dealing with local flat morphisms

$A \longrightarrow B$ rather than localizations. But their validity has been fully investigated only in the μ-series case - in fact, with a positive answer (see, for example, [7; Chapter 2, Proposition 4.1] and [3], or [2; Theorems 10.10, 10.31]). Here we shall discuss the two other cases.

Definitions. Let $(Q,\underline{m})$ be a local ring with residue field k. The *Hilbert series* H_Q of Q is given by $H_Q(T) = \sum_{n=0}^{\infty} \dim_k(\underline{m}^n/\underline{m}^{n+1})T^n$. The *deviation series* D_Q is defined by $D_Q(T) = \sum_{n=1}^{\infty} e_n(Q)T^n$ where the $e_n(Q)$ are non-negative integers determined by the following equation (whose left member is known as the *Poincaré series* of Q):

$$\sum_{n=0}^{\infty} \dim_k(\mathrm{Tor}_n^Q(k,k))T^n = \prod_{n=1}^{\infty} (1-(-T)^n)^{(-1)^{n+1}e_n(Q)}.$$

(The non-negativity of the $e_n(Q)$ is a consequence of the Tate-Gulliksen analysis of $\mathrm{Tor}_*^Q(k,k)$.)

PROBLEMS. Decide if the statements (1) - (4) below are generally valid under the assumptions that $\underline{p}$ is a prime ideal of coheight d in the local ring Q and that A and B are local rings of equal Krull dimension for which there exists a local flat ring homomorphism $A \longrightarrow B$.

(1) $H_{Q_{\underline{p}}}(T)(1-T)^{-d} \le H_Q(T)$.

(2) $D_{Q_{\underline{p}}}(T) + d.T \le D_Q(T)$.

(3) $H_A(T)(1-T)^{-1} \le H_B(T)(1-T)^{-1}$.

(4) $D_A(T) \le D_B(T)$.

(The proposed inequalities are coefficientwise inequalities between formal power series.)

Comments. The statements (1) and (2) have actually been established when Q is a sufficiently "geometric" ring. What remains to be proved is perhaps better isolated in the (respectively) equivalent statements (3) and (4). (For the equivalence, see [6].)

In what follows, let us concentrate on (3) and (4). The status of our knowledge is markedly different in these two cases.

In respect of (4) one is not too far from having an exact relation between the two members, giving the desired inequality as a trivial consequence. For it has been proved (Avramov, André; see [1]) that $D_B - D_A$ is equal to D_C, where C is the fibre of $A \longrightarrow B$, minus a polynomial (with non-negative coefficients). This polynomial cannot vanish in general, as is shown by the simple example $A = B = K[X]_{(X)}$, $X \mapsto X^2$, but it might always have the form $a(T + T^2)$ with $a \in \mathbb{Z}$ (as conjectured by André, Roos ...). The situation can be said to be even more straight in the μ-series case. But for Hilbert series nothing similar appears to have been conceived; see however [5] where the relation $H_B = H_A H_C$ is studied as a condition on local morphisms $A \longrightarrow B$ (with fibre C).

Very little is known about (3). The simplest undecided instance is probably the conclusion, for A,B of Krull dimension 3, that the multiplicity of A should not exceed the multiplicity of B. No counter-example to the stronger assertion that $H_A \le H_B$ seems to be known, but it would of course already be a great thing if one could prove that there always exists a natural number n, that may depend on the morphism $A \longrightarrow B$, such that

$$H_A(T)(1 - T)^{-n} \le H_B(T)(1 - T)^{-n}.$$

Let us finally remark that there is a relation between the Hilbert and Poincaré series of a local ring which might make it possible to attack Hilbert series problems by means of Poincaré series (cf. the formula (19) of [8], which can be generalized).

References (further references can be found in [6])

1. M. André, "Le caractère additif des déviations des anneaux locaux", preprint, École Polytechnique Fédérale de Lausanne, 1982.

2. H.-B. Foxby, "A homological theory of complexes of modules", Preprint Series 1981, No. 19, Københavns Universitet, Matematisk Institut, 1981.

3. H.-B. Foxby and A. Thorup, "Minimal injective resolutions under flat base change", Proc. Amer. Math. Soc., 67 (1977), 27-31.

4. B. Herzog, "Effect of certain flat local extensions on the local Hilbert functions", Seminar D. Eisenbud-B. Singh-W.Vogel, Vol. 1, Teubner Texte zur Math. Bd.29 (Teubner, Leipzig, 1980), pp. 78-93.

5. B. Herzog, "On a relation between the Hilbert functions belonging to a local homomorphism", J. London Math. Soc. (2), 25 (1982), 458-466.

6. T. Larfeldt and C. Lech, "Analytic ramifications and flat couples of local rings", Acta Math., 146 (1981), 201-208.

7. P. Roberts, Homological invariants of modules over commutative rings, Séminaire de mathématiques supérieures (Les Presses de l'Université de Montréal, Montréal, 1980).

8. J.-E. Roos, "Relations between the Poincaré-Betti series of loop spaces and of local rings", Séminaire d'algèbre Paul Dubreil, Proceedings, Paris 1977-78, Lecture Notes in Mathematics 740 (Springer, Berlin, Heidelberg, New York, 1979), pp. 285-322.

Department of Mathematics,
University of Stockholm,
Box 6701,
S-113 85 Stockholm, Sweden.

2. *Problems on asymptotic prime divisors (submitted by L.J. Ratliff, Jr. and received on 8 September, 1981)*

Throughout, I is an ideal in a Noetherian ring R, $A^*(I) = \text{Ass } R/I^n$ and $\hat{A}^*(I) = \text{Ass } R/(I^n)_a$ for all large n, where I_a denotes the integral closure of I in R. (It is shown in [1] that the sets Ass R/I^n are equal for all large n, and a similar result holds for Ass $R/(I^n)_a$, by [6].)

PROBLEM 2.1 It is known [6] that the sets Ass $R/(I^k)_a$ are increasing; that is, if P is a prime divisor of $(I^k)_a$ for some $k \geq 1$, then $P \in \hat{A}^*(I)$ and in fact P is a prime divisor of $(I^{k+i})_a$ for all $i \geq 1$. The first part of this is not true for the sets Ass R/I^k, by [1]. Is it true that if $P \in A^*(I)$ and P is a prime divisor of I^k for some $k \geq 1$, then P is a prime divisor of I^{k+i} for all $i \geq 1$?

PROBLEM 2.2. Given a finite set S of positive integers, do there exist a Noetherian ring R, an ideal I in R, and $P \in \text{Spec } R$ such that P is a prime divisor of I^k if and only if $k \in S$?

PROBLEM 2.3. It is shown in [1] that the sets Ass (I^n/I^{n+1}) are equal for all large n. Let $B^*(I)$ denote this set. Then it is shown in [2; Proposition 10 and Corollary 13] that $A^*(I) = B^*(I) \cup \{P: I \subseteq P \in \text{Ass } R\}$. Characterize $B^*(I) \cap \{P: I \subseteq P \in \text{Ass } R\}$. (This was asked in [2; p. 75].)

PROBLEM 2.4. It is shown in [4; (16)] that a local domain (R,M) of altitude n+1 > 1 is a Cohen-Macaulay ring if and only if R is quasi-unmixed and there exists an integrally closed ideal of the principal class of height n in $R(X_1,\dots,X_n) = R[X_1,\dots,X_n]_{MR[X_1,\dots,X_n]}$. (An ideal I is of the principal class in case I can be generated by h = height I elements.) Is the quasi-unmixedness assumption essential? (This was asked in [4; (18)].)

PROBLEM 2.5. In [3; Theorem 4], the set $\mathscr{A}$ = {R: R is a Noetherian domain such that $\hat{A}^*(I) = A^*(I)$ for all ideals I in R} was essentially characterized. However, it was left open in [3] if there exists an integrally closed local domain R of altitude three in $\mathscr{A}$, but it is shown there that any such R cannot satisfy the altitude formula. Does there exist such an integral domain in $\mathscr{A}$? (The only known example of such a ring is the recent example of T. Ogoma that shows that the strongest of the catenary chain conjectures, the Chain Conjecture (the integral closure of a local domain is catenary), is false.) It is shown in [6] that if there exists a local UFD of altitude three which does not satisfy the altitude formula, then it must be in $\mathscr{A}$. Does there exist such a ring? (If so, then the weakest of the catenary chain conjectures, the Normal Chain Conjecture (if the integral closure of a local domain R satisfies the first chain condition for prime ideals, then every integral extension domain of R does), is false.)

PROBLEM 2.6. In [5], it is shown that asymptotic sequences in R are an excellent analogue of R-sequences in general Noetherian rings. (Elements $b_1,\dots,b_g$ in R are an asymptotic sequence in case $B_g \neq R$ and $(B_{i-1}{}^n)_a : b_iR = (B_{i-1}{}^n)_a$ for i = 1,...,g and for all large n, where $B_i = (b_1,\dots,b_i)R$ (i = 0,1,...,g).) It is also shown in [5] that the asymptotic grade of an ideal I in R satisfies most of the basic properties of the usual grade of an ideal. (The asymptotic grade of I is defined as the maximum of the lengths of asymptotic sequences contained in I.) The literature is crowded with useful, interesting, and important results concerning R-sequences and grade(I). To what extent are the asymptotic versions of these results valid?

References

1. M. Brodmann, "Asymptotic stability of Ass(M/IM)", Proc. Amer. Math. Soc., 74 (1979), 16-18.

2. S. McAdam and P. Eakin, "The asymptotic Ass", J. Algebra, 61 (1979), 71-81.

3. L.J. Ratliff, Jr., "Integrally closed ideals and asymptotic prime divisors", Pacific J. Math., 91 (1980), 445-456.

4. L.J. Ratliff, Jr., "New characterizations of quasi-unmixed, unmixed, and Macaulay local domains", J. Algebra, to appear.

5. L.J. Ratliff, Jr., "Asymptotic sequences", preprint, University of California at Riverside, 1981.

6. L.J. Ratliff, Jr., "Asymptotic prime divisors", in preparation.

Department of Mathematics,
University of California,
Riverside,
California 92521, U.S.A.

3. *Miscellaneous problems*

PROBLEM 3.1 *(submitted by Winfried Bruns and received on 19 April, 1982).* Let R be a noetherian commutative ring and let n be a positive integer. What can be said about the set

$$V^{(n)}(\underline{a}) := \{\underline{p} \in \text{Spec } R: \underline{a} \subset \underline{p}^{(n)}\}$$

for an ideal $\underline{a}$ of R?

PROBLEM 3.2 *(submitted by E.G. Evans and received on 24 July, 1981).* Let $(R,\underline{m},k)$ be a regular local ring for which k is infinite, and let M be a finitely generated R-module. For $x \in M$, let

$$O_M(x) = \{f(x) \mid f \in \text{Hom}_R(M,R)\}.$$

Does there exist an $x \in M - \underline{m}M$ for which

$$\text{height } O_M(x) \leq \text{rank } M?$$

PROBLEM 3.3 *(submitted by E.G. Evans and received on 24 July, 1981).* In the situation of Problem 3.2, can one put reasonable assumptions on M and then predict the behavior of some $O_M(x)$ (for example, can one predict their primeness or otherwise, their depth, etc.)? The hope would be to generalize to height greater than 2 the results obtained in [EG].

[EG] E.G. Evans and P. Griffith, "Local cohomology modules for normal domains", J. London Math. Soc. (2), 19 (1979), 277-284.

PROBLEM 3.4 *(submitted by E.G. Evans and received on 24 July, 1981).* In the situation of Problems 3.2 and 3.3, if M is reflexive and one knows all the $O_M(x)$ for $x \in M - \underline{m}M$ (for example, that they all have height $\geq r$), then what can one say about M (for example, about its depth, its syzygyness, etc.)?

PROBLEM 3.5 *(submitted by Jan-Erik Roos and received on 5 February, 1982).* Let $(R,\underline{m})$ be a commutative noetherian local ring. Is λ-dim R α I(k) $< \infty$? (The notation is the same as that on p.191.)

This problem is equivalent to Problem 3.6 below.

PROBLEM 3.6 *(submitted by Jan-Erik Roos and received on 5 February, 1982).* Let $(R,\underline{m})$ be a commutative noetherian local ring. Does there exist an integer $\alpha(R)$ such that, if M is a finitely generated R-module with $\mathrm{Ext}_R^i(M,R)$ of finite length for $1 \leq i \leq \alpha(R)$, then we have that $\mathrm{Ext}_R^i(M,R)$ is of finite length for all $i \geq 1$?

PROBLEM 3.7 *(submitted by Jan-Erik Roos and received on 5 February, 1982).* Let $(R,\underline{m})$ be a commutative noetherian local ring. Does there exist an integer $\beta(R)$ such that, if, for a finitely generated R-module M, we have $\mathrm{Ext}_R^i(M,R) = 0$ for $1 \leq i \leq \beta(R)$, it follows that $\mathrm{Ext}_R^i(M,R) = 0$ for all $i \geq 1$?

PROBLEM 3.8 *(submitted by Jan-Erik Roos and received on 5 February, 1982).* Let $(R,\underline{m})$ be a commutative noetherian local ring. Does there exist an integer $\gamma(R)$ such that, if a finitely generated R-module is a $\gamma(R)$-th syzygy, then it is a $(\gamma(R) + 1)$-st syzygy?

PROBLEM 3.9 *(submitted by Judith D. Sally and received on 2 June, 1982).* Is there a d-dimensional Cohen-Macaulay local ring $(R,\underline{m})$ having minimal reductions $x_1,\ldots,x_d$ and $y_1,\ldots,y_d$ such that $\underline{m}^{r+1} = (x_1,\ldots,x_d)\underline{m}^r$ but $\underline{m}^{r+1} \neq (y_1,\ldots,y_d)\underline{m}^r$?

PROBLEM 3.10 *(submitted by R.Y. Sharp).* If a commutative Noetherian ring A possesses a dualizing complex, must A be a homomorphic image of a finite-dimensional Gorenstein ring?

PROBLEM 3.11 *(submitted by R.Y. Sharp).* Let A be a commutative Noetherian local ring and let M be a balanced big Cohen-Macaulay A-module (this terminology is explained on p. 74). Let $\underline{p}$ be an associated prime ideal of $M/(a_1,\ldots,a_r)M$ for some M-sequence

$a_1,\ldots,a_r$. Must it be the case that $M_{\underline{p}}$ is a balanced big Cohen-Macaulay $A_{\underline{p}}$-module?

This question has an affirmative answer when A is a catenary local domain: see [S; Theorem 4.3].

[S] R.Y. Sharp, "A Cousin complex characterization of balanced big Cohen-Macaulay modules", Quart. J. Math. Oxford (2), to appear.